ENVIRONMENTAL SCIENCE DEMYSTIFIED

Demystified Series

Advanced Statistics Demystified
Algebra Demystified
Anatomy Demystified
Astronomy Demystified
Biology Demystified
Biotechnology Demystified
Business Statistics Demystified
Calculus Demystified
Chemistry Demystified
College Algebra Demystified
Differential Equations Demystified
Digital Electronics Demystified
Earth Science Demystified
Electricity Demystified
Electronics Demystified
Environmental Science Demystified
Everyday Math Demystified
Geometry Demystified
Math Proofs Demystified
Math Word Problems Demystified
Microbiology Demystified
Physics Demystified
Physiology Demystified
Pre-Algebra Demystified
Precalculus Demystified
Probability Demystified
Project Management Demystified
Quantum Mechanics Demystified
Relativity Demystified
Robotics Demystified
Statistics Demystified
Trigonometry Demystified

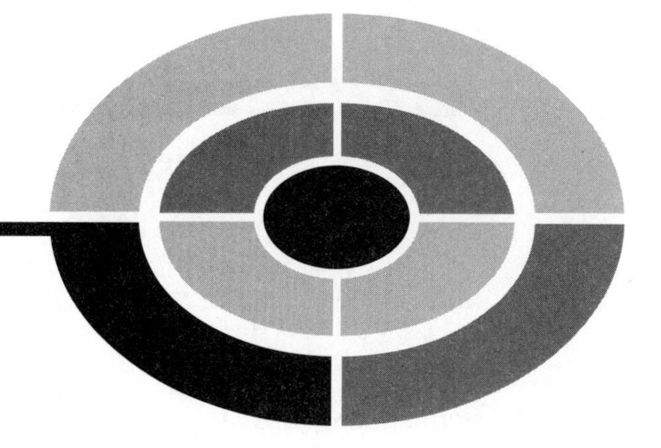

ENVIRONMENTAL SCIENCE DEMYSTIFIED

LINDA D. WILLIAMS

McGRAW-HILL
New York Chicago San Francisco Lisbon London
Madrid Mexico City Milan New Delhi San Juan
Seoul Singapore Sydney Toronto

The McGraw-Hill Companies

Library of Congress Cataloging-in-Publication Data

Williams, Linda D.
Environmental science demystified / Linda D. Williams.
p. cm.
ISBN 0-07-145319-9 (acid-free paper)
1. Environmental sciences—Popular works. I. Title.

GE110.W55 2005
628—dc22 2005047985

1 2 3 4 5 6 7 8 9 0 DOC/DOC 0 1 0 9 8 7 6 5

ISBN 0-07-145319-9

The sponsoring editor for this book was Judy Bass and the production supervisor was Pamela A. Pelton. It was set in Times Roman by Fine Composition. The art director for the cover was Margaret Webster-Shapiro; the cover designer was Handel Low.

Printed and bound by RR Donnelley.

This book is printed on recycled, acid-free paper containing a minimum of 50% recycled, de-inked fiber.

This book is dedicated to the environmental heroes
of the past 200 years, who had the vision, courage,
and quiet persistence to preserve pristine forests,
open lands, and endangered species,
as well as bring pollution issues into general view.
Because of their efforts, we have a good chance
of surviving our global growing pains.
Thank you.

Linda D. Williams

CONTENTS

	Preface	ix
	Acknowledgments	xiii
PART ONE:	ATMOSPHERE	1
CHAPTER 1	Our Planet Earth	3
CHAPTER 2	Ecosystems and Biodiversity	21
CHAPTER 3	Atmosphere	41
CHAPTER 4	Greenhouse Effect and Global Warming	73
	Part One Test	91
PART TWO:	WATER	99
CHAPTER 5	The Hydrologic Cycle	101
CHAPTER 6	Oceans and Fisheries	125
CHAPTER 7	Glaciers	147
CHAPTER 8	Water Pollution and Treatment	171
	Part Two Test	193

PART THREE: LAND 201

CHAPTER 9 Weathering and Erosion 203

CHAPTER 10 Deserts 221

CHAPTER 11 Geochemical Cycling 243

CHAPTER 12 Solid and Hazardous Waste 261

Part Three Test 277

PART FOUR: WHAT CAN BE DONE 285

CHAPTER 13 Fossil Fuels 287

CHAPTER 14 Nuclear and Solar Energy 303

CHAPTER 15 Wind, Hydroelectric and Geothermal Energy 319

CHAPTER 16 Future Policy and Alternatives 335

Part Four Test 351

Final Exam 359

Answers to Quiz, Test, and Exam Questions 377

APPENDIX I Terms and Organizations 383

APPENDIX II Conversion Factors 393

References 395

Index 401

PREFACE

This book is for anyone with an interest in Environmental Science who wants to learn more outside of a formal classroom setting. It can also be used by home-schooled students, tutored students, and those people wishing to change careers. The material is presented in an easy-to-follow way and can be best understood when read from beginning to end. However, if you just want more information on specific topics like greenhouse gases, geothermal energy, or glaciers, then you can review those chapters individually as well.

You will notice through the course of this book that I have mentioned milestone theories and accomplishments of geologists and ecologists along with national and international organizations making a difference. I've highlighted these innovative people and agencies to give you an idea of how the questions and strong love of nature have motivated individuals and countries to take action.

Science is all about curiosity and the desire to find out how something happens. Nobel prize winners were once students who daydreamed about new ways of doing things. They knew answers had to be there and they were stubborn enough to dig for them. The Nobel prize for Science has been awarded over 475 times since 1901.

In 1863, Alfred Nobel experienced a tragic loss in an experiment with nitroglycerine that destroyed two wings of the family mansion and killed his younger brother and four others. Nobel had discovered the most powerful weapon of that time—dynamite.

By the end of his life, Nobel had 355 patents for various inventions. After his death in 1896, Nobel's will described the establishment of a foundation to create five prizes of equal value "for those who, in the previous year, have contributed best towards the benefits for humankind," in the areas of Earth Science, Physics, Physiology/Medicine, Literature, and Peace. Nobel wanted to recognize the heroes of science and encourage others in their quest for knowledge. Perhaps the simple ideas that changed our understanding of the Earth, ecosystems, and biodiversity will encourage you to use your own creative ideas in tackling important Environmental Science concerns.

This book provides a general overview of Environmental Science with chapters on all the main areas you'll find in an Environmental Science classroom or individual study of the subject. The basics are covered to familiarize you with the terms and concepts most common in the experimental sciences, of which Environmental Science is one. Additionally, I've listed helpful Internet sites with up-to-date information on global warming, atmospheric factors, and energy alternatives, to name a few.

Throughout the text, I've supplied lots of everyday examples and illustrations of natural events to help you visualize what is happening beneath, on, or above the Earth's surface. There are also quiz, test, and exam questions throughout. All the questions are multiple-choice and a lot like those used in standardized tests. There is a short quiz at the end of each chapter. These quizzes are "open book." You shouldn't have any trouble with them. You can look back at the chapter text to refresh your memory or check the details of a natural process. Write your answers down and have a friend or parent check your score with the answers in the back of the book. You may want to linger in a chapter until you have a good handle on the material and get most of the answers right before moving on.

This book is divided into four major parts. A multiple-choice test follows each of these parts. When you have completed a part, go ahead and take the part test. Take the tests "closed book" when you are confident about your skills on the individual quizzes. Try not to look back at the text material when you are taking them. The questions are no more difficult than the quizzes, but serve as a more complete review. I have thrown in lots of wacky answers to keep you awake and make the tests fun. A good score is three-quarters of the answers right. Remember, all answers are in the back of the book.

The final exam at the end of the course is made up of easier questions than those in the quizzes and part tests. Take the exam when you have finished all the chapter quizzes and part tests and feel comfortable with the material as a whole. A good score on the exam is at least 75% of correct answers.

With all the quizzes, part tests, and the final exam, you may want to have your friend or parent give you your score without telling you which of the questions you missed. Then you will be tempted not to memorize the answers to the missed questions, but instead to go back and see if you missed the point of the idea. When your scores are where you'd like them to be, go back and check the individual questions to confirm your strengths and any areas that need more study.

Try going through one chapter a week. An hour a day or so will allow you to take in the information slowly. Don't rush. Environmental Science is not difficult, but does take some thought to get the big picture. Just plow through at a steady rate. If you're really interested in deserts, spend more time on Chapter 10. If you want to learn the latest about the oceans and fisheries, allow more time for Chapter 6. At a steady pace, you'll complete the course in a few months. After completing the course, you will have become a geologist-in-training. This book can then serve as a ready reference guide, with its comprehensive index, appendix, and many examples of cloud structures, energy types, erosion, and geochemical cycling.

Suggestions for future editions are welcome.

LINDA D. WILLIAMS

ACKNOWLEDGMENTS

Illustrations in this book were generated with CorelDRAW and Microsoft PowerPoint and Microsoft Visio, courtesy of the Corel and Microsoft Corporations, respectively.

National Oceanographic and Atmospheric Administration (NOAA), Environmental Protection Agency (EPA), and United States Geological Survey (USGS) statistics and forecasts were used where indicated.

A very special thanks to Dr. Karen Duston of Rice University for the technical review of this book.

Many thanks to Judy Bass at McGraw-Hill for her unfailing confidence and assistance.

Thank you also to Rice University's staff and faculty for their friendship, support, and flexibility in the completion of this work.

To my children, grandchildren, and great-grandchildren who will inherit the Earth that is left to them.

ENVIRONMENTAL SCIENCE DEMYSTIFIED

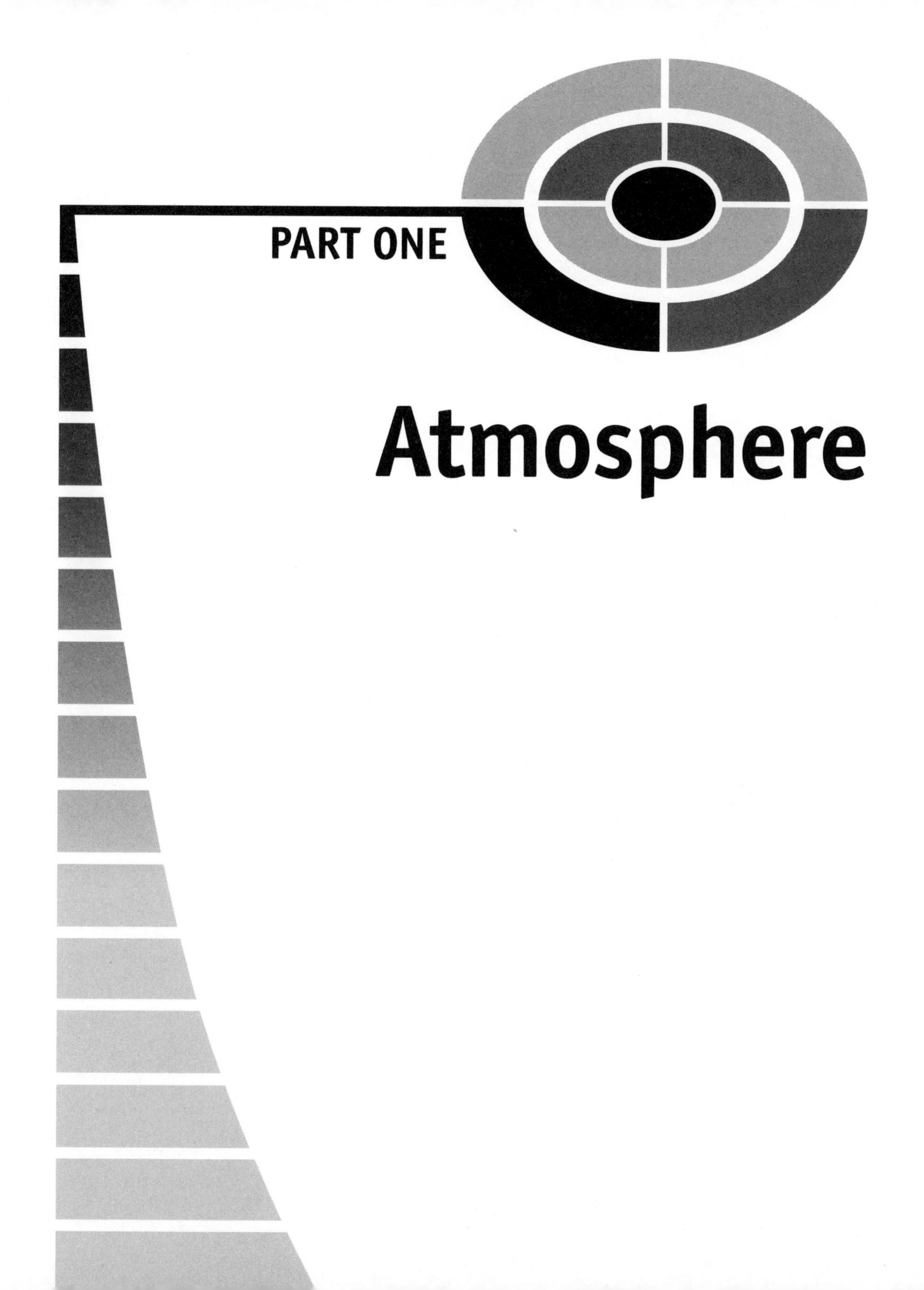

PART ONE

Atmosphere

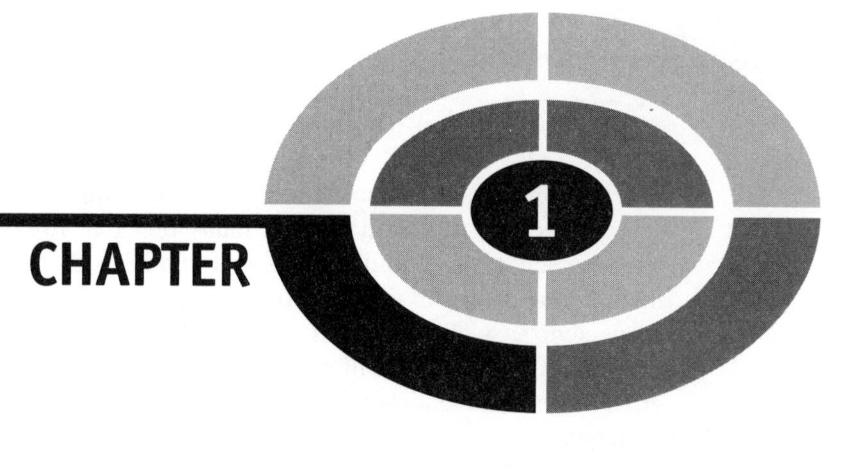

CHAPTER

Our Planet Earth

Native peoples, completely dependent on nature for everything in their lives, worshipped Earth as a nurturing mother. The soil sprouted plants and trees that provided food, clothing, and shelter. The rivers and seas gave up fish and shellfish for food, trade articles, and tools. From the atmosphere came rain, snow, and wind to water crops and adjust the seasons. Earth, never stagnant or dull, provided abundantly for early stewards of her resources. Ancient peoples thought that Mother Earth worked together with Father Sun to provide for those who honored her.

In early Greek mythology, the Earth goddess, Gaia, mother of the Titans, was honored as an all-nourishing deity. When Gaia was happy, crops flourished, fishermen and hunters were successful, and everything thrived.

Today, astronauts who orbit Earth in space ships and scientific laboratories marvel at her beauty while working toward her care. Other scientists, engineers, and test pilots have communicated their wonder and appreciation for our fragile world through environmental efforts that address global issues. Any study of the environment includes many facets of this planet we call home. *Environmental science* encompasses worldwide environmental factors like air, light, moisture, temperature, wind, soil, and other living organisms.

> **Environmental biology** includes all the external factors that affect an organism or community and that influence its development or existence.

The Earth's almost limitless beauty and complexity provide broad areas for scientific study. Researchers from many different fields are focusing their skills on the mechanisms and interactions of hundreds of environmental factors. These natural and industrial factors affect the environment in ways that are known and suspected, as well as those totally unidentified. Although some changes have been taking place for millions of years, some appear to be accelerating. Today, environmental scientists are sorting through tons of data in order to increase their understanding of the impacts of modern processes on all environmental aspects. Table 1-1 lists a sampling of the various environmental fields of scientific study.

Size and Shape

The size and shape of the earth was a mystery for thousands of years. Most people thought the land and seas were flat. They were afraid that if they traveled too far in one direction, they would fall off the edge. Explorers who sailed to the limits of known navigation were considered crazy and on paths to destruction. Since many early ships did not return from long voyages, people thought they had been sunk by storms, eaten by sea monsters, or just went too far and fell off.

It wasn't until the Greek philosopher Aristotle (384–322 BC), who noticed that Earth's shadow on the moon was curved, that people began to question the "flat earth" idea. It was another 1500 years, however, before the earth's round shape was well understood.

Compared to the sun, which is over 332,000 times the mass of the Earth, our home planet is tiny—a bit like a human compared to an ant. The sun is 1,391,000 kilometers in diameter compared to the Earth, which is approximately 12,756 km in diameter. That means the diameter of the sun is over 100 times that of the Earth. To picture the size difference, imagine that the sun is the size of a basketball. By comparison, the earth would be about the size of this "o."

Earth's Formation

In 1755, Immanuel Kant offered the idea that the solar system was formed from a rotating cloud of gas and thin dust. In the years since, this idea has become known as the *nebular hypothesis*. The clouds that Kant described could be seen

Table 1-1 Fields of study.

Environmental science	Area of interest
Agrology	Analysis and management of usable land for growth of food crops
Bioengineering	Designing or reconstructing sustainable ecosystems
Botany	Characterization, growth, and distribution of plants
Conservation biology	Preserve, manage, or restore endangered areas or species
Ecology	Study of relationships between living organisms and their environment
Environmental geology	Conservation of resources and future planning
Exploration geophysics	Crustal composition to find resources (e.g. oil, gold)
Forestry	Characterization, growth, distribution, and planting of trees
Geochemistry	Chemical composition of rocks and their changes
Geomorphology	Nature, origin, development, and surface of land forms
Geophysics	Earth's magnetism, gravity, electrical properties, and radioactivity
Glaciology	Formation, movement, and makeup of current glaciers
Hydrology	Composition and flow of water over the earth
Mineralogy	Natural and synthetic minerals with a crystalline structure
Oceanography	Water makeup, currents, boundaries, topography, marine life
Pedology	Origin, treatment, character, and utilization of soil
Petrology	Origins, composition, alteration, and decay of rock
Structural geology	Rock changes and distortions within the earth's layers
Volcanology	Formation, activity, temperature, and explosions of volcanoes
Wildlife biology	Characterization and distribution of animal communities
Zoology	Characterization, growth, and distribution of animals

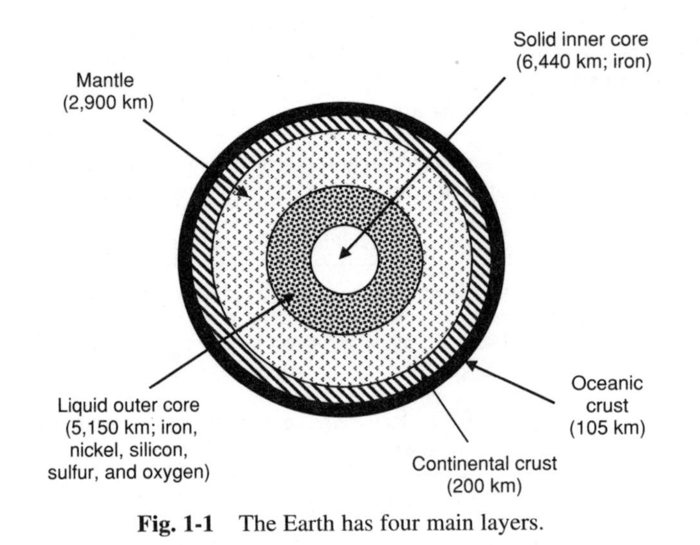

Fig. 1-1 The Earth has four main layers.

by powerful telescopes. The NASA Hubble space telescope has sent back images of many of these beautiful formations, called *nebulae*.

It is likely that when the Earth was first forming in our young solar neighborhood, it was a molten mass of rock and metals simmering at about 3,600° Fahrenheit (2,000° Celsius). The main cloud elements included hydrogen, helium, carbon, nitrogen, oxygen, silicon, iron, nickel, phosphorus, sulfur, and others. As the sphere (Earth) cooled, the heavier metals like iron and nickel sunk deeper into the molten core, while the lighter elements like silicon rose to the surface, cooled a bit, and began to form a thin crust. Fig. 1-1 shows the way that early elements formed into a multilayer crust. This crust floated on a sea of molten rock approximately four billion years ago, sputtering volcanic gases and steam from the impact of visitors like ice comets. Millions more years passed like this while an atmosphere gradually formed. Rain condensed and poured down, cooling the crust into one large chunk and gathering into low spots and cracks, forming oceans, seas, lakes, rivers, and streams.

Our Place in the Galaxy

Even though the sun seems to be the center of our universe, it is really just one of many kids on the block. Our solar system is found on one of the spiral arms, Orion, of the galaxy known as the Milky Way.

The **Milky Way** is one of millions of galaxies in the universe. The *Andromeda* galaxy is the nearest major galaxy to the Milky Way.

Think of the Milky Way galaxy as one "continent" among billions of other continents in a world called the universe. Its spiraling arms or "countries" are called Centaurus, Sagittarius, Orion, Perseus, and Cygnus. The Milky Way galaxy is 80,000 to 120,000 light years across (a light year is a measure of distance equal to or more than 9 trillion km, or 9.46×10^{12} km). The center of the Milky Way is made up of a dense molecular cloud that rotates slowly clockwise, throwing off solar systems and cosmic debris. It contains roughly 200 billion (2×10^{12}) stars.

Although Andromeda is the closest full-size galaxy to the Milky Way, the Sagittarius Dwarf, discovered in 1994, is the closest galaxy. It is 80,000 light years away, or nearly 24 kiloparsecs. A parsec is a unit of measurement equal to 3.26 light years.

Atmosphere

Earth's *atmosphere* is the key to the development of life on this planet. Other planets in our solar system contain various levels of hydrogen, methane, and ammonia atmospheres (Jupiter, Neptune), carbon dioxide and nitrogen (Venus, Mars), or hardly any atmosphere at all (Mercury, Pluto).

The atmosphere of the Earth, belched from prehistoric volcanoes, extends nearly 563 km (350 miles) out from the solid surface of our planet. It is made up of a mixture of gases that combine to allow life to exist. In the lower atmosphere, nitrogen is found in the greatest amounts, 78%, followed by oxygen at 21%. Carbon dioxide, vital to the growth of plants, is present in trace levels of atmospheric gases, along with argon and a sprinkling of neon and other minor gases. Table 1-2 compares the earth's atmosphere with the atmosphere of neighboring planets.

Oxygen, critical to human life, developed as microscopic plants and algae began using carbon dioxide and photosynthesis to make food. From that process, oxygen is the most important byproduct.

The mixture of gases that we call air penetrates the ground and most openings in the earth not already filled with water. The atmosphere is perhaps the most active of the different environmental components. To people around the world, it has a constantly changing personality.

Table 1-2 Planets, atmospheric gases.

Planets	Atmospheric gases
Sun	Hydrogen, nitrogen
Mercury	None
Venus	None
Earth	Oxygen, nitrogen, hydrogen
Mars	None
Jupiter	Methane, hydrogen
Saturn	Hydrogen, methane
Uranus	None
Neptune	None
Pluto	None

We will study these atmospheric factors that favor life on Earth when we take a closer look at the atmosphere in Chapter 3.

Gaia Hypothesis

In 1974, James Lovelock explained that the Earth existed as a single living organism in his *Gaia hypothesis.* He described how organic and inorganic components interact through complex reactions to balance an environment where life can exist. Lovelock's Gaia theory considers the evolution of a tightly integrated system made up of all living things and the physical environment: the atmosphere, oceans, and land. Natural regulation of important factors, like climate and chemical composition, is a result of intricate evolutionary development. Like many living organisms and closed-loop self-regulating systems, Lovelock considered Gaia as one system in which the whole is greater than the sum of its parts.

An active, adaptive control process, able to maintain the Earth in overall balance, is known as the **Gaia hypothesis.**

Today, there is heated debate about which global environmental problem is the most crucial. Depending on a person's geographical and economic position, it can be pollution, overpopulation, ozone depletion, deforestation, habitat destruction, global warming, overfishing, drought, radioactive waste storage—or all or none of these. Some scientists believe that these environmental impacts will be overcome in the long run when Gaia makes the corrections needed to bring the Earth back into equilibrium.

The problem with these balancing forces is that they can be sudden and violent (think earthquakes and volcanic eruptions). A powerful, natural environmental adjustment often brings disaster upon all inhabitants unfortunate enough to be in the vicinity.

In 1979, Lovelock further described his theory in *GAIA: A New Look at Life on Earth.* Since then, many scientists and environmentalists have begun to study global changes within the context of the Gaia idea, although not everyone agrees with the theory.

The updated Gaia hypothesis proposes that Earth's atmosphere, oceans, and land masses are held in equilibrium by the living inhabitants of the planet, which includes millions of species besides humans. The Gaia concept suggests that this living world keeps itself in worldwide environmental balance.

One example of this balancing act takes place in the oceans. Salts are constantly added to the oceans by physical and chemical processes, raising salinity. Eventually, affected seas (like the Dead Sea) reach an uninhabitable salinity level. According to the Gaia hypothesis, the sea's salinity is controlled biologically through the mutual action of ocean organisms. In fact, living sea creatures, primarily algae and protozoa, have processed and removed salt throughout geological time, balancing salinity levels that allow life to thrive.

Actually, that is fairly straightforward. Salt is removed from ocean waters when it piles up on the bottom. This happens following the death of microorganisms that sink to the ocean floor. As ocean salinity rises, plankton that include salt into their outer coverings die and sink to the ocean depths, lowering salt levels. In this way, the ocean's salinity stays in equilibrium.

The Gaia theory can also be applied to the balancing of atmospheric gases in fairly constant proportions needed to support life. Without the ongoing biological creation of oxygen and methane, for example, the balance of critical atmospheric elements would be severely altered. Organisms all over the world work together to create a breathable atmosphere, not just in one habitat or location.

Scientists questioning the Gaia concept think that evolutionary changes account for the adjustments needed for life to exist. The argument goes that when ocean salinity increases, oxygen levels change or global temperatures increase. Only evolved organisms are able to able to survive new conditions, and their genetically stronger offspring are then able to thrive in a changed environment.

Lovelock counters this by stressing that environmental conditions can't operate independently of living world processes. Table 1-3 lists the diverse Earth energy resources and processes to be considered.

In the Gaia concept, humans are seen as one species among millions, with no special rights. Whether humans were here or not makes little difference to Gaia's survival, which eventually adjusts for overpopulation, global warming, or habitat destruction. Some might even argue that Gaia would function

Table 1-3 Energy resources.

Energy resources
Solar energy
Fossil fuels
Atmospheric absorption
Surface heating
Wind energy
Heat from the earth's core
Shortwave and longwave radiation
Flowing rivers
Hydrologic cycle and precipitation
Tidal energy
Gravitational energy
Tectonic energy
Energy absorption in the earth's crust

more effectively without us here at all. However, whether humans are responsible for environmental problems or not, global balancing may severely impact our future.

To understand the Earth as one living system (whether called Gaia or something else) we need to understand more about all the parts that make it unique.

Biosphere

All plants and animals on the earth live in the *biosphere,* which is measured from the ocean floor to the top of the atmosphere. It includes all living things, large (whales) and small (bacteria), grouped into species or separate types. The main compounds that make up the biosphere contain carbon, hydrogen, and oxygen. These elements also interact with other earth systems.

> The **biosphere** includes the hydrosphere, crust, and atmosphere. It is located above the deeper layers of the earth.

Surprisingly, life is found in many hostile environments on this planet. Very hot temperatures (5,000°C) near volcanic spouts rising from the ocean floor, and polar, subzero temperatures (–84°C) are at the extreme ends of the temperature range. The earth's biodiversity is truly amazing. Everything from exotic and fearsome deep-ocean creatures to sightless fish in underground lakes exists as part of the earth's diverse inhabitants. There are sulfur-fixing bacteria that thrive in sulfur-rich, boiling geothermal pools and frogs that dry out and remain barely alive in desert soils until the rare rains bring them back to life. This makes environmental study fascinating to people of all cultures, geographies, and interests.

However, the large majority of biosphere organisms that grow, reproduce, and die are found in a much more narrow range. In fact, most of the Earth's species live in a thin slice of the biosphere. This slice is located at temperatures above zero (most of the year) and in upper ocean depths where sunlight can penetrate.

The vertical range of the biosphere is roughly 20,000 meters, but the section most populated with living species is only a fraction of that. It includes a section measured from just below the ocean's surface to about 1,000 meters above it. Most living plants and animals live in this narrow layer of the biosphere. The biosphere and the impacts of today's world will be described in greater detail in Chapter 2.

Hydrosphere

The global ocean, the Earth's most noticeable feature from space, makes up the largest single part of the planet's total covering. The Pacific Ocean, the largest ocean, is so big that the land mass of all the continents could fit into it.

The combined water of all of the oceans makes up nearly 97% of the earth's water. These oceans are much deeper on average than the land is high, and make up what is known as the *hydrosphere*.

> The **hydrosphere** describes the ever-changing total water cycle that is part of the closed environment of the earth.

The hydrosphere is never still. It encompasses the evaporation of oceans into the atmosphere, the raining of this water back onto land, the run-off into streams and rivers, and finally the flow back into the oceans. The hydrosphere also contains the water in underground aquifers, lakes, and streams.

The *cryosphere* is a subset of the hydrosphere. It includes all of the Earth's frozen water found in colder latitudes and higher elevations in the form of snow and ice. At the poles, continental ice sheets and glaciers cover vast wilderness areas of barren rock that have hardly any plant life. Antarctica is a continent two times the size of Australia and contains the world's largest ice sheet. In Chapter 13 we will learn much more about the cryosphere, its beauty, and its hazards, when we study glaciers.

Lithosphere

The crust and very top part of the mantle are collectively known as the *lithosphere* (*lithos* is Greek for "stone"). This layer of the crust is rigid and brittle, acting as an insulator over the active mantle layers below. It is the coolest of all the Earth's land layers and thought to float or glide over the layers beneath it. Table 1-4 lists the amounts of different elements in the earth's crust.

> The **lithosphere** is about 65 to 100 km thick and covers the entire Earth.

Table 1-4 Elements in Earth's crust.

Element	Percentage of Earth's crust
Oxygen	47
Silica	28
Aluminum	8
Iron	5
Calcium	4
Sodium	3
Potassium	3
Magnesium	2

Scientists have determined that around 250 million years ago, all the land mass was in one big chunk or continent. They called the solid land mass *Pangaea,* meaning "all earth." The huge surrounding ocean was called *Panthalassa,* which means "all seas." By nearly 65 million years ago, things had gradually broken apart to form the continental land masses we know and love today, separated by huge distances of water.

Crust

The Earth's *crust* is the hard, outermost covering of the planet. This is the layer exposed to weathering like wind, rain, freezing snow, hurricanes, tornadoes, earthquakes, meteor impacts, volcano eruptions, and everything in between. It has all the wrinkles, scars, colorations, and shapes that make nature interesting. Just as everyone is different, with diverse ideas and histories depending on their experiences, the land varies widely around the globe. Lush and green in the tropics to dry and inhospitable in the deep Sahara to fields of frozen ice pack in the Arctic, the crust has many faces.

CONTINENTAL CRUST

The land mass of the crust is thin compared to the rest of the Earth's layers. It makes up only about 1% of the earth's total mass, but the continental crust can be as much as 70 km thick. The land crust with mountain ranges and high peaks is thicker in places than the crust found under the oceans and seas, but the ocean's crust, about 7 km thick, is denser.

The continents are the pieces of land that sit above the level of ocean basins, the deepest levels of land within the crust. Continents have broken up into six major land masses: Africa, Antarctica, Australia, Eurasia, North America, and South America. This hard continental crust forms about 29% of the Earth's surface.

Beside dry land, continents include undersea *continental shelves* that extend the land mass even further, like the crust around the edge of a pie. A continental shelf provides a base for the deposit of sand, mud, clay, shells, and minerals washed down from the land mass.

> A **continental shelf** is the thinner, extended edges of a continental land mass that are found below sea level.

A continental shelf can extend beyond the shoreline from 16 to 320 km, depending on location. The water above a continental shelf is fairly shallow (between 60 and 180 km deep) compared to the greater depths at the slope and below. There is a drop-off, called the *continental slope*, that slips away suddenly to the ocean floor. Here, the water reaches depths of up to 5 km to reach the average level of the seafloor.

"Land" or "dry" crust has more variety than its undersea brother, the *oceanic crust,* because of weathering and environmental conditions. The continental crust is thicker, especially under mountains, but less dense than the "wet crust" found under the oceans. Commonly, the continental crust is 32 km thick, but can be up to 80 km thick from the top of a mountain.

OCEANIC CRUST

The land below the levels of the seas is known as the *oceanic crust.* This "wet" crust is much thicker than the continental crust. The average elevation of the continents above sea level is 840 meters. The average depth of the oceans is about

3,800 meters, or $4^1/_2$ times greater. The oceanic crust is roughly 7 to 10 km thick below the bottom of the oceans.

Though not pounded by wind and rain like the continental crust, the oceanic crust is far from dull. It experiences the effects of the intense heat and pressures of the mantle more than the continental crust, because the oceanic crust covers more area.

Even slow processes like sediment collection can trigger important geological events. This happens when the build-up of heavy sediments onto a continental shelf by ocean currents causes pieces to crack off and slide toward the ocean floor in a rush. When this happens, the shift can roar downward at speeds of between 50 and 80 km per hour, smashing everything in its path. Delicate ocean communities are as affected by these types of undersea events as land animals would be after a mudslide or earthquake. The sudden water movement causes intense *turbidity currents* that can slice deep canyons along the ocean floor. These currents cause disruptive undersea avalanches that change the underwater seascape and affect its many inhabitants.

The winds from the Northern and Southern Hemispheres also keep the oceans churning and recycling. Their pushing movement, along with the Earth's rotation, keeps ocean currents moving until they hit a land mass and are deflected. A large, circular rotation pattern in the subtropical ocean is called a *gyre.* The circulation of gyres in the Northern Hemisphere is clockwise, while the circulation in the Southern Hemisphere is counterclockwise. Fig. 1-2 shows the flow of these global gyres.

These are just some of the currents that circulate the oceans. In Chapter 6, we'll learn more about the currents and constant motion of the earth's oceans.

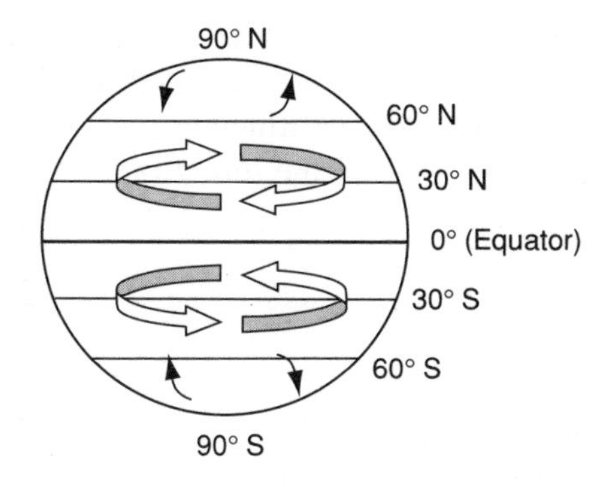

Fig. 1-2 Northern and Southern Hemisphere currents are constantly circulating.

Mantle

The mantle is the next layer in the Earth's crust. It is located just below the lithosphere. The mantle makes up over 80% of the earth's volume. It is estimated to be about 2,900 km thick. The mantle is not the same all the way through. It is divided into two layers: the upper mantle, or *asthenosphere* (*asthenes* is Greek for "weak"), and the lower mantle. These layers are not the same. They contain rock of different density and makeup.

> The highest level of the mantle is called the **asthenosphere,** or **upper mantle.** It is located just below the lithosphere.

The asthenosphere is solid, but found at much greater depths than the lithosphere. Compared to the crust, this layer is hot, near the melting point of rock. Think of it as something like oatmeal: When it is hot, it is fairly liquid, but if you leave it to cool on the table for a few hours, it turns to stone and is nearly impossible to get out of the bowl!

Heat and pressure create malleability within the lithosphere. This acts like a series of ball bearings under the chunks of the lithosphere. Mantle layers move and glide on this moldable, creeping underlayer. This allows a lot of activity to take place.

The heated materials of the asthenosphere become less dense and rise, while cooler material sinks. This works very much like it did when the planet originally formed. Dense matter sank to form a core, while lighter materials shifted upward. The lower part of the mantle, or *mesosphere,* measures roughly 660 km from the Earth's molten outer core to the bottom of the asthenosphere.

Different amounts of heating in the upper and lower parts of the mantle cause extremely slow currents to form and allow solid rock to creep along one atom at a time in a flow direction. The continental and oceanic crusts are pulled down and moved around depending on the direction of these deep currents.

Core

The distance from the Earth's surface to its core is nearly 6,500 km. Like the Hollywood movie *The Core,* the deeper you go, the hotter it gets. The Earth's outer peel, the crust, is 5 to 55 km thick and insulates the surface from its hot interior.

The temperature gradient (rise in temperature with regard to depth) of the earth's crust is 17 to 30°C per kilometer of depth. The mantle has temperatures between 650 and 1250°C. At the earth's core (liquid outer and solid inner) temperatures are between 4000 and 7000°C.

Because heat moves from hotter areas to colder areas, the earth's heat moves from its fiery center toward the surface. This outward heat flow creates a convective mantle movement that drives plate tectonics. At spots where the plates slide apart, magma rises up into the rift, forming new crust.

Where continental plates collide, one plate is forced under the other, a process called *subduction*. As a subducted plate is forced downward into areas of extreme heat, it is forced by increasing pressure, temperature, and water content to melt, becoming magma (lava). Hot magma columns rise and force their way up through the crust, transferring huge amounts of heat.

This very center of the earth (core) is made up mostly of iron with a smattering of nickel and other elements. The core, which is under extreme pressure, makes up around 30% of the Earth's total mass. It is divided into an inner and outer core. Look back to Fig. 1-1 to see the elemental makeup of the core.

Earthquake wave measurements have suggested that the outer core is fluid and made of iron, while the inner core is solid iron and nickel. The solid center, under extremely high pressure, is unable to flow at all.

MAGNETISM

The earth acts as a giant magnet with lines of north/south magnetic force looping from the North Pole to the South Pole. Ancient sailors noticed and used this magnetism to chart and steer a course. The magnetic field around the earth is formed by the rotation of the inner core as a solid ball, the different currents in the liquid outer core, and the much slower movement of the mantle. The earth's magnetic field is supported by this circulation of molten metals in the outer core.

The Earth's magnetic field further shows that the core must be made of a conducting substance (metal). Iron is thought to be the only element that is abundant enough and conductive at the extreme pressures and temperatures typical of the core.

The global magnetic environment is a fairly big unknown. Geologists have discovered that the directional pattern of magnetism in rocks, at the time they were formed, provides an accurate record of the Earth's magnetic profile over geologic time.

When volcanic rock bubbles from volcanic vents, its elements are aligned with the magnetic pole at the time of its formation and locked into that structure as the molten rock cools. By studying ancient volcanic rock, scientists found that the earth's magnetic environment changes in a cyclic manner. In fact, by studying rock magnetism, geologists know that the North and South Poles have actually flipped positions. Because this takes place over the entire globe, scientists are trying to understand the atmospheric, geologic, and oceanic impacts of such a huge magnetic switch. What would it mean to today's cities and land masses? Would hurricanes, tornadoes, and earthquakes increase? Would there be more and more droughts? How are the polar ice caps affected? Can species adapt? No one really knows.

We have surveyed the birth and characteristics of our home planet, but the story doesn't stop there. Now, let's study the factors and forces that have shaped the Earth since the beginning. In Chapter 2, we'll learn more about the complex global ecosystems that support life on the Earth's crust.

Quiz

1. What percentage of water covers the earth's surface?
   (a) 40%
   (b) 50%
   (c) 70%
   (d) 80%

2. Aristotle was the first person to notice that
   (a) the moon was round
   (b) mice always live near grain barns
   (c) bubbles appear in fermenting liquids
   (d) the earth's shadow on the moon is curved

3. What is the nearest major galaxy to the Milky Way?
   (a) Orion
   (b) Draco
   (c) Andromeda
   (d) Cirrus

4. A large, circular rotation pattern in the subtropical ocean is called a
   (a) plateau
   (b) gyre
   (c) mantle
   (d) hydrosphere

5. The magnetic pole is
   (a) kept moving by outer core currents
   (b) located exactly at the geographical pole
   (c) only observed in the Southern Hemisphere
   (d) based on observations of the tides

6. The lithosphere is
   (a) located below the ionosphere
   (b) the crust and very top part of the mantle
   (c) roughly 5 to 20 km thick
   (d) fluid and soft in all areas

7. An active, adaptive control process that is able to maintain the Earth in overall balance is known as the
   (a) Geary hypothesis
   (b) Gaia hypothesis
   (c) Miller hypothesis
   (d) Gladiator hypothesis

8. The diameter of the sun is over how many times the diameter of the Earth?
   (a) 50
   (b) 75
   (c) 100
   (d) 125

9. The biosphere includes the
   (a) hydrosphere, crust, and atmosphere
   (b) oceans and trenches
   (c) crust, mantle layer, and inner core
   (d) hydrosphere and lithosphere

10. The extremely slow atom-by-atom movement and deformation of rock under pressure is known as
    (a) commuting
    (b) sedimentation
    (c) lithification
    (d) creep

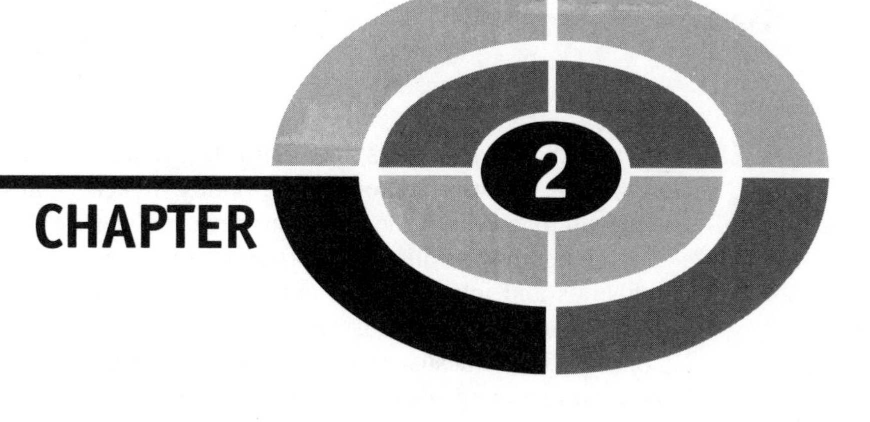

CHAPTER 2

Ecosystems and Biodiversity

Although humans are primarily land dwellers, the Earth's surface is largely water. The world's oceans make up 99% of the planet's biosphere and contain the greatest diversity of life. Even the most biologically rich tropical rain forests cannot match the biodiversity (measured by the number of species) found in a coral reef community.

Rain forests, deserts, coral reefs, grasslands, and a rotting log are all examples of *ecosystems*.

> An **ecosystem** is a complex community of plants, animals, and microorganisms linked by energy and nutrient flows that interact with each other and their environment.

Since the oceans seemed limitless for thousands of years, it's hard to understand pollution's heavy impact on plant and animal marine species and ecosystems. Within the last 30 years, population increases, new technology, increased seafood demand, and many other factors have impacted marine ecosystems in ways unknown 100 years ago. With the planet's population having now passed six billion, scientists, economists, policy makers, and the public are becoming increasingly aware of the strain on the ocean's natural ecosystems and resources.

Climate impacts on coral reefs and forest ecosystems have affected associated industries and jobs (lumber and fishing). Public policy in many countries has begun to address climate issues at the regional, national, and international levels. Conservation and sustainable biodiversity activities are becoming more common with a strong interest toward *sustainable use.*

> **Sustainable use** affects a species or environment and protects its numbers and complexity without causing long-term loss.

Some of the biologically diverse areas currently under study include marine and coastal, island, forest, agricultural, and inland waters, as well as dry, subhumid, and mountain regions. Scientists are initiating research programs that address basic principles, key issues, potential output, timetables, and future goals of single and overlapping systems.

Biosphere

We learned in Chapter 1 that the part of the Earth system that directly supports life, including the oceans, atmosphere, land, and soil, is the *biosphere.* All the Earth's plants and animals live in this layer, which is measured from the ocean floor to the top of the atmosphere. All living things, large and small, are grouped into *species,* or separate types. The main compounds that make up the biosphere contain carbon, hydrogen, and oxygen. These elements interact with other Earth systems.

> The **biosphere** includes the hydrosphere, crust, and atmosphere. It is located above the deeper layers of the earth.

The vertical range that contains the biosphere is roughly 20,000 meters high. The section most populated with living species is only a fraction of that. It includes a section measured from just below the ocean's surface to about 1,000 meters above it. Most living plants and animals live in this narrow layer of the biosphere. Fig. 2-1 gives an idea of the depth of the biosphere.

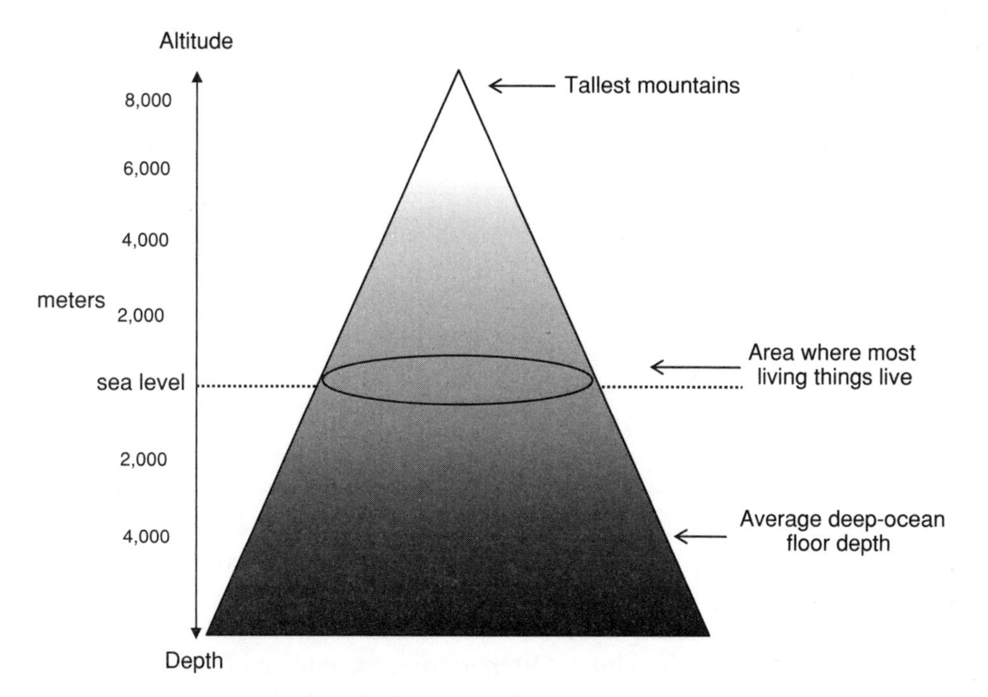

Fig. 2-1 Most of the Earth's life is found in a small wedge.

BIODIVERSITY

The idea of a biologically diverse environment is easy to imagine in the middle of the tropical rain forest, where there are living organisms all around you, but what about the desert? A lot of sand, cactus, scrubby plants, and stunted trees may not seem important, but they are. Every member of a particular environment or ecosystem has a specific purpose, or *ecological niche.*

> An animal or plant with a specific relationship to its habitat or other species, filled by that organism alone, exists within an **ecological niche.**

The interrelationships between the ecological niches make up a complex ecosystem. Whenever a major overlap exists between species or a foreign species is introduced, the ecological balance is upset and things get out of whack. A new ecosystem balance must be achieved for the natural system to work smoothly. If *biodiversity* is unbalanced and a species eliminated, then niches within the system must adjust. Some adjustments are minor, but more often a domino effect takes place, with all members of the ecosystem rebalancing. The groups that cannot make the change die out.

Biodiversity is a measure of the number of different individuals, species, and ecosystems within an environment.

Endemic Species

Plants and animals are scattered all across the globe. Some, like humans and cockroaches, are widespread, while others with very restricted territories are found only on a single river, lake, island, or mountaintop. These highly specialized organisms are called *endemic species* because they are unique to a specific region. Consequently, endemic species are usually studied within their geographic location.

In order for a species to survive for thousands or millions of years, the organisms must adapt to their habitat. Often, during the course of their development, they obtain ecological characteristics that help them thrive.

Dogs and cats live in many habitats (mostly around humans), but even in the wild, they are widely distributed. However, most species are restricted to certain areas because their ecological requirements are only found in a limited location. They might be able to do well in another region, but not if they have to travel huge distances (like across an ocean) to enjoy a better climate and food supply.

Endemic species, naturally occurring in only one area or region, are unique to that specific region.

You don't find polar bears in Arizona because they are endemic to polar regions. Plants and animals that need warmer climates or a longer growing season are restricted by environmental conditions like temperature and rainfall.

A species' geographical range often reaches across broad areas, depending on the environmental conditions. As long as the core habitat needs of a species are met, its members can survive.

The total area in which a plant, animal, insect, or other organism may travel in its lifetime is called its **range.**

The range of the once limitless American bison (which numbered in the millions) has been reduced to a tiny fraction of what it once was. Range loss and the massive slaughter that took place during the nineteenth century construction of the North American east-west railroad across their territory took a heavy toll on the buffalo.

HABITAT

The area in which an animal, plant, or microorganism lives and finds nutrients, water, sunlight, shelter, living space, and other essentials is called its *habitat.* Habitat loss, which includes the destruction, degradation, and fragmentation of habitats, is the primary cause of biodiversity loss.

Loss of habitat is perhaps the most important factor that affects species. Think of when a tornado or hurricane levels a town. Not only are homes and businesses destroyed, but water supplies, food crops, communications, and transportation methods may be lost. The area may become unlivable. Without the necessities that humans require to live or adapt to an environment, they have to find some place else to live.

When a species is continually crowded out of its habitat or its habitat is destroyed, it cannot reproduce, and its numbers drop. When this happens, a species is said to be *endangered.* Table 2-1 lists the top species in the world on the World Wildlife Fund's *Endangered Species List.*

Table 2-1 Endangered species.

Giant pandas	Atlantic salmon
Tigers	Pikas
Whales	Polar bears
Dolphins	Snow leopards
Coral	Rhinos
Marine turtles	Elephants
Great apes	Common sturgeon

Endangered species are those species threatened with extinction (like the Florida panther and the California condor).

Sometimes habitat loss is so severe or happens so quickly that it results in a species being eliminated from the planet. This happened to the dinosaurs. Scientists are still trying to decide what caused the mass extinction. There are a lot of theories, but except as seem in the occasional Hollywood movie, huge dinosaurs no longer roam the earth.

A species that is no longer living anywhere on the earth is said to be **extinct.**

Extinction takes place naturally, because for some species to succeed, others must fail. Since life began, about 99% of the earth's species have disappeared and, on several occasions, huge numbers died out fairly quickly. The most recent of these mass extinctions, about 65 million years ago, swept away the dinosaurs and many other forms of life. Though not extinct as a result of human actions, the dinosaurs are a good example of a large number of species that could not adapt to environmental changes.

Local extinction takes place when every member of a specific population in a specific area has died. Table 2-2 shows the number of species evaluated and those placed on the Endangered Species List in 2004 by the World Conservation Union (IUCN).

For the past forty years, the World Conservation Union's Species Survival Commission (SSC) has been ranking the conservation status of species, subspecies, varieties, and selected subpopulations worldwide in order to pinpoint groups threatened with extinction. To promote their conservation efforts, the SSC provides the most current, objective, scientifically-based information available on the status of globally threatened biodiversity. The groups assessed for the IUCN Red List of Threatened Species possess genetic diversity and provide the foundation for ecosystems. The collected data on species rank and distribution gives policy makers solid information with which to make informed decisions on preserving biodiversity at all levels.

A few species that have either approached or have completely gone extinct include *Gorilla beringei beringei* (African mountain gorilla), *Pyrenean ibex* (European goat), *Canis rufus floridianus* (Florida wolf), and *Hippopotamus madagascariensis* (Madagascan hippo). Global extinction happens when every member of a species has died. The passenger pigeon and the dodo are examples of globally extinct birds. Extinction is forever.

Table 2-2 Endangered species numbers.

	No. of species	No. of species evaluated (2004)	No. of threatened species (2004)
Mammals	5,416	4,853	1,101
Birds	9,917	9,917	1,213
Reptiles	8,163	499	304
Amphibians	5,743	5,743	1,770
Fishes	28,500	1,721	800
Insects	950,000	771	559
Mollusks	70,000	2,163	974
Crustaceans	40,000	498	429
Others	130,200	55	30
Mosses	15,000	93	80
Ferns	13,025	210	140
Gymnosperms	980	907	305
Dicotyledons	199,350	9,473	7,025
Monocotyledons	59,300	1,141	771
Lichens	10,000	2	2
TOTAL	**1,545,594**	**38,046**	**15,503**

Wetlands

One area that contains diverse populations is the nation's wetlands. Once considered unimportant wasteland, wetlands are now known to support important and extensive ecosystems. Wetland plants convert sunlight into plant material or biomass, which provides food to many different kinds of aquatic and land animals, supporting the aquatic food chain. Wetlands, often protected, also provide moisture and nutrients needed by plants and animals alike.

Wetlands are low, soggy places where land is constantly or seasonally soaked, or even partly underwater.

Wetlands are the transitional areas between land and marine areas. The water table is above, even with, or near the land's surface. Wetland soils hold large amounts of water and their plants are tolerant of occasional flooding. Wetlands can be coastal (estuaries or mangroves) or inland.

Wetlands can be swamps, bogs, peat lands, fens, marshes, and swamp forests. Two large wetland areas in the United States are the Florida Everglades and the Okefenokee Swamp in Georgia.

Wetlands are also habitats for many types of fish and wildlife. About 60% of the United States' major commercial fisheries use estuaries and coastal marshes as nurseries or spawning sites. Migratory waterfowl and other birds also rely on wetlands for homes, stopovers, and food.

Some wetlands form, over time, from lakes that have filled in, becoming wetlands or forest. This provides breeding and nursery areas for thousands of fish, shellfish, microorganisms, amphibians, reptiles, insects, invertebrates (like worms), and birds. Many species are forced into extinction as wetlands disappear. Table 2-3 lists the many ways wetlands are impacted by human activities.

Table 2-3 Wetland impacts.

Source	Wetland impact
Agriculture	Plowed for farmland; pesticide and fertilizer runoff pollutes wetlands
Cities	Filled for commercial development like homes, businesses, resorts, airports, and roads
River pollution	Carry wastewater, spilled oil, chemicals and industrial waste
Canals, ports, and harbors	Destroyed by dredging to widen areas for ships
Oil	Spilled and offshore oil drilling waste often washed back into coastal estuaries
Aquaculture	Mangroves and other areas destroyed to grow fish, shellfish, or shrimp
Dams	Restrict freshwater, nutrient, and sediment flow; also divert water for irrigation
Air pollution	Burning, industrial and automobile exhaust
Non-native species	Alien plant species push out native species, slow water flow and suffocate freshwater ponds and marshes

Some wetlands are created artificially. These are used for aquaculture or for filtering and purifying sludge and sewage from cities and industrial processes. Although these can serve as environmental buffers for plants and animals, they just don't support the same variety of life found in natural ecosystems.

Wetlands also play an important role in flood control by absorbing high flows and releasing water slowly. Along the coast, wetlands decrease storm surges from hurricanes and tropical storms. When wetlands are filled in for development or residential use, heavy rainfall flows off the land, taking rich topsoil and agricultural chemicals with it.

Wetlands serve as natural filters, allowing water to trickle down to the water table and underground reservoirs. Refilled and purified, groundwater is then stored in deep underground aquifers. When wetlands are lost, groundwater resupply or recharge may not take place. This causes drought in spring-fed areas. Loss of wetlands also increases the chance of groundwater contamination by removing the natural detoxification process.

Wetlands are home to more than 600 animal and 5,000 plant species. In the United States, nearly 50% of the animals and 25% of the plants on the endangered species list live in or rely on wetlands. One-half of the United States migratory birds are dependent on wetlands.

Internationally, wetlands are taking a hit as well. In Canada, which contains one-quarter of the world's wetlands, 15% of the wetlands have been lost. Germany and the Netherlands have lost over 50% of their wetlands.

Europe's largest wetland, the Danube Delta, stretches across 2,860 km in nine nations. Severe yearly erosion and pollution by nitrogen, phosphorus, pesticides, and other chemicals have caused a 50% drop in fish harvests in the past 20 years.

Mangroves in Africa, Asia, and Central and South America have been impacted or destroyed for firewood, rice fields, and aquaculture. India, the Philippines, and Thailand have lost over 80% of their mangrove forests. Bangladesh, Ghana, Pakistan, Somalia, Tanzania, Kenya, and Mozambique have all lost over 60% of their mangroves.

DEFORESTATION

From the time when early peoples switched from being hunters and gatherers to settling down and growing crops, humans have had a larger and larger impact upon the land's surface. The clearing of trees is commonly done to grow food crops. Because this has taken place over several thousand years, there are few forested areas untouched by change today.

Deforestation is the large-scale destruction of trees in a land area by disease, cutting, burning, flooding, erosion, pollution, or volcanic activity.

One of Earth's wonders is the diversity of plant life, including ancient forests. Although forests are complex ecosystems in themselves, they are also divided according to climate.

Forests are mainly divided into *temperate* (moderate climate) or *tropical* regions. Temperate forests are grouped into *conifers* (needle-leaf trees) like pine, spruce, redwood, cedar, fir, sequoia, and hemlock, while tropical forests contain flat-leaf trees. Old growth forests contain mostly conifers. Second- and third-growth forests contain trees of the same age and size as some of the younger old-growth trees, but have far fewer plant and animal species.

Temperature and rainfall are the major determiners of forest type. Some of these include:

- Temperate rain forest
- Tropical dry forest
- Tropical rain forest

A *temperate rain forest* is found in only a few special places around the world, such as the Pacific temperate rain forest on the west coast of North America. These temperate forests are often dominated by conifer trees adapted to wet climates and cool temperatures.

Located near the equator, a *tropical dry forest* has distinct rainy and dry seasons. Most tropical dry forest plants have adapted to withstand high temperatures and seasonal droughts.

The third major forest type receives and contains a lot of moisture. This *tropical rain forest*, found near the equator, harbors the richest diversity of terrestrial plant and animal species.

Today, the largest and most severe species loss is taking place in the tropical rain forests near the equator. Trees are being cut for grazing, farming, timber, and fuel. Ecologists have found that over 50% of the earth's original rainforests had been cleared by 1990. The remaining forests are lost by around 1.8 percent per year.

Old Growth Forests

Old-growth forests are found primarily in northern climates, although there are small caches of untouched trees in remote locations like Tasmania.

Old-growth forests are made up of trees that are often several thousand years old and take at least 100 years to regrow to maturity if cut down.

Old-growth forests are those that have never been harvested. They contain a variety of trees that are between 200 and 2000 years old. Forest floor leaf litter and fallen logs provide habitat for a complex mix of interdependent animals, birds, amphibians, insects, bacteria, and fungi that have adapted to each other over geological time. When the forests are cleared, the animals, birds, and insects that live under their protective cover are displaced or destroyed as well. Genetic uniqueness (biodiversity) of these affected species is permanently lost.

Redwood National Park and three California state parks contain some of the world's tallest trees: old-growth coastal *redwoods.* Living to be as much as 2,000 years old, they grow to over 91 meters tall. Spruce, hemlock, Douglas fir, and sword ferns create a multilevel tree canopy that towers above the forest floor. The local shrubs include huckleberry, blackberry, salmonberry, and thimbleberry. Park animals include raccoons, skunks, Roosevelt elk, deer, squirrels, minks, weasels, and black bears.

Redwoods live so long because they are particularly resistant to insects and fire. Scientists have discovered that giant redwoods contain high levels of bark tannins that protect them from disease. Additionally, redwoods can grow either from seeds or new sprouts from a fallen tree's root system, stumps, or burls. The economic clearing of fallen trees therefore limits new growth.

The most important factor in redwood survival is their biodiversity. Forest floor soils play a big role in tree growth. A healthy redwood forest includes a variety of tree species, as well as ferns, mosses, and mushrooms. These are important to soil regeneration. Fog from the nearby Pacific provides cooling and moisture for the trees.

The redwoods, however hardy, have lost a lot of ground. Of the original 1,950,00 acres of redwood forests growing in California, only 86,000 acres remain today. Three percent of these acres are preserved in public lands and 1 percent is privately owned and managed.

It comes as no surprise that when trees are lost through deforestation, biodiversity drops. When forests or grasslands are cleared and planted as a single cash or food crop, the number of species drops to one, plus a few weeds. But this is only part of the problem. Since forests support animal species with food and shelter, these species are also eliminated. Often, new species replace the originals, but generally the number of species goes down. When plant cover is removed, other area populations (mammals, birds, and insects) are greatly affected.

If a species cannot find new habitat or adapt to changed land use, it often becomes **extinct**. Its unique genetic information and position in the ecosystem are lost.

A species that has survived, while other similar ones have gone extinct, is called a *relict species.* A relict species, like the European white elm tree in western Siberia, may have had a wider range originally, but is now found only in particular areas. Other relict species, like horseshoe crabs or cockroaches, have survived unchanged since prehistoric times, even as other species become extinct.

HOTSPOTS

In 1988, British ecologist Norman Myers described the biodiversity *hotspot* idea. Although the tropical rain forests have the highest extinction rates, they aren't the only places at risk. Myers wanted to point out the resource problem facing ecologists. Since they were unable to save everything at once, they needed a way to identify areas with endangered species.

Globally, there are hundreds of species facing extinction because of habitat destruction and loss. Myers identified 18 high-priority areas where habitat cover had already been reduced to less than 10% of its original area or would be within 20 to 30 years. These regions make up only 0.5% of the earth's land surface, but provide habitats for 20% of the world's plant species facing extinction. Table 2-4 lists the world's top 10 hotspots as designated by *Conservation International.*

Two factors are weighed heavily in identifying a hotspot: (1) high diversity of endemic species; and (2) significant habitat impact and alteration from human activities.

Plant diversity is the biological basis for hotspot designation. In order to qualify as a hotspot, a region must support 1500 endemic plant species, 0.5% of the global total. Existing natural vegetation is used to assess human impact in a region.

An ecological region that has lost more than 70% of its original habitat is considered a **hotspot.**

Since plants provide food and shelter for other species, they are used in rating an area as a hotspot. Commonly, the diversity of endemic birds, reptiles, and

Table 2-4 Hotspots.

Hotspot	Endemic plants	Endemic vertebrates	Endemic plants per 100 km^2	Endemic vertebrates per 100 km^2	% Remaining natural vegetation
Madagascar and Indian Ocean Islands	9,704	771	16.4	1.3	9.9
Philippines	5,832	518	64.7	5.7	3.0
Sundaland	15,000	701	12.0	0.6	7.8
Atlantic Forest	8,000	654	8.7	0.6	7.5
Caribbean	7,000	779	23.5	2.6	11.3
Indo-Burma	7,000	528	7.0	0.5	4.9
Western Ghats and Sri Lanka	2,180	355	17.5	2.9	6.8
Eastern Arc Mountains & Coastal Forests	1,500	121	75.0	6.1	6.7

animals in hotspot areas is also extremely high. Hotspot animal species are found only within the boundaries of the hotspot, since they are often specifically adapted to endemic plant species as their main food source.

In recent hotspot designations by world conservation agencies, 25 biodiversity hotspots—containing 44% of all plant species and 35% all terrestrial vertebrate species in only 1.4% of the planet's land area—were listed. Hotspots target regions where the extinction threat is the greatest to the greatest number of species. This allows biologists to focus cost-effective efforts on critical species.

Endemic species have been isolated over a long period of geologic time. Islands, surrounded by water, have the most endemic species. In fact, many of the world's hotspots are islands. Mild environments, like Mediterranean regions, give shelter to the greatest diversity of species. Topographically different, mild environments, like mountain ranges, allow the greatest ecosystem diversity.

Several hotspots are tropical island archipelagos, like the Caribbean and the Philippines, or big islands, like New Caledonia. However, other hotspots are continental islands isolated by surrounding deserts, mountain ranges, and seas.

Peninsulas are key regions for hotspots. They are similar to islands and some, like Mesoamerica, Indo-Burma, and the Western Ghats in India, were islands at some time in the past. Other hotspots are landlocked islands isolated between high mountains and the sea. The Andes Mountains, which separate South America from north to south, are an impassible barrier to many species. On the western coast, the lowlands form a thin, isolated ecosystem from the eastern side of the continent.

The Cape Floristic Province in South Africa is isolated by the extreme dryness of the Kalahari, Karoo, and Namib deserts, and large rivers like the Zambezi and the Limpopo.

Why Are Hotspots Fragile?

Island ecosystems are particularly fragile because they are never or are only rarely exposed to outside influences. Ecologists have found that most extinct species were island species and not widely spread. They were confined to certain habitat areas, whether an island or an isolated part of a continent, that supported their existence. Once a one-of-a-kind population is gone, the species is lost forever.

Isolated species lose their defenses over time, because they are only exposed to limited number of specific species. When they have to compete with new, previously unknown species, they can't adapt fast enough. This is especially true if the new species is highly competitive and adaptable.

For example, large extinct birds like the moa and dodo, which had no predators on the remote Australian continent, lost their ability to fly. When humans and other predators arrived, these birds were easy targets and quickly dropped in numbers.

Since many of the global hotspots are beautiful and unique, humans have been drawn to their natural diversity throughout history. Ecosystems and landscapes were changed, first by hunter-gatherers, then by farmers and herdsmen, and most extensively by the global growth and sale of agricultural crops. During the past 500 years, many species were harvested to the last individual. Currently, growing human populations in world hotspots add to their decline by the introduction of nonnative species, illegal trade in endangered species, industrial logging, slash-and-burn agricultural practices, mining, and the construction of highways, dams, and oil wells. Eleven hotspots have lost at least 90% of their original natural vegetation, and three of these have lost 95%.

Today, the world's regions considered the "hottest of the hot spots" are Madagascar and the Indian Ocean Islands, the Philippines, Sundaland, the Atlantic Forest, and the Caribbean. These five hotspots have the most unique biodiversity to lose, and are at extreme risk of losing it without immediate and effective conservation.

Conservation

Since hotspots possess the highest concentrations of distinctive biodiversity on the planet, they are also at the greatest risk. It is essential to conserve hotspot species in order to prevent a domino effect of species extinction. A strategy to protect global biodiversity, especially in hotspots, is an important next step. In order to have a positive impact on currently threatened species in hotspots, the following characteristics shown in Fig. 2-2 must be identified and considered. For conservation capability to be realized in various threatened areas, the knowledge and tools needed to conserve the hotspots must be present. Steady monitoring is important since the political, social, and biological environments of hotspots change. For this reason, an early warning system is under development by the Center for Applied Biodiversity Science at Conservation International (CI).

The Biodiversity Early Warning System is supported by several monitoring programs that are being done by CI, its partner countries and organizations, and hotspot experts. These programs assess species, habitats, and socioeconomic factors to identify problem regions before they are too far gone.

Species data are being collected and rapid biological assessments of poorly known terrestrial, freshwater, and marine ecosystems made. Sponsored expeditions send small teams of international and host-country biologists to the field for three to four weeks to conduct the hotspot assessments. Additionally, a network of field stations in all of the world's major tropical ecoregions is being established to monitor biodiversity.

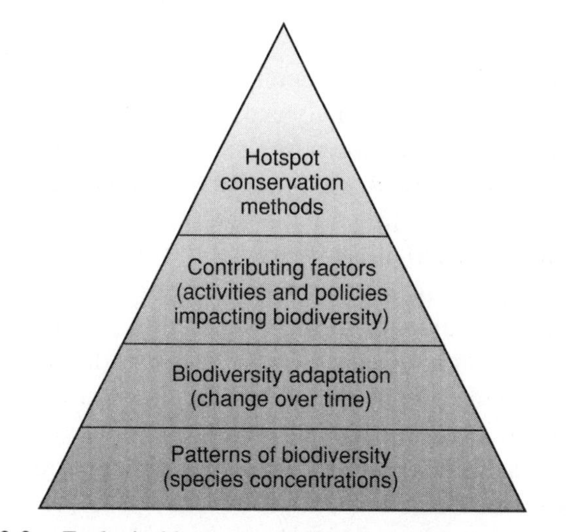

Fig. 2-2. Ecological hotspots contain unique and diverse species.

Remote Sensing

When environmentalists study nature by making careful observations and measurement, collecting data directly or in place, it is called *in situ* data collection. This is the same thing as when children collect fallen acorns, rocks, or colorful autumn leaves. It is sometimes called *field sampling*. Field sample collection and analysis are done at the sample site. However, when too many people collect samples in the same place, the area can get stomped down or otherwise disturbed. This is why national parks and forests ask visitors to stay in approved areas and on trails. They are trying to limit human contact with and disturbance of protected areas.

When scientists want to study an area and avoid disturbing the environment, they use *remote sensing*, such as aircraft, high-altitude balloons, and satellites. Much of this technology was developed as stealth imaging during wartime or for space exploration by NASA. Today, the environmental benefits from this sophisticated technology, some of which can accurately picture an object to within a meter, are diverse.

> The measurement or gathering of information about some characteristic of an object, area, or event by a distant recording device (not in contact with the study sample) is known as **remote sensing.**

One example of a remote sensing instrument is the ASTER (Advanced Spaceborne Thermal Emission and Reflection Radiometer). This imaging equipment is flying on *Terra,* a satellite launched in December 1999 as part of NASA's Earth Observing System (EOS). ASTER is a cooperative effort between NASA and the Japanese Ministry of Economy, Trade, and Industry (METI) and the Earth Remote Sensing Data Analysis Center. ASTER data is used to draw detailed maps of land surface temperature, emissivity, reflectance, and elevation. The EOS platforms are part of NASA's Earth Science Enterprise, whose goal is to obtain a better understanding of the interactions between the biosphere, hydrosphere, lithosphere, and atmosphere.

Various instruments involved in remote sensing can gather information on many different characteristics including, temperature, chemistry, photosynthetic ability, moisture content, and location. Table 2-5 provides a sampling of the geographical and ecological characteristics that can be studied with different types of remote sensing.

Habitat hotspots are also similarly studied with satellite imagery and aerial photography. These methods help to produce maps of existing vegetation or changes in vegetation over time. This noninvasive imaging is particularly impor-

Table 2-5 Remote sensing.

Biological and physical characteristics	Remote sensing systems
x, *y* Location	Aerial photography, Landsat, SPOT HRV, Space Imaging IKONOS, ASTER, Radarsat, ERS-1,2 microwave, ATLAS
z Topographic / depth measurement	Aerial photography, TM, SPOT, IKONOS, ASTER, Radarsat, LIDAR systems, ETM
Vegetation *(chlorophyll concentrations, biomass, water content, absorbed photosynthetic radiation, phytoplankton)*	Aerial photography, TM, SPOT, IKONOS, ETM, Radarsat, TM Mid-IR, SeaWiFS, AVHRR, IRS-1CD
Surface temperature	GOES, SeaWiFS, AVHRR, TM, Daedalus, ATLAS, ETM, ASTER
Soil moisture	ALMAZ, TM, ERS-1,2 Radarsat, Intemap Star 3i, IKONOS, ASTER
Evapotranspiration	AVHRR, TM, SPOT, CASI, ETM, MODIS, ASTER
Atmosphere *(chemistry, temperature, water vapor, wind speed/direction, energy input, precipitation, clouds, and particulates*	GOES, UARS, ATREM, MODIS, MISR,CERES
Reflectance	MODIS, MISR, CERES
Ocean *(color, biochemistry, phytoplankton, depth)*	POPEX/POSEIDON, Sea WiFS, ETM, IKONOS, MODIS, MISR, ASTER, CERES
Snow and Sea ice *(distribution and characteristics)*	Aerial photography, AVHRR, TM, SPOT, Radarsat, SeaWiFS, ICONOS, ETM, MODIS, ASTER
Volcanoes *(temperature, gases, eruption characteristics)*	ATLAS, MODIS, MISR, ASTER
Land Use	Aerial photography, AVHRR, TM, SPOT, IRS-ICD, Radarsat, Star 3i, IKONOS, MODIS

tant in places that are already in trouble. When a sensor or satellite passively records the electromagnetic energy reflected or emitted from vegetation, it gives information without the need for physical samples. For greater detail of what's happening, like potato blight or pine bark beetle invasion, local farmers and scientists can take samples of the affected plants or trees. Remote sensing results can help direct their efforts and guide them straight to problem areas.

Using images taken at different times, maps are made that show at-risk areas and ongoing deforestation. These maps are important in that they show the location and size of remaining habitats.

Activities and other regional biodiversity threats, such as forest fires, illegal logging, and construction development, can also be detected with remote sensing. These assessments make it possible for researchers and policy makers to develop models that help conservationists predict and prevent potential mass extinctions.

Identifying biodiversity hotspots requires identification of unique species and areas that face the greatest risk of extinction. In order to set targets and avoid extinctions, scientific evaluations of the threatened species' status must be made and trends identified. Also, critical regions must be protected. A solid foundation of hotspot conservation and scientific research requires three parts: (1) remote sensing; (2) laboratory analysis; and (3) field (*in situ*) sampling.

Solutions

A variety of conservation methods are needed to protect hotspots' biodiversity. These range from the creation of protected areas to the establishment of alternatives like *ecotourism.* Hotspot conservation also requires educating people at the local and national levels through conservation policies and awareness programs. It includes working with international corporations to improve business practices that protect against even more biodiversity loss.

Strengthening existing conservation efforts not only helps lessen potential climate destabilization, but also offers greater resilience against weather disasters that threaten both people and habitat. Creating protected areas and conservation corridors, as well as establishing and improving the management of roughly 55 million acres of parks and protected areas in biodiversity hotspots and high-biodiversity wilderness areas, is critical to ensuring biodiversity.

However, species' habitat ranges adjust in response to climate change. This impacts environmentalists' ability to protect them in existing parks. Shifting of range boundaries due to temperature increases has been happening for the past 75 years. To reduce the extinction risk associated with global warming, conservation methods must solve this problem.

The challenge of conserving biodiversity in hotspots and throughout the world is so great that no one country or organization can do it alone. Everyone has to work together at all levels, from working with a single expert to protect an endangered species to aid for countries like Madagascar to help fund national conservation projects.

Many medicines are derived from plants and fungi. Little known species in today's hotspots may hold the key to treatments for human disorders like arthritis and cancer. It is important to keep these potential cures from slipping away due to environmental neglect.

As the world's population continues to climb, environmental issues will become critical for more and more species, including our own. Remember, extinction is forever.

Quiz

1. The world's oceans make up what percentage of the planet's biosphere?
   (a) 45 percent
   (b) 60 percent
   (c) 75 percent
   (d) 99 percent

2. Which diverse ecosystem also helps in flood control by absorbing high flow and releasing water slowly?
   (a) High plains
   (b) Wetlands
   (c) Arctic
   (d) Rocky Mountains

3. The vertical range that contains the biosphere is roughly
   (a) 5,000 meters high
   (b) 10,000 meters high
   (c) 20,000 meters high
   (d) 40,000 meters high

4. The effect upon a species or environment that protects its numbers without causing long-term decline is known as
   (a) extinction
   (b) sustainable use
   (c) clear cutting
   (d) evolutionary use

5. The total area in which a plant, animal, insect, or other organism travels in its lifetime determines its
   (a) itinerary
   (b) life span
   (c) personality type
   (d) range

6. A hotspot is an ecological region that has lost
   (a) 20 percent of its original habitat
   (b) 45 percent of its original habitat
   (c) 60 percent of its original habitat
   (d) more than 70 percent of its original habitat

7. When a species like the dodo bird becomes extinct, it is
   (a) hibernating for the winter
   (b) gone forever
   (c) in remission
   (d) gone for 10 years, then returns

8. Madagascar and the Indian Ocean Islands, the Philippines, Sundaland, the Atlantic Forest, and the Caribbean are all considered
   (a) hotspots
   (b) expensive vacation spots
   (c) sustainable use areas
   (d) arid regions

9. A species that has survived while other similar ones have gone extinct is called a
   (a) relict species
   (b) barren species
   (c) anthropogenic species
   (d) gravimetric species

10. Aircraft, high-altitude balloons, and satellites are all used in
    (a) birthday parties
    (b) acrobatic air shows
    (c) remote sensing
    (d) field testing

CHAPTER 3

Atmosphere

Next to the ground we walk on, the *atmosphere* is the easiest to identify. Just look up. The atmosphere is the realm of kites, hot air balloons, and spaceships. It starts at ground level and then goes straight up, getting thinner and colder, until it finally ends at the edge of space. The atmosphere protects us and makes our world livable. Without it, the Earth would be no better off than Venus or Mars.

Most people think of the weather when they hear the word atmosphere. The weather affects crops, habitat, recreation, and outdoor planning of all kinds. It affects the rush hour commute as well as the weekend beach trip. The weather makes life pleasant or miserable, depending on conditions.

The atmosphere also provides the air (oxygen) we breathe. Humans can survive for about 28 days without food and 3 days without water—but only three to four minutes without air. For this reason, it is the single most important resource we have. All other environmental concerns must tie into the preservation of our atmosphere.

Atmospheric variables include temperature, pressure, and water vapor. The gradients and interactions of these variables and how they change over time are also important.

> A **meteorologist** is a person who studies the weather and its atmospheric patterns.

Along the Texas Gulf Coast, the common saying is, "If you don't like the weather, just wait 15 minutes and it will change!" True of many places, the weather can change suddenly, especially at the turn of the seasons. Temperature drops of 30°F in the two hours preceding a cold front are possible.

What is the atmosphere made of? Air? Water? Smoke? The answer depends on what is happening at that moment. Most of the time you can't even see the atmosphere unless there is fog, rain, snow, clouds, wind, or some other atmospheric player.

Most, if not all, of the earth's atmosphere was missing at the beginning. The original atmosphere was a lot like a solar nebula and similar to the gaseous planets. That atmosphere was gradually replaced by compounds emitted from the crust or the impacts of meteoroids and comets loaded with volatile matter.

The atmosphere contains oxygen produced almost solely by algae and plants. The atmosphere's present composition is nitrogen, oxygen, and other miscellaneous gases. While our current atmosphere is an oxidizing atmosphere, the primordial atmosphere was a reducing atmosphere. That early atmosphere had little, if any, oxygen. It was more a product of volcanic blasts.

Today, there are several layers that make up our protective atmosphere. The lower layers have a larger percentage of total oxygen, while upper layers have much less.

Composition

The atmospheric gases blanketing the Earth exist in a mixture. This mixture is made up (by volume) of about 79% nitrogen, 20% oxygen, 0.036% carbon dioxide, and trace amounts of other gases.

The atmosphere is divided into four layers according to the mixing of gases and their chemical properties, as well as temperature. The layer nearest the earth is the *troposphere,* which reaches an altitude of about 8 km in polar regions and up to 17 km around the equator. The layer above the troposphere is the *stratosphere,* which reaches to an altitude of around 50 km. The *mesosphere* reaches up to approximately 90 km and lies above the *stratosphere.* Finally, the *thermosphere,* or *ionosphere,* is still further out and eventually fades to black in outer space. There is very slight mixing of gases between the layers.

The four layers of the earth's atmosphere are illustrated in Fig. 3-1. The location of the ozone layer is also shown.

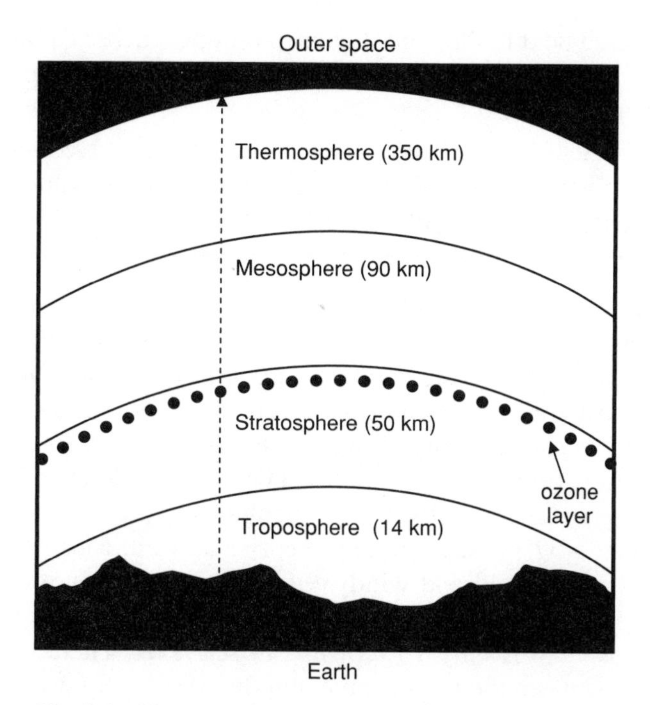

Fig. 3-1 The atmosphere is divided into four main layers.

TROPOSPHERE

The lowest of the atmospheric layers, the troposphere, extends from the earth's surface up to about 14 km in altitude. Virtually all human activities occur in the troposphere. Mt. Everest, the tallest mountain on the planet, is only about 9 km high.

Nitrogen and oxygen make up the majority of the Earth's gases, even in the higher altitudes. But it's the atmospheric level closest to the Earth where everything is perfect to support life. At this level, living organisms are protected from harmful cosmic radiation showers that constantly assault the earth's atmosphere.

This active layer is called the *troposphere*. If you have ever survived a hurricane or tornado, you know that the troposphere is an active place. It is the atmospheric layer where all the weather we experience takes place. Rising and falling temperatures, as well as circulating masses of air, keep things lively. Air pressure also adds to the mix.

When measured next to the other layers, the troposphere is a fairly slim layer, extending only 14 km up from the Earth's surface. Located within this thin layer, weather alert material is born.

The **troposphere** is where all the local temperature, pressure, wind, and precipitation changes take place.

The warmest portions of the troposphere are found at the lowest altitudes. This is because the earth's surface absorbs the sun's heat and radiates it back into the atmosphere. Commonly, temperature decreases as altitude increases.

However, there are some exceptions. Depending on wind currents and the like, mountain ranges can cause lower areas in the troposphere to have just the opposite effect. Temperatures actually increase with altitude. This is called a *temperature inversion.* Generally, the temperatures at the top of the troposphere have lows around –57°C. The wind speeds rise as well, causing the upper tropospheric limits to be cold and windy. Of course, there is not enough oxygen to breath at those heights, so it doesn't really affect us. The air pressure at the top of the troposphere is only 10% of that at sea level. There is a thin "shock absorber" zone between the troposphere and the next layer (stratosphere) called the *tropopause.* This is a gradual mixer zone between the two layers.

STRATOSPHERE

Above the troposphere is the *stratosphere,* where air flow is mostly sideways. There is a gradual change from the troposphere to the stratosphere, which starts at around 14 km in altitude. The stratosphere extends from 14 km to around 50 km. Most commercial aircraft travel takes place in the lower part of the stratosphere. Military aircraft travel at much higher altitudes: Some classified stealth aircraft are thought to graze the boundary of the mesosphere and beyond. NASA's Space Shuttle generally travels to altitudes between 160 and 500 km.

Although the temperature in the lower stratosphere is cold and constant, hovering around at –57°C, there are strong winds in this layer that are part of specific circulation patterns. Extremely high and wispy clouds can form in the lower stratosphere. In general, there are no major weather formations that take place regularly in the stratosphere.

The stratosphere has an interesting feature from midlevel on up. Its temperature jumps up suddenly with an increase in altitude. Instead of a frosty –57°C, the temperature jumps up to a warm 18°C around 40 km in altitude in the upper stratosphere. This temperature change is due to increasing ozone concentrations, which absorb ultraviolet radiation.

The melding of the stratosphere upward into the mesosphere is called the *stratopause.*

Ozone

Ozone is one of our atmospheric bodyguards. Even small amounts have an important role in protecting planetary life. Concentrated in a thin layer in the upper stratosphere, atmospheric ozone is an exceptionally reactive form of oxygen. It is found in the stratospheric layer, around 15 to 30 km above the Earth's surface. The ozone layer is largely responsible for absorbing most of the sun's ultraviolet (UV) radiation. Most importantly, it absorbs the fraction of ultraviolet light called UVB.

Ultraviolet radiation with a wavelength between 200 and 400 nanometers (nm) is usually divided into three main ranges of the spectrum. Table 3-1 shows these different ultraviolet categories and their characteristics.

Ultraviolet radiation is a bad, bad thing! It causes breaks in the body's nuclear proteins, leaving the door open for cancers and other health issues to get a foothold. UVB has been connected with many serious health problems, like different kinds of skin cancer and cataracts. It is also harmful to certain crops, materials, and marine organisms.

Ozone is much less widespread than normal oxygen. The formation of the ozone layer is a tricky matter. Out of every 10 million air molecules, about 2 million are normal oxygen and only three are ozone molecules. Instead of two atoms of oxygen like normal oxygen molecules (O_2), ozone (O_3) contains three oxygen atoms. Ozone has a distinctive odor and is blue in color. Regular oxygen has no odor or color.

Only through the production of atmospheric oxygen can ozone form to block ultraviolet radiation from reaching the earth's surface and the plants and animals

Table 3-1 UV radiation types.

UV radiation type	Wavelengths (nm)	Effects on life
UVA	400–320	Fairly safe: tanning but not burning
UVB	320–290	Harmful: sunburn; skin cancer; other problems
UVC	290–200	Very harmful, but mostly absorbed by ozone

living there. In the past 30 years, there has been intense concern over decreased ozone levels. This big problem must be solved if we want to go on enjoying the outdoors and growing food in the centuries to come!

Ozone molecules are constantly created and destroyed in the stratosphere. The amount is relatively constant and only affected naturally by sunspots, seasons, and latitude. Atmospheric scientists have studied and recorded these annual and geographical fluctuations for years. There are usually yearly cyclic downturns in ozone levels, followed by a recovery. However, as our population increases along with industrialization, global atmospheric changes are taking place as well.

There are other bit players in the stratosphere. These are collections of droplets called *polar stratospheric clouds.* These unique cloud-like condensations of trace chemicals condense in the cold (–80° C or below) Southern Hemisphere's winter. During this time, the atmospheric mass above Antarctica is kept cut off from exchanges with midlatitude air by prevailing winds known as the *polar vortex.* This leads to very low temperatures, and in the cold and continuous darkness of the season, polar stratospheric clouds are formed that contain chlorine. These clouds are often a combination of water and nitric acid. Although very weak, they affect the chemistry of the lower stratosphere by interacting with nitrogen and chlorine and providing surface area for other reactions to take place.

Chlorine and nitrogen interact in the atmosphere in ways that form chlorine nitrate. This compound does not interact with ozone or atomic oxygen, so it serves as a storage tank for chlorine in the environment. When polar stratospheric clouds form, they tie up stratospheric nitrogen so that it is not available to bind extra chlorine. This interference allows atomic chlorine to interact with ozone and destroy it.

The Antarctic winter (May to September) and the many months of very cold temperatures maintain this interference long enough for ozone levels to drop steeply. As warmer spring temperatures arrive, the combination of returning sunlight and the presence of polar stratospheric clouds leads to the splitting of chlorine into highly ozone-reactive radicals that break ozone apart into individual oxygen molecules. A single molecule of chlorine can break down thousands of molecules of ozone.

Ozone Depletion

For the past 50 years, *chlorofluorocarbons* (CFCs) held the answer to lots of material problems. They were stable, nonflammable, not too toxic, and cheap to produce. They had a variety of uses including applications as refrigerants, solvents, and foam-blowing agents.

Chlorine has been used for everything from disinfecting water to serving as solvents (methyl chloroform and carbon tetrachloride) in chemistry labs.

Unfortunately, these compounds are not so good for the atmosphere. They don't just break down and disappear. They hang around. This lingering characteristic allows them to be carried by winds into the stratosphere. The net effect is to destroy ozone faster than it is naturally created. Roughly 84% of stratospheric chlorine comes from manmade sources, while only 16% comes from natural sources.

Unfortunately, CFCs break down only by exposure to strong UV radiation. When that happens, CFCs release chlorine. Scientists have found that one atom of chlorine can destroy over 100,000 ozone molecules. As CFCs decay, they release chlorine and damage the ozone layer.

Thirty years ago, researchers started looking at the effects of different chemicals on the ozone layer. They looked at chlorine and its surface origins. Chlorine from swimming pools, industrial use, sea salt, and volcanic eruptions were found to be minor factors in ozone depletion. Because they mixed with atmospheric water first and quickly precipitated out of the troposphere, they didn't reached the stratosphere.

Stable CFCs, however, act differently. There wasn't anything in the lower atmosphere to cause them to break down. The outcry from scientists and environmentalists over ozone depletion led to a 1978 ban on the use of aerosol CFCs in several countries, including the United States.

In 1985, since other types of chlorine compounds were still being used, the policy of the Vienna Convention was adopted to gather international cooperation and reduce the number of all CFCs by half. It's important to remember that just because CFCs were banned doesn't mean that long-lived chemicals will disappear immediately from the atmosphere. Until CFCs degrade to negligible levels, the annual South Polar ozone hole will keep appearing for many years to come.

The annual "hole" or thinning of the ozone layer over Antarctica was first noticed in 1985. This area of extremely low ozone levels showed drops of over 60% during bad years. No corresponding hole was found over the Arctic.

The European Space Agency's Envisat Earth observation satellite records the arrival of the annual opening of the hole in the Earth's ozone layer. Since it first appeared, satellites have been tracking its arrival and shape for years, and scientists have gotten good at predicting the conditions that create the opening.

The ozone hole usually shows up around the first or second week of September and then closes up again in November or December. When higher temperatures around the South Pole mix ozone-rich air into the region—causing the winds surrounding the South Pole to weaken—ozone-poor air inside the vortex is mixed with ozone-rich air outside it.

Envisat contains an instrument called the Scanning Imaging Absorption Spectrometer for Atmospheric Cartography that provides new atmospheric data on the ozone layer every day.

This information presents a good way of eventually identifying long-term ozone trends.

Subsequent research found that some ozone thinning (though not as severe) also took place over the latitudes of North America, Europe, Asia, Australia, South America, and much of Africa. It became obvious that ozone decreases were a global concern.

In 1992, with new information on the ever-shrinking ozone layer, developed countries decided to totally stop production of halons by 1994 and CFCs by 1996. *Halons* are compounds in which hydrogen atoms of a hydrocarbon are replaced by bromine or fluorine. The halons are used as fire-extinguishing agents, both in built-in systems and in handheld fire extinguishers.

Halons cause ozone depletion because they contain bromine, which is a lot stronger than CFCs in destroying ozone. Halons are also very stable and break down slowly once formed. The United States' halon production was stopped by December 31, 1993, because of its contribution to ozone depletion.

This course of action turned out to be what turned the tide in falling ozone levels. Levels of inorganic chlorine in the atmosphere stopped increasing in 1997–1998, and stratospheric chlorine levels peaked and are no longer rising. If nothing happens to change this trend, natural ozone recovery should mend the ozone layer in about 50 years.

However, we need to hold off on celebrations until we see whether this actually takes place. Newly developed industrial chemicals must be watched as well. Whether the ozone layer has begun to recover is a hotly debated subject; scientists will know for sure only with time and continued monitoring.

MESOSPHERE

Above the stratosphere is the *mesosphere,* a middle layer separating the lower stratosphere from the inhospitable thermosphere. Extending from 80 to 90 km and with temperatures around –101°C, the mesosphere is the intermediary of the earth's atmosphere layers.

THERMOSPHERE

The changeover from the mesosphere to the *thermosphere* layer begins at a height of approximately 80 km. The thermosphere is named because of the return to rising temperatures that can reach an amazing 1,982°C. The different temperature ranges in the thermosphere are affected by high or low sun spot and solar flare activity. The more active the sun is, the higher the heat generated in the thermosphere.

Extreme thermospheric temperatures are a result of UV radiation absorption. This radiation enters the upper atmosphere, grabbing atoms from electrons and creating positively charged ions. This ionization gives the thermosphere its other name: the *ionosphere.* Because of ionization, the lowest area of the thermosphere absorbs radio waves, while other areas reflect radio waves. Since this area drops and disappears at night, radio waves bounce off the thermosphere. This is why far distant radio waves can often be received at night.

Electrically charged atoms build up to form layers within the thermosphere. Before modern satellite use, this thermospheric deflection was important for long distance radio communication. Today, radio frequencies that can pass through the ionosphere unchanged are selected for satellite communication.

The thermosphere is where the *aurora* resides. The *Aurora Borealis* and *Aurora Australis,* known as the Northern Lights and Southern Lights, respectively, are found in the thermosphere. When solar flares impact the magnetosphere and pull electrons from their atoms, they cause magnetic storms near the poles. Dazzling red and green lights are emitted when scattered electrons reunite with atoms, returning them to their original state.

Even higher, above the auroras and the ionosphere, the gases of this final atmospheric layer begin to dissipate. Finally, several hundred kilometers above the earth, they fade off into the vastness of space.

Jet Stream

When watching the evening weather report, chances are good that you will hear something about the *jet stream.* This speedy current is commonly thousands of kilometers long, a few hundred kilometers wide, and only a few kilometers thick. Jet streams are usually found somewhere between 10 to 14 km above the earth's surface in the troposphere. They blow across a continent at speeds of 240 km per hour, usually from west to east, but can dip northward and southward depending on atmospheric conditions.

The **jet stream** is a long, narrow current of fast-moving air found in the upper levels of the atmosphere.

Air temperature differences cause jet streams. The bigger the temperature differences, the stronger the pressure differences between warm and cold air. Stronger pressure differences create stronger winds. This is why jet streams fluctuate so much in speed.

During the winter months, polar and equatorial air masses form a sharp temperature contrast, causing an intense jet stream. Stronger jet streams push farther south in the winter. However, during summer months, when the surface temperature changes are less, the jet stream winds are weaker. The jet stream blows farther north.

Airline companies like the jet stream since it helps to push aircraft along. It's like walking with the wind instead of against it. Going with the flow is easier.

Pressure

Bakers that live in the mountains have to consider the pressure of air when creating light cakes and soufflés. The decrease in pressure at high altitudes (over 6,000 meters) changes the baking process from that of sea-level baking. That is why some cake mixes give different directions for high-altitude baking, to make up for the difference of pressure on the rising cake.

Air pressure is the force applied on you by the weight of air molecules.

Although air is invisible, it still has weight and takes up space. Since air molecules float freely in the vastness of the atmosphere, they become pressurized when crowded into a small volume. The downward force of gravity gives the atmosphere a pressure or a force per unit area. The Earth's atmosphere presses down on every surface with a force of 1 kilogram per square centimeter.

Weather scientists measure air pressure with a *barometer*. Barometers are used to measure air pressure at a particular site in centimeters of mercury or *millibars*. A measurement of 76 centimeters of mercury is equivalent to 1013.25 millibars.

Air pressure can tell us a lot about the weather. If a *high-pressure* system is coming, there will be cooler temperatures and sunny skies. If a *low-pressure system* is moving in, then look for warmer temperatures and thunder showers.

Atmospheric pressure falls with increasing altitude. A pillar of air in cross section, measured from sea level to the top of the atmosphere, would weigh approximately 14.7 pounds per square inch (psi). The standard value for atmospheric pressure at sea level is equal to:

1 atm = 760 mm Hg (millimeters of mercury) = 1013 millibars =
14.7 psi (pounds force per square inch) = 1013.25 hPa (hectopascals)

On weather maps, changes in atmospheric pressure are shown by lines called *isobars.* An isobar is a line connecting areas of the same atmospheric pressure. It's very similar to the lines connecting equal elevations on a topographical map of the earth's surface.

Wind

Winds are a product of atmospheric pressure. Pressure differences cause air to move. Like fluids, air flows from areas of high pressure to areas of low pressure. Meteorologists predict winds by looking at the location and strength of regional high- and low-pressure air masses. If the changes are small, the day is calm. However, if pressure differences are high and close together, then strong winds whip up.

In 1806, Admiral Sir Francis Beaufort of the British Navy came up with a way of describing wind effects on the amount of canvas carried by a fully rigged frigate. This scale, named the *Beaufort wind scale,* has been updated for modern use. Wind speeds are described according to their effects on nature and surface structures. Table 3-2 lists the different wind effects using Beaufort numbers.

> The **wind chill factor** measures the rate of heat loss from exposed skin to that of surrounding air temperatures.

Wind chill happens when winter winds cool objects down to the temperature of the surrounding area—the stronger the wind, the faster the rate of cooling. For example, the human body is usually around 36°C in temperature, a lot higher than a cool Montana day in November. Our body's heat loss is controlled by a thin insulating layer of warm air held in place above the skin's surface by friction. If there is no wind, the layer is undisturbed and we feel comfortable. However, if a sudden wind gust sweeps by, we feel chilled. Our protective warm

Table 3-2 Beaufort scale.

Beaufort scale no.	Wind speed (kilometers per hour)	Wind	Sign
0	<1	Calm	Smoke rises vertically
1	1–3	Light air	Smoke drifts
2	6–11	Light breeze	Leaves rustle
3	12–19	Gentle breeze	Small twigs rustle
4	20–29	Moderate breeze	Small branches move
5	30–38	Fresh breeze	Small trees move
6	39–50	Strong breeze	Large branches move
7	51–61	Moderate gale	Whole trees move
8	62–74	Fresh gale	Twigs break off trees
9	75–86	Strong gale	Branches break
10	87–101	Whole gale	Some trees uprooted
11	102–119	Storm	Widespread damage
12	>120	Hurricane	Severe damage

air cushion is blown away and has to be reheated by the body. See Table 3-3 to get an idea of the wind chill equivalent temperatures at different wind speeds.

Relative Humidity

In the hot, sticky, suffocating days of summer, you always hear people saying, "It's not the heat; it's the humidity." What is humidity, anyway?

Humidity is the amount of water vapor in the air. When it has been raining and the air is saturated, there is 99 to 100 percent humidity.

> **Relative humidity** is the relationship between the air's temperature and the amount of water vapor it contains.

Table 3-3 Wind chill.

Wind speed (kph)	Temperature (°Celsius)											
	–15°C	–10°C	–5°C	0°C	5°C	10°C	15°C	20°C	25°C	30°C	35°C	40°C
0	–15	–10	–5	0	5	10	15	20	25	30	35	40
5	–18	–13	–7	–2	3	9	14	19	25	31	36	41
10	–20	–14	–8	–3	2	8	13	19	25	31	37	42
30	–24	–18	–12	–6	1	7	12	18	25	32	38	43
50	–29	–21	–14	–7	0	6	12	18	25	32	38	44
70	–35	–24	–15	–8	–1	6	12	18	25	32	38	44
90	–41	–30	–19	–9	–2	5	12	18	25	32	38	45

At any specific temperature, there is a maximum amount of moisture that air can hold. For example, when the humidity level is forecast to be 75%, it means that the air contains three-quarters of the amount of water it can hold at that temperature. When the air is completely saturated and can't hold any more water (100% humidity), it rains.

The air's ability to hold water is dependent on temperature. Hotter air holds more moisture. So you get the double whammy of heat *and* humidity. It's like adding sugar to a hot drink compared to a cold drink. The sugar dissolves quickly in the hot drink, but more slowly in the cold drink. The cold drink usually has undissolved sugar crystals in the bottom of the glass. Heat a cold drink and the sugar dissolves.

This temperature-dependent, moisture-holding capacity of air contributes to the formation of all kinds of clouds and weather patterns.

Convection

In atmospheric studies, convection refers to vertical atmospheric movement. As the earth is heated by the sun, different surfaces absorb different amounts of energy, convection occurs, and surfaces are quickly heated.

Hot air rises. This is why it's warmer in the upper floors of a house or building in the summer. The same is true of the atmosphere. As the Earth's surface warms, it heats the overlying air, which gradually becomes less dense than the surrounding air and begins to rise.

A **thermal** is the mass of warm air that rises from the surface upward.

Thunderstorms are the result of convection. When warm air rises in some areas and sinks in others, convection is established. Thunderstorms feed on unstable atmosphere conditions. They like warm air on the ground with cold air above. This is the reason why more thunderstorms happen in the spring and summer. During those seasons, the ground is warmed by the sun, followed by cold winter air blowing in on top of it. In the fall and winter, the air and surface are either cooling off or are already cold.

The concern over global warming includes its impact on convection and the formation of severe weather. Fronts form when thermal heating forces air currents to circulate.

FRONTS

Atmospheric fronts are conflicts between air masses. Depending on the air masses involved and which way the fronts move, fronts can be warm, cold, stationary, or occluded.

In the case of a *cold front,* a colder, denser air mass picks up the warm, moist air ahead of it. As the air rises, it cools and its moisture condenses to form clouds and rain. Since cold fronts have a steep slope face, strong uplifting is created, leading to the development of showers and severe thunderstorms. Fig. 3-2 illustrates the air flow of a cold front.

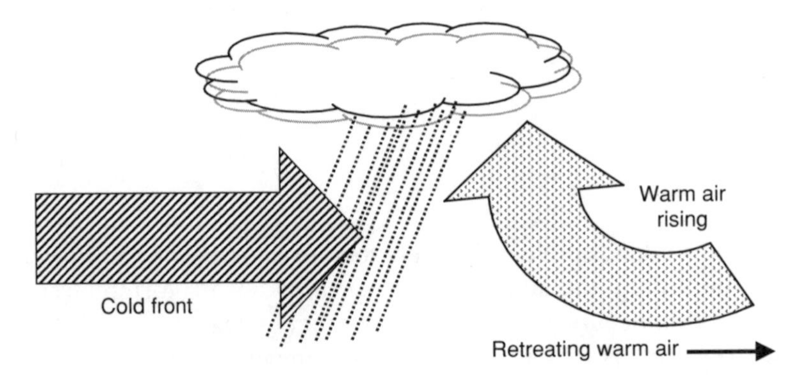

Fig. 3-2 As warm, moist air rises before a cold front, it is pushed sharply out of the way.

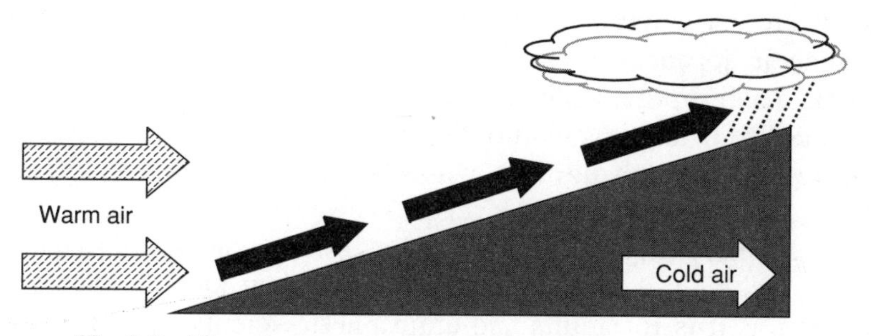

Fig. 3-3 Warm fronts approach more slowly than cold fronts.

In the case of a *warm front,* the warm, less dense air slips up and over the colder air ahead of the front. Like cold fronts, the air cools as it rises and its moisture condenses to produce clouds and rain. Warm fronts, with gentler slope faces, usually move more slowly than cold fronts, so the forward vertical motion along warm fronts is much more gradual. Rainfall that develops in front of a warm front is often constant and more wide-ranging than rain connected with a cold front. Fig. 3-3 shows the gradual air flow angle and approach of a warm front.

When neither the cold front nor the warm front is pressing forward, the fronts are known as *stationary fronts.* They are at a standoff. Broad bands of clouds form on both sides of the stalled front edge.

With an *occluded front* or *occlusion,* it's a combination. Cold, warm, and cool air meet, forming boundaries above the ground and at the surface. When this happens, atmospheric scientists say that a cold front has caught up with a warm front. Occluded-front rainfall is a mixture of the different types found within cold and warm fronts.

Clouds

These visible wonders of the troposphere are found in a variety of shapes and sizes. While some clouds are happy just to be, others come with precipitation like mist, rain, sleet, hail, and snow.

> A **cloud** is a combination of tiny water droplets and/or ice crystals suspended in the atmosphere.

Clouds are classified and often named using Latin prefixes and suffixes to describe their appearance. Table 3-4 lists a few of these Latin root words. Additional detail is provided by measuring a cloud's base height or altitude from the ground. For example, cloud names containing the prefix *cirr-,* as in *cirrus* clouds, are found at high altitudes, while cloud names with the prefix *alto-,* as in *altostratus,* are found at middle levels.

Clouds are extremely varied, coming in all sizes, colors, and shapes. They can also change within minutes. This variety makes their identification as much fun for children as it is for adults and atmospheric scientists. Table 3-5 provides some of the common characteristics of different cloud types.

HIGH-LEVEL CLOUDS

High-level clouds form above 6,000 meters. The temperatures at these high elevations are cold, so high-level clouds are mostly made up of ice crystals. High-level clouds are generally thin and white in appearance, but can appear in a terrific variety of colors when the sun is setting on the horizon.

The most common forms of high-level clouds are thin and wispy *cirrus* clouds. Typically found at heights greater than 6,000 meters, cirrus clouds are formed out of ice crystals that come from frozen water droplets. Cirrus clouds usually form in fair weather and point in the direction of the air flow at their altitude.

Cirrostratus are sheet-like, high-level clouds made of ice crystals. Although cirrostratus can blanket the sky and be many thousands of meters thick, they are fairly transparent. The sun or the moon is usually seen through cirrostratus. These high-level clouds form when a wide air layer is lifted by large-scale fronts.

Table 3-4 Latin root clouds.

Latin root word	Atmospheric meaning
cirrus	curl
cumulus	pile
nimbus	rain
stratus	layer

Sometimes cirrostratus are so filmy, the only sign of their existence is a halo around the sun or the moon. *Halos* are formed when light refracts off a cloud's ice crystals. Since cirrostratus clouds get thicker when warm fronts advance, a halo gradually fades or disappears from around the sun or moon as the weather changes. Fig. 3-4 illustrates high-level, cirrus-type clouds (top) and midlevel altocumulus clouds.

Table 3-5 Cloud types.

Cloud	Altitude (meters)	Shape	Composition
Cumulus	12,000	Vertical, fluffy, defined edges and flat bases	Condensed water vapor
Cumulonimbus	12,000+	Massive, dark, vertical towers	Water droplets and ice crystals
Cirrus	6,000+	Thin and wispy	Ice crystals
Cirrostratus	6,000	Sheet-like, almost transparent	Ice crystals
Altocumulus	2,000 – 6,000	Parallel bands or rounded masses	High humidity and water droplets
Nimbostratus	< 2,000	Dark, low	Water or snow
Stratocumulus	< 2,000	Light to dark gray, low, lumpy masses, and rolls	Weak rainfall with clear sky breaks in between
Contrail (condensation trail)	6,000 – 12,000+	Long thin lines following a jet's exhaust path	Water droplets freeze to ice crystals
Orographic	2,000 – 6,000+	Fluffy, circling mountain peaks	Condensed water vapor
Mammatus	2,000 – 6,000	Light to dark gray	Water droplets
Billow	2,000 – 6,000	Horizontal eddies	Condensed water vapor

Fig. 3-4 High altitude cirrus clouds are wispy, while mid-level altostratus are flat bottomed.

Mid-Level Clouds

The bases of midlevel clouds usually form around 2,000 to 6,000 meters. Found at lower altitudes, they are usually warm enough that their water droplets don't freeze, but they can contain ice crystals when temperatures are cold enough.

Altocumulus are found as parallel bands or rounded masses. Commonly a portion of an altocumulus cloud is shadowed, which helps you tell them apart from high-level cirrocumulus. Altocumulus clouds often form by convection in an unstable upper air layer. This can be caused by the gradual lifting of air before a cold front. When you see altocumulus clouds on a warm, humid summer morning, there will often be thunderstorms later that day.

Low-Level Clouds

Low clouds are mostly made up of water droplets, since their bases sit below 2,000 meters. However, when temperatures are cold enough, they can pick up ice particles and snow.

Nimbostratus are dark, low-level clouds accompanied by light-to-medium rainfall. However, when temperatures are cold enough, these clouds can contain ice particles and snow.

Stratocumulus clouds generally appear as low, lumpy, layered clouds that come with weak rainfall. Stratocumulus vary in color from dark gray to light gray and may appear as rounded masses, rolls, and so on, with breaks of clear sky in between. Fig. 3-5 shows the medium-to-dark, lumpy, low-level-clouds seen on dreary, rainy days.

VERTICAL CLOUDS

Probably the most familiar of the basic cloud shapes is the *cumulus cloud.* Formed by either thermal convection or frontal lifting, cumulus clouds reach altitudes of more than 15,000 meters. Additionally, huge amounts of energy are freed through the condensation of water vapor within the cloud itself. Fig. 3-6 shows the puffy, cotton-like nature of cumulus clouds.

Fair-weather cumulus clouds look like floating cotton balls to most people. Identified by their flat bases and fluffy outlines, fair-weather cumulus show little

Fig. 3-5 Low level clouds can sheet the sky or form dark, overcast clouds.

Fig. 3-6 Fluffy air weather cumulus clouds are easy to identify.

vertical growth, with cloud tops at the limit of the rising air. Given frontal action, though, fair weather cumulus become tigers, turning into gigantic cumulonimbus clouds—the citadel of violent thunderstorms.

Supplied by rising pockets of air, or *thermals*, and lifting vertically from the earth's surface, fair-weather cumulus water vapor cools and condenses to create cloud droplets. Newly formed fair-weather cumulus clouds have sharply defined margins and bases, while older cumulus edges are rougher, showing cloud erosion. Evaporation around a cloud's edges cools the surrounding air, making it heavier and causing it to drop outside the cloud.

Cumulonimbus clouds are much bigger and taller than fair-weather cumulus. They either build vertically as separate soaring towers or form a line of structures known as a *squall line.* Fed by intense updrafts, cumulonimbus clouds are easily the giants of the cloud forms, reaching 15,000 meters or higher.

Lower cumulonimbus clouds are made up of water droplets, but at higher altitudes and temperatures below freezing, ice crystals take over. When everything is just right for a thunderstorm, happy, fair-weather cumulus clouds quickly turn into large, nasty cumulonimbus clouds linked to awesome, towering thunderstorms called *supercells.*

SPECIALTY CLOUDS

Some clouds are very specialized. They form when specific events take place. For example, some of these clouds are formed by aircraft, earlier storms, and the presence of mountain peaks. A few of these specialty clouds are shown in Fig. 3-7.

Fig. 3-7 Contrails, mammatus, and orographic clouds are very distinct.

A *contrail,* short for *condensation trail,* is a cirrus-like trail of condensed water vapor that looks like the tail of a kite or a wide piece of yarn. Contrails are created at high altitudes when the heat coming from a high-performance jet engine hits the extremely cold atmosphere, condensing water vapor into a miniature cloud, which forms a line behind the engine. If the engines are close together, the contrail meshes into a single line. If they are farther apart, a distinct condensation trail forms behind each one. The formation of contrails is a function of the atmosphere's water content at the aircraft's altitude and the heat characteristics of the aircraft's engines. At very high altitudes and in polar regions, there usually isn't enough water content for contrails to form, but it does happen.

Mammatus clouds are shaped like hanging fruit and are formed in sinking air. Although fairly scary in appearance, mammatus clouds are safe and don't signal the imminent appearance of tornadoes (though it is a common idea). Actually, mammatus clouds usually appear after a thunderstorm has done its worse and passed.

Air is also lifted by the land's topography. When air flow runs into a mountain range, it is forced to rise up and over the mountains. If it is lifted high enough, water vapor condenses and produces rain. *Orographic* clouds are formed in response to air that is uplifted by the shape of the earth. Mountains create a barrier to air currents and weather fronts. Orographic clouds are formed when rising air is cooled and water vapor condenses. In the United States, prevailing winds often blow from west to east, so most orographic clouds form on the western side of a mountain. These clouds are also seen around mountain peaks.

Billows clouds are created from the instability connected with vertical shear air flows and weak thermal layering. These clouds, formed by winds blowing at different speeds in different air layers, are often seen at air mass margins of different densities (warm air layered over cold air). They look a lot like breaking waves.

Lenticular clouds are shaped like lenses or flying saucers. These flat clouds are found in windy areas where rising air is cooling. After the water vapor condenses into cloud droplets, it is pulled along by the blowing wind. As the droplets warm and begin to drop, they gradually turn back into vapor and disappear. These clouds are often seen downwind of mountain ranges. Fig. 3-8 shows the characteristic shape of billows and lenticular clouds.

Tornadoes

Tornadoes are the children of severe thunderstorms. As speeding cold fronts smash into warm humid air, a convection of temperature and wind is formed. Winds can easily reach speeds of over 250 km per hour. Large tornadoes stir up the fastest winds ever measured on the earth's surface. They have been measured at over 480 km per hour (kph).

In April 1974, a huge weather system blowing cold air down from the Rocky Mountains hit rising warm, humid air from the Gulf of Mexico and caused a terrible series of storms, later referred to as the Super Outbreak. The storm stretched from Indiana, where it began, to Alabama, Ohio, West Virginia, and

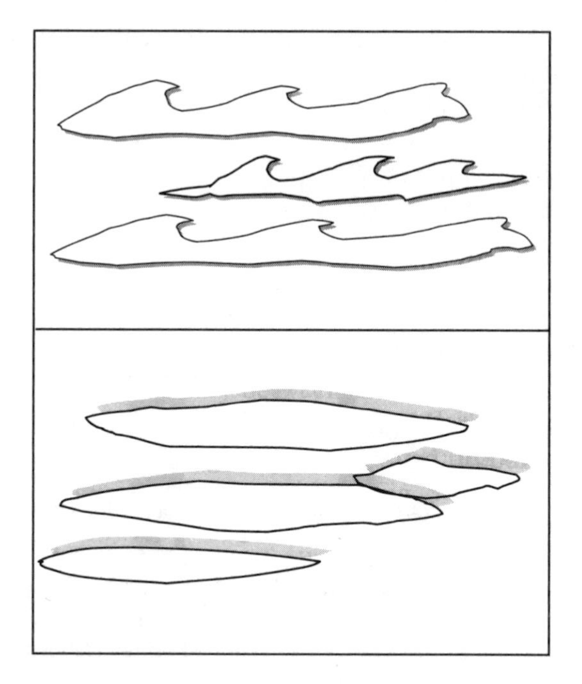

Fig. 3-8 Winds have a major effect on the shape of billows and lenticular clouds.

Virginia. Winds were recorded over 420 kph spawning 127 tornadoes! Three hundred fifteen people were killed in 11 states and 6,142 were injured.

Tornadoes are usually classified into one of the following three different levels:

1. Weak tornadoes (F0/F1) make up roughly 75% of all tornadoes. They cause around 5% of all tornado deaths and last approximately 1–10 minutes, with wind speeds of less than 180 kph.
2. Strong tornadoes (F2/F3) make up most of the remaining 24% of all tornadoes. They cause nearly 30% of all tornado deaths and last 20 minutes or longer, with wind speeds between 180 and 330 kph.
3. Violent tornadoes (F4/F5) are rare and account for less than 2% of all tornadoes, but cause nearly 65% of all tornado deaths in the United States. They have been known to last for one to several hours, with extreme wind speeds of 330 to 500 kph.

In the late 1960s, University of Chicago atmospheric scientist T. Theodore Fujita realized that tornado damage patterns could be predicted according to certain wind speeds. He described his observations in a table called the *Fujita wind damage scale*. Table 3-6 shows the Fujita scale with its corresponding wind speeds and surface damage.

The big problem with tornadoes is that they are unpredictable. Weather forecasters can tell when conditions are ripe for tornadoes, but cannot predict if or where they will strike. Think of a sleeping rattlesnake: if irritated, it will probably strike, but when and where is the question.

Table 3-6 Tornado ratings: Fujita scale.

Tornado rating	Type	Speed	Damage
F0	Gale	64–116 kph (40–72 mph)	Light damage: some damage to chimneys, tree branches break, shallow-rooted trees tip over, and sign boards damaged
F1	Moderate	117–180 kph (73–112 mph)	Moderate damage: beginning of hurricane wind speeds, roofs peeled, mobile homes moved off foundations or overturned, and moving cars shoved off roads
F2	Significant	181–251 kph (113–157 mph)	Considerable damage: roofs peeled, mobile homes smashed, boxcars pushed over, large trees snapped or uprooted, and heavy cars lifted off ground and thrown
F3	Severe	252–330 kph (158–206 mph)	Severe damage: roofs and walls torn off well-made houses, trains overturned, most trees in forest uprooted, and heavy cars lifted off ground and thrown
F4	Devastating	331–416 kph (207–260 mph)	Devastating damage: well-made houses leveled, structures blown off weak foundations, and cars and other large objects thrown around
F5	Incredible	417–509 kph (261–318 mph)	Incredible damage: strong frame houses are lifted off foundations and carried a considerable distance and disintegrated, car-sized missiles fly through the air in excess of 100 meters, and trees debarked
F6	Inconceivable	510–606 kph (319–379 mph)	The maximum wind speed of tornadoes is not expected to reach the F6 wind speeds

Hurricanes

Hurricanes are formed from groups of thunderstorms. The ocean's temperature and atmospheric pressure also play important roles. To start, ocean water must be warmer than 26.5°C. The heat and water vapor from this warm water serves as a hurricane's fuel source.

A hurricane can exist for as long as two or three weeks. It begins as a series of thunderstorms over tropical ocean waters. The first phase in the formation of a hurricane is the lowering of barometric pressure. This is called a *tropical depression.*

Then to get to the next phase, the intensification of the storm to a *tropical storm* must take place. Favorable atmospheric and oceanic conditions affect the speed of the hurricane's development.

Relatively high humidity in the lower and middle troposphere is also needed for hurricane development. This high humidity decreases the amount of cloud evaporation and increases the heat released through increased rainfall. The concentration of heat is critical to driving the system.

Vertical wind shear is also a factor in a hurricane's development. When wind shear is weak, a hurricane grows vertically and condensation heat is released into the air directly above the storm. This causes it to build.

> **Wind shear** describes the sudden change in the wind's direction or speed with increasing altitude.

When wind shear is intense, a hurricane becomes slanted, with heat being released and distributed over a large area. Atmospheric pressure and wind speed change across the diameter of a hurricane. Barometric pressure falls more quickly as the wind speed increases.

Hurricanes have winds over 64 knots and rotate counter-clockwise about their centers in the Northern Hemisphere and clockwise in the Southern Hemisphere. Fig. 3-9 shows the circular wind motion that signals a hurricane.

Hurricanes produce dangerous winds, torrential rains, storm surges, flooding, riptides, and tornadoes that often result in huge amounts of property damage, deaths, and injuries in coastal locations. In August and September of 2004, hurricanes Charlie, Francis, Ivan, and Jean pounded the Caribbean and the United States in rapid succession. They were responsible for nearly 60 deaths and more than $20 billion in property damage over 10 states.

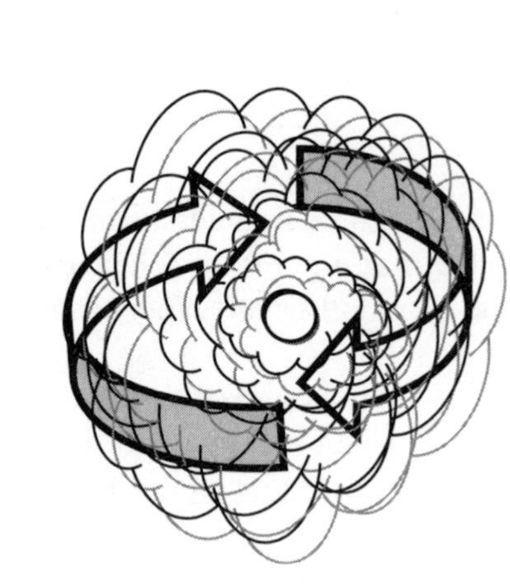

Fig. 3-9 Circular wind motion that signals a hurricane.

A hurricane's course depends upon its location. A hurricane beginning in the eastern tropical Atlantic is driven westward by easterly trade winds. These storms turn northwest around a subtropical high pressure system and move into higher latitudes. As a result, the Gulf of Mexico and the East Coast of the United States are at risk for hurricanes yearly. Table 3-7 lists the different hurricane force categories.

Hurricanes get their energy from the warm surface water of the tropics, which explains why they fizzle quickly after moving over cold water or continental land masses. However, if the oceans continue to heat and remain warmer for longer periods of time—for example, through global warming—the chance of violent hurricanes increases. The destructive hurricanes of 2004 may be just a glimpse of this.

Severe storms are known by different names in the different parts of the world. For example, the name *hurricane* is given to systems that form over the Atlantic or the eastern Pacific Oceans. In the northwestern Pacific Ocean and Philippines, these systems are called *typhoons,* while in the Indian and the south Pacific Ocean they are known as *cyclones.*

In 1953, the Tropical Prediction Center created a list of hurricanes names. When a tropical depression turns into a tropical storm, it is given the next name available on the list. Listed in alphabetical order, the names are alternately male and female. If a hurricane has been particularly vicious, the name is never used again. For example, hurricanes Alicia, Andrew, Betsy, Camille, Carmen, Gilbert, Hugo, and Roxanne have been retired from use, to name just a few.

Table 3-7 Hurricane category.

Hurricane category	Strength	Winds and storm surge
1	Weak	65–82 knot winds 1.2–1.7 meters above normal storm surge
2	Moderate	83–95 knot winds 1.8–2.6 meters above normal
3	Strong	96–113 knot winds 2.7–3.8 meters above normal
4	Very strong	114–135 knot winds 3.9–5.5 meters above normal
5	Near total devastation	> 135 knot winds > 5.5 meters above normal

The good news is that weather prediction is much more sophisticated than it was 100 years ago. Now, there is usually time to issue tornado warnings or to define a hurricane's strength and find an evacuation path, keeping any human casualties to a minimum.

Air Pollution

The atmosphere has many faces. Polluting compounds present another area for concern. For the great majority of most industrialized nations, the very air we breathe is laced with smoke, particulates, and an array of toxic chemicals from human activities and industrial processes. This is a major problem.

The most common air pollutants are:

- Carbon monoxide
- Chlorofluorocarbons (CFCs)
- Heavy metals (arsenic, chromium, cadmium, lead, mercury, zinc)
- Hydrocarbons

- Nitrogen oxides
- Organic chemicals [volatile organic compounds (VOCs), dioxins]
- Sulfur dioxide
- Particulates

Most of the time, people depend on local, state, and national agencies to serve as watchdogs for these pollutants, since they are often colorless, odorless, and hard to detect. However, we can't ignore what our eyes and nose tell us is true; sometimes the air is brown and it smells! Table 3-8 lists the major sources of air pollution and their effects.

Atmospheric pollution affects everyone all around the world. When a volcano erupts in one part of the world, dust and particles are not limited to one area, but instead encircle the globe. In 1982, after Mount Pinatubo in the Philippines erupted, global temperature drops were recorded for several years.

The dust from the Pinatubo eruption was blown 32 km above the earth. Satellites observed the plume of volcanic ash as it circled the globe at around 120 kph. In fact, a month after the eruption, which killed 350 people, a 4800-km-long ash and sulfur cloud still ringed the earth. Satellite temperature measurements found that dust shaded the Earth's surface from the sun's rays, causing the average global temperature to dip about 0.3°C.

Shooting more than 20 million tons of dust and ash into the atmosphere, altering its heat balance, and speeding ozone depletion across the globe, Pinatubo has become the focus of several in-depth studies. Environmental biologists use the term *Pinatubo effect* to describe how volcanic ash and debris, if sent high enough into the atmosphere, can affect temperature and weather for years. By studying the Pinatubo effect, scientists hope to understand how and to what extent manufactured pollution may cause long-term and irreversible climatic change.

Damaging effects from air pollution and volcanic eruption on vegetation and tropical and temperate forests have been recorded in the United States, Germany, Sweden, and Czechoslovakia, among others. Acid rain, heavy metals, nitrogen, and other pollutants act singly and in combination to produce plant stress and injury.

Amazingly, the air quality indoors isn't much better. With all the chemicals used in our indoor environments, the United States Environmental Protection Agency (EPA) has discovered that indoor air is 5 to 10 times as toxic as outdoor air since tightly sealed buildings allow chemical levels to concentrate.

We will discuss the various alternatives and solutions to air pollutants and other environmental issues in greater detail in Chapter 4.

Table 3-8 Pollution sources.

Pollution source	Composition of pollutant
Automobiles	Burning of oil and gas produces carbon monoxide, VOCs, hydrocarbons, nitrogen oxides, peroxyacetyl nitrate, benzene, and lead
Utility power plants	Burning of coal, oil, and gas produces nitrogen oxides, heavy metals, sulfur dioxide, and particulates
Industry	Particulates, sulfur dioxide, nitrogen oxides, heavy metals, fluoride, CFCs, and dioxins
Incineration	Carbon monoxide, nitrogen oxides, particulates, dioxins, and heavy metals
Biomass burning	Burning of grasslands, crop stubble, agricultural waste, organic fuel, and forests produces sulfur, methane, radon, carbon dioxide, nitrogen oxides, carbon monoxides, and particulates
Small engines	Mowers, blowers, trimmers, chain saws, and other machines produce nitrogen oxides and hydrocarbons
Disasters	Radiation leaks, chemical leaks, burning of oil wells produce radiation, nitrogen oxides, carbon monoxide, sulfur, heavy metals, and particulates
Mining	Rock breakdown and processing produces nitrogen oxides, heavy metals, radiation particles, and particulates
Erosion	Road work and farm work produces dust, particulates, dried pesticides, and fertilizers
Indoor air pollution	Carpeting, cooking, and other indoor products and activities produce formaldehyde, lead and asbestos dust, radon, and other incorporated chemicals
Nature	Volcanic eruptions and forest fires produce dust and particles, sulfur dioxide, carbon monoxide, carbon dioxide, chlorine, nitrogen oxides, heavy metals, radon, and particulates

Quiz

1. Which of the following is a layer of the atmosphere?
   (a) Angiosphere
   (b) Gymnosphere
   (c) Stratosphere
   (d) Pycnosphere

2. What is it called when a NASA T-38 jet zips through the sky and leaves a trace?
   (a) Mammatus cloud
   (b) Contrail
   (c) Lenticular cloud
   (d) Really cool

3. Which two gases make up the majority of the Earth's gases?
   (a) Oxygen and methane
   (b) Oxygen and propane
   (c) Nitrogen and carbon dioxide
   (d) Nitrogen and oxygen

4. The relationship between air temperature and the amount of water vapor it contains is known as
   (a) indistinct humidity
   (b) point source humidity
   (c) relative humidity
   (d) aridity

5. High-level clouds form above
   (a) 1,000 meters
   (b) 2,000 meters
   (c) 6,000 meters
   (d) 20,000 meters

6. A long, narrow, upper atmosphere current of fast-moving air is known as a
   (a) contrail
   (b) jet stream
   (c) thermophile
   (d) typhoon

7. Which atmospheric layer is largely responsible for absorbing most of the sun's ultraviolet (UV) radiation?
   (a) Troposphere
   (b) Cumulus cloud
   (c) Stratonimbus
   (d) Ozone

8. Conflicts between air masses are called
   (a) fossilization
   (b) fronts
   (c) precipitation
   (d) auroras

9. A sudden change in the wind's direction or speed with increasing altitude is called
   (a) wind shear
   (b) wind stop
   (c) wind chill
   (d) wind rear

10. Carbon monoxide, chlorofluorocarbons (CFCs), heavy metals, hydrocarbons, and nitrogen oxides are all
    (a) found in automobile tires
    (b) common air pollutants
    (c) good for the environment
    (d) organic compounds

CHAPTER 4

Greenhouse Effect and Global Warming

Have you ever been in a greenhouse? Greenhouses are special buildings usually made from glass and steel. They are used to grow plants that need humidity, tropical temperatures, and constant growing conditions. Gardeners also use greenhouses to protect plants from freezing in the winter.

Greenhouses, sometimes called *hothouses,* work by trapping the sun's heat. A greenhouse's glass sides and roof let sunlight in, but keep heat from escaping. This causes the greenhouse to heat up, like a vehicle parked outside in the summer sun with the windows rolled up. A greenhouse offers plants a warm and humid environment, even if the outside weather is dry, windy, or cold. For a warm climate plant like the tropical orchid, a balmy greenhouse is the perfect environment.

In Chapter 3, we saw how the earth's atmosphere surrounds our planet like a blanket. It protects us from harmful cosmic radiation, regulates temperature and humidity, and controls the weather. The atmosphere is critical to life on this planet and provides the air we breathe.

Greenhouse gases are a natural part of the atmosphere. They trap the sun's warmth, and preserve the earth's surface temperature at a median level needed to support life. Atmospheric greenhouse gases act like the glass windows of a greenhouse. Sunlight enters the Earth's atmosphere, passing through greenhouse gases that act like a lens. Then, as it reaches the Earth's surface, the land, water, and biosphere absorb the sun's energy. Once absorbed, this energy gets recycled into the atmosphere. A portion of the heat is reflected into space, but a lot of it stays locked in the atmosphere by greenhouse gases, causing the Earth to heat. Fig. 4-1 shows how these greenhouse gases act in the atmosphere to trap energy.

The **greenhouse effect** describes how atmospheric gases prevent heat from being released back into space, allowing it to build up in the Earth's atmosphere.

The more gases there are, the more the Earth heats up. The greenhouse effect is important. Without it, the Earth would not be warm enough for most living things to survive. However, if the greenhouse effect gets too strong, it can make

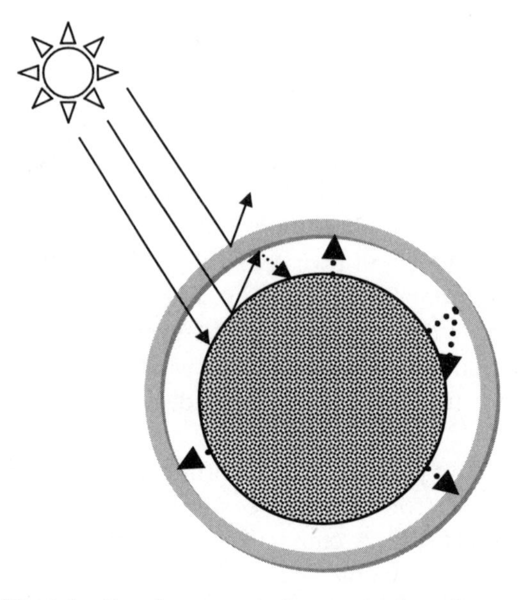

Fig. 4-1 Greenhouse gases trap energy from the sun in the atmosphere, causing it to heat up.

the Earth warmer than normal. The problem is that even a little more heat (a few degrees higher) creates problems for people, plants, and animals.

Greenhouse Gases

We know that the earth is surrounded by a mixture of gases. The Earth's atmosphere consists of roughly 79.1% nitrogen, 20.9% oxygen, 0.03% carbon dioxide, and trace amounts of other gases.

Greenhouse gases are a natural part of the atmosphere. Greenhouse gases include water vapor, carbon dioxide, methane, nitrous oxide, halogenated fluorocarbons, ozone, perfluorinated carbons, and hydrofluorocarbons. Water vapor is the most important greenhouse gas, but human activity doesn't have much direct impact on its amount in the atmosphere.

Global warming is caused by an increase in the levels of these gases brought about by human activity. The greatest impact on the greenhouse effect has come from industrialization and increases in the amounts of carbon dioxide, methane, and nitrous oxide. The clearing of land and burning of fossil fuels, for example, have raised atmospheric gas concentrations of soot and other *aerosols* (fine particles in the air).

Manufactured greenhouse gases and particles, rather than the occasional volcanic eruption, now account for higher gas concentrations. The planet has begun to warm at a steep rate, and future temperature increases are predicted by climatic models programmed with the volumes of gases released yearly into the atmosphere. Some scientists are already seeing the consequences of global warming, such as the melting of the polar ice sheets and rising sea levels. Since 1991, the National Academy of Sciences has found clear evidence of global warming and recommended immediate reductions in greenhouse gases. Depending on whether or not changes are made, they have predicted temperature increases of between 3° and 9° Fahrenheit in the next 100 to 200 years, with sea level increases of 3 to 25 feet.

Since these greenhouse gases are long-lasting, even if everyone *stopped* using their cars today, greenhouse effects would continue for another 150 years. Think of it like a speeding train, even when the engineer hits the brakes hard, the total speed combined with the train's mass causes it to take a long time to stop. For this reason, it's important to accelerate the search for alternative energy sources and to curb and/or stop the release of greenhouse gases into the atmosphere. Global warming is likely to be the subject of intense scientific study and growing concern from now on, as the general public is just becoming aware of the consequences of industrial progress.

CARBON DIOXIDE

Carbon dioxide (CO_2) is a natural greenhouse gas and also the biggest human-supplied gas to the greenhouse effect (about 70%). A heavy, colorless gas, carbon dioxide is the main gas we exhale during breathing. It dissolves in water to form carbonic acid, is formed in animal respiration, and comes from the decay or combustion of plant/animal matter. Carbon dioxide is absorbed from the air by plants in photosynthesis and is also used to carbonate drinks.

You may be thinking that the Earth's inhabitants don't have the option to stop breathing. You're right. However, the amount of carbon dioxide in the atmosphere, unfortunately, is about 30% higher now than it was at the beginning of the 1800s. Fig. 4-2 shows carbon dioxide concentration trends over the past 250 years.

The industrial revolution is responsible for this jump. Ever since fossil fuels such as oil, coal, and natural gas were first burned to create energy for electricity and transportation fuel, carbon dioxide levels have climbed. Additionally, when farmers clear and burn weeds and crop stubble, carbon dioxide is also produced.

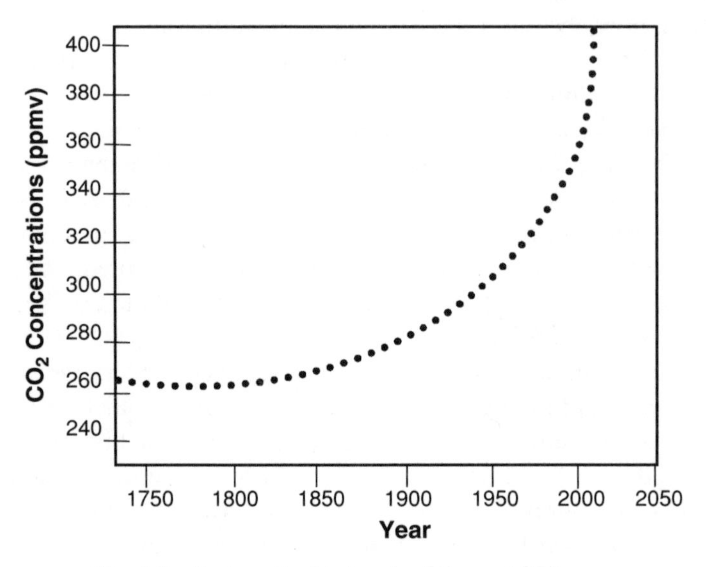

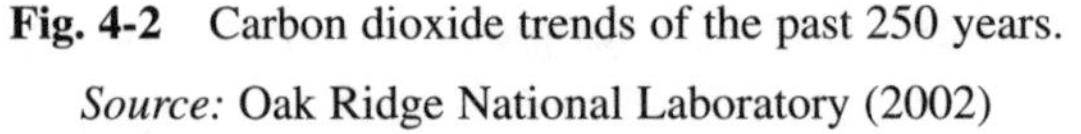

Fig. 4-2 Carbon dioxide trends of the past 250 years.

Source: Oak Ridge National Laboratory (2002)

Table 4-1 Volcanic gases.

Volcanic gas	% of total volcanic gas (average)
Water vapor (steam) and carbon dioxide	90–95
Sulfur dioxide	< 1
Nitrogen	< 1
Hydrogen	< 1
Carbon monoxide	< 1
Sulfur	< 0.5
Chlorine	< 0.2

Carbon dioxide gases also come from the earth. When volcanoes explode, about 90 to 95% of the spewed gases are made of water vapor and carbon dioxide. Table 4-1 lists the different gases released into the atmosphere during a volcanic eruption.

NITROGEN OXIDES

The colorless gas known as nitrous oxide is an atmospheric pollutant produced by combustion. It is also one of the greenhouse gases.

There are several ways that nitrogen and oxygen team up in the atmosphere, including *nitrogen dioxide, nitric oxide,* and *nitrous oxide.* Nitrogen oxides are stable gases and do not break down quickly. For this reason, they build up in the atmosphere in greater and greater concentrations. Nitrogen dioxide and other gases and particulates gives the sky that yellow-brown look that many people call *smog.*

Nitrogen combines with moisture in the atmosphere to form nitric acid. This comes down as rain and acidifies lakes and soils, killing fish, and small animal populations and damaging forests. Acid particulates are also precipitated, along with the leaching of heavy metals, into water supplies. Scientists believe that this increase comes from the burning of crops, industrial releases, and the effects of nitrogen fertilizers used in agriculture.

Nitrous oxide is also used in dentistry to put patients to sleep during dental procedures and is sometimes called *laughing gas.* The amount of nitrous oxide in the atmosphere is about 15% higher now than it was in the 1800s.

Besides the burning of fossil fuels, nitrogen oxides are also produced by kerosene heaters, gas ranges and ovens, incinerators, deforestation, leaf burning, aircraft engines, and cigarettes. Lightning and natural soil sources also produce nitrogen oxides. Scientists estimate that vehicles produce 40% and electric utilities and factories produce 50% of industrial nitrogen oxide emissions. The remaining 10% comes from other sources.

At high altitudes, nitrogen oxides are responsible for some ozone depletion. When ozone is thin or gone in places, the amount of solar ultraviolet radiation that reaches the ground is increased. This causes plant damage and injury to animals and humans in the form of skin cancers and other problems.

As a greenhouse gas, nitrogen oxides trap heat much more efficiently than carbon dioxide.

METHANE

Another greenhouse gas, *methane,* is a colorless, odorless, flammable hydrocarbon that is released by the breakdown of organic matter and the carbonization of coal. This gas is the second biggest additive, after carbon dioxide, to the greenhouse effect at around 20%.

Methane is the main component of natural gas, which is found in deposits, like oil, in the earth's crust. Methane is a byproduct of the production, transportation, and use of natural gas. Underwater decaying plants create methane known as *marsh* or *swamp gas.*

One of the best known sources of methane in rural populations is that of belching farm animals. Cows have complicated digestive systems, and release large amounts of methane in satisfying belches. It sounds funny, but when you consider herds of hundreds of animals, it adds up!

The U.S. National Aeronautics and Space Administration (NASA) reports that methane is increasing in the atmosphere three times faster than carbon dioxide; atmospheric methane levels have tripled in the past 30 years. Over 500 million tons of methane are emitted yearly from bacterial decomposition and fossil fuel burning.

The amount of methane in the atmosphere is about 145% higher now than it was in the 1800s. The major causes of this increase are thought to include:

- Digestive gases of sheep and cattle
- Growth and cultivation of rice
- Geologic release of natural gas
- Decomposition of garbage and landfill waste

Methane's interaction in the atmosphere is fairly complex. Naturally occurring hydroxyl radicals in the atmosphere combine with and pull methane from the atmospheric mix. Unfortunately, methane and carbon monoxide (from car emissions) lower hydroxyl levels; as hydroxyl concentration drops and methane emissions continue, methane concentration increases. Naturally, methane interacts in the stratosphere as a sink for chlorine (helping maintain ozone levels), with its ups and downs affected by its breakdown to hydrogen and water vapor.

HALOCARBONS

As we learned earlier, halocarbons levels dropped since being banned in the 1990s. The phasing out of chlorofluorocarbons has removed a lot of the ozone threat and is allowing the protective ozone layer to recover. However, other problem gases, like perfluorocarbons and sulfur hexafluoride, affect the atmosphere and are given off during aluminum smelting, production of electricity, magnesium processing, and semiconductor manufacturing. These can be limited through different manufacturing methods, but it's a matter of economics. Until environmental concerns on excess halocarbon release are widely understood, manufacturers have little incentive to change.

Formation of Greenhouse Gases

The burning of fossil fuels for energy creates most greenhouse gases. When oil, gas, or coal burns, carbon in the fuel mixes with atmospheric oxygen to form carbon dioxide. Methane is produced from coal mining and certain natural gas pipelines. Animals, especially sheep and cattle, produce methane as food breaks down within their stomachs. Some fertilizers release nitrous oxide. Rice production in paddy fields generates methane under water.

When organic matter such as table scraps, garden waste, and paper is left in landfills, its decomposition forms methane and carbon dioxide. Sewage and water treatment plants also release these gases in the process of breaking down wastes.

Many industrial processes create greenhouse gases. Cement production, used for everything from building roads to laying the foundations of homes and businesses, requires chemical processes that produce an assortment of greenhouse gases.

OZONE

Ozone is another major air pollutant that the Environmental Protection Agency (EPA) regulates, but it is not emitted directly from specific sources. It's formed in the atmosphere from nitrogen oxides and volatile organic compounds (VOCs). Sources of VOCs include:

1. Combustion products from motor vehicles and machinery that burns fossil fuels
2. Gasoline vapors from cars and fueling stations
3. Refineries and petroleum storage tanks
4. Chemical solvent vapors from dry-cleaning processes
5. Solid waste facilities
6. Metal-surface paints

The fumes from internal combustion engines contain many VOCs that, when released into the atmosphere, interact with other gases in the presence of sunlight to create the ozone part of smog. The EPA has targeted VOC reduction as an important control mechanism for reducing high levels of ozone-containing smog in cities.

Since the reactions that form ozone are affected by sunlight, high ozone levels usually take place in the summer months when the air is hot and not moving much. In the summer, more people are also traveling in addition to their daily commute. So vehicle emissions rise.

Ozone greatly irritates the mucous membranes of the nose, throat, and lungs. The extent of irritation depends on the amount of ozone in the air and the frequency and duration of exposure. Health effects related to ozone have been studied and when ozone concentrations are up, illness and hospital admissions go up.

NATURE'S PART

Green plants use the sun's energy and carbon dioxide from the air as part of photosynthesis. This is a good thing, because they soak up carbon dioxide in the process. Plants are considered to be carbon dioxide storehouses. During the photosynthetic cycle, they form carbohydrates, which make up the foundation of the food chain.

Maybe this is why people talk to houseplants. The carbon dioxide that people exhale during breathing is a critical part of plants' energy process. The more you talk to them, the more carbon dioxide you provide.

Forests absorb carbon dioxide in a big way. Over time, forests build up a significant supply of stored carbon in their tree trunks, roots, stems, and leaves. Then, when the land is cleared, this stored carbon is converted back to carbon dioxide by burning or decomposition.

Another big player in this process of carbon dioxide absorption is the oceans. The oceans absorb carbon dioxide from the atmosphere. They also act as a moderating influence on temperature ranges. We will look at the details of the earth's carbon cycle more carefully in Chapter 11.

GREENHOUSE EFFECT AND OZONE

The difference between the greenhouse effect and ozone reduction can be confusing. Both are important environmental topics and are affected by molecules released into the atmosphere by human activities.

The greenhouse effect refers to the ability of the greenhouse gases to trap heat within the atmosphere. These gases include water vapor, carbon dioxide, methane, and nitrous oxide. Without them, life on Earth would not be possible. A problem occurs when greenhouse gases increase, and affect the atmosphere and climate.

Ozone reduction is a global issue since it makes the earth vulnerable to solar radiation. When the protective ozone layer is thin or absent, the sun's harmful UV radiation is able to get through the atmosphere and reach the Earth's surface. Exposure to this radiation causes skin cancer, eye damage, and other health problems.

We learned in Chapter 3 that ozone depletion is strongly affected by chlorofluorocarbons and halons that make their way into the stratosphere. Ozone drops in the stratosphere are thought to have allowed the upper part of the atmosphere to cool and attend ozone distribution.

Ozone is also an atmospheric pollutant, especially in urban areas. Ozone pollution is believed to cause planet-wide heating in addition to its bad health effects. To slow ozone's contribution to global warming, exceptional international cooperation is crucial.

ENHANCED GREENHOUSE EFFECT

Solar cycles and changes in the sun's radiation increase the Earth's temperature naturally. They affect local climate and allow the sun's energy to reach the Earth's surface, keeping heat from escaping. The Earth gets slowly hotter.

The **enhanced greenhouse effect** brought about by the burning of fossil fuels (oil, coal, and natural gas) creates global warming and changes the environment.

Industrial activity produces greenhouse gases that serve as additional blankets to heat the Earth even more.

Global warming effects differ around the world and make it hard to predict exactly how the climate may change. Temperature increases are expected to be higher in polar areas than around the equator. Land temperatures might be higher than those over oceans. Rainfall might be heavier in some areas and lower in others.

A major climatic change would greatly affect local weather through the frequency and intensity of storms. Some scientists fear the succession of Atlantic hurricanes in 2004 may be just the beginning of severe climate changes. Ranching, farming of crops, number of pests and diseases, ocean levels, and the populations of native plants and animals would all be impacted.

However, as global warming has increased many people in different communities, states, industries, and countries around the world have started working toward developing "clean" energy alternatives.

Carbon

Carbon is essential to life, but increasingly cast as a villain in today's environmental discussions and research. Worldwide burning of fossil fuels has noticeably increased carbon dioxide in the atmosphere. However, unless you live in a polluted city or near industrial plants, it's hard to believe the earth's immense atmospheric layers can't handle industrial pollutants. In fact, global warming is often dismissed as a vivid dream of knee-jerk environmentalists. It would be great if this were true.

Greenhouse gases are measurable. Global warming happens when heat is retained near the earth. This is science fact, not fiction, but there is one good bit of news. The Earth is a living, recycling, and changing system. It is an open system and allows adjustments.

In the summer of 2004, researchers reported in *Science* that the ocean had taken up nearly half of the carbon dioxide gas that humans had released into the air since the 1800s. This was good news: less atmospheric greenhouse gas, means less global warming. However, there was a downside to the news. Too much stored carbon will eventually limit the ocean's ability to soak up carbon dioxide. It will one day reach its maximum.

The carbon balancing act is a natural, interwoven process. It starts with plankton. These tiny organisms, drifting along on ocean currents, perform photosynthesis to produce energy and draw carbon out of the atmosphere. While building intricate calcium carbonate shells, they bind up carbon compounds. This circle of life is well understood.

Eventually, these organisms die and sink to the bottom of the deep ocean. A portion come to rest in the siliceous material that covers the abyssal plains, dropping out of the carbon cycle, while the rest dissolve in the low-calcium carbonate conditions of deep waters. Either way, the carbon from these organisms doesn't return to the atmosphere for a very long time.

A growing problem is the amount of human-produced carbon dioxide being absorbed into the ocean. If the levels of CO_2 get high enough, the ocean's top layer will become more acidic, reducing calcium carbonate's availability to plankton. Then, as human-produced carbon dioxide works its way down into the ocean, it may speed up the dissolution of calcium carbonate.

This becomes especially important when shell-making is impacted. In the last decade, scientists have found that even small decreases in the amount of calcium carbonate in seawater can limit plankton and coral's ability to build exoskeletons.

To better understand this situation, oceanographer Richard Feely (Pacific Marine Environmental Laboratory in Seattle) and others mapped the ocean's carbon chemistry. They compared what the ocean looked like before the industrial revolution (subtracting out carbon from fossil fuels) to current carbon values. Their findings, reported in the July 16, 2004, issue of *Science,* show that where human-produced carbon dioxide has sunk deep enough, the layer of carbonate-dissolving ocean water is now roughly 200 meters closer to the surface. If this continues, over several more centuries, the ability of organisms to create shells may be compromised. If sinking shells from dead organisms dissolve in this shallower ocean water and their carbon returns into the atmosphere a lot sooner, the greenhouse effect will speed up dramatically.

The take-home message is that the ocean may become less effective in the coming decades to serve as a sink for human-produced carbon dioxide. This will accelerate the atmospheric CO_2 buildup greatly, and add fuel to the changing climate fire.

Climate Change

Climate change includes temperature increases (global warming), rise in sea-levels, rainfall pattern changes, and increased incidence of extreme weather events. Scientific data discussed at the 2002 World Summit on Sustainable Development and the 2004 United Nations' Convention on Biological Diversity suggests that global warming is causing shifts in species' habitat and migrations that average 6.1 km per decade towards the poles. This shift, predicted by climate change models, notes that spring arrives 2.3 days earlier per decade, on average, in temperate latitudes. Entire boreal and polar ecosystems are also showing the effects of global warming.

> **Climate change** describes the difference in either the average state of the climate or its variability over an extended period of time.

Past changes in the global climate resulted in major shifts in species ranges and huge reorganization of biological communities, landscapes, and biomes during the last 1.8 million years. These changes occurred in a much simpler world than today's and with little or no pressure from human activities. Species' biodiversity is impacted by climate changes as well as human pressure and adaptation.

In a 2003, National Science Foundation (NSF) Report on Global Warming, Anthony Leiserowitz at the University of Oregon Survey Research Laboratory asked Americans their opinions on global warming. Some of the survey results showed that:

- Of the 92% of Americans who had heard of global warming, over 90% think the United States should reduce its greenhouse gas emissions;
- 77% support government regulation of carbon dioxide as a pollutant and investment in renewable energy (71%); and
- 88% support the Kyoto Protocol and 76% want the United States to reduce greenhouse gas emissions regardless of what other countries do.

Global warming is not just a theory anymore. Data confirms that it is definitely happening, and our biggest enemy is the human race. Environmental biologists and chemists have calculated that in terms of equivalent units of carbon dioxide, humans are releasing roughly the same amount of greenhouse gases into the atmosphere every two days as that of a Mount St. Helen's eruption. Not a good thing.

Global scientific agreement within the past 5 years is that the growing levels of heat-trapping gases are definitely affecting global climate and regional weather patterns. Those changes are causing a domino effect and impacting biodiversity.

Scientists have already seen regular and extensive effects on many species and ecosystems. In the past 30 years alone, climate change has resulted in large shifts in the distribution and abundance of many species. These impacts range from the disappearance of toads in Costa Rica's cloud forests to the death of coral reefs throughout the planet's tropical marine environments. This will only get worse in future decades.

Greenhouse Gas Inventories

We've seen how the combustion of coal, oil, and natural gas causes most global CO_2 levels to rise. Less well known is that trees absorb CO_2 during photosynthesis and then release it when cut down. The 34 million acres of tropical forests destroyed annually, an area the size of the state of New York, release between 20 and 25% of total global CO_2 emissions.

In 1992, at the United Nations Earth Summit in Rio de Janeiro, the international community first acknowledged the threat of climate instability. Over 185 nations agreed to reduce greenhouse gas emissions to their 1990 levels by 2012. More importantly, the participants agreed to stabilize atmospheric concentrations of greenhouse gases to prevent dangerous human interference with the global climate system.

The EPA's Clean Air Markets Division developed the annual Inventory of U.S. Greenhouse Gas Emissions and Sinks. This EPA atmospheric inventory estimates, documents, and evaluates greenhouse gas emissions and sinks for all source categories. To update the report, the Inventory Program polls dozens of federal agencies, academic institutions, industry associations, consultants, and environmental organizations for up-to-date information. It also gets data from a network of continuous carbon dioxide emission monitors installed at most U.S. electric power plants.

In 2002, the EPA reported the following greenhouse trends during the years 1990–2000:

- Total greenhouse gas emissions increased 14.2%;
- Carbon dioxide (CO_2), mostly from the burning of fossil fuels, was the dominant greenhouse gas;
- Methane emissions decreased by 5.6%;
- Nitrous oxide emissions increased by 9.8%; and
- Hydrofluorocarbon, perfluorocarbon, and sulfur hexafluoride emissions increased by over 29.6%.

If business-as-usual industrial output doesn't change, global CO_2 levels will double by the end of the twenty-first century.

Under the umbrella of the Intergovernmental Panel on Climate Change (IPCC), over 200 scientists and national experts worked together to develop a set of methodologies and guidelines to help countries create similar atmospheric inventories across international borders. Since then, scientists have determined that stabilizing atmospheric levels of carbon dioxide will mean reducing CO_2 emissions and other heat-trapping gases to 80% of the 1990 global levels. According to some models, this means decreasing or stopping the release of 1.2 trillion tons of CO_2 by 2050. This isn't an impossible mission. Energy efficiency improvements throughout the global economy could prevent one-third or more of these emissions while also cutting energy costs.

Greenhouse gas emission inventories are important for many reasons. Scientists use inventories of natural and industrial emissions as tools when developing atmospheric models. Policy makers use inventories to test policy compliance and progress. Regulatory agencies and corporations depend on inventories to confirm compliance with permissible emission rates. Businesses and the general public use inventories to better understand greenhouse gas sources and emission trends.

Most inventories contain the following information:

- Chemical and physical identity and properties of pollutants
- Geographical area affected
- Time period when emissions were generated
- Types of activities that cause emissions
- Description of methods used
- Data collected

Levels of greenhouse gases as well as global warming change in naturally occurring hotter and cooler cycles, independent of human activities. This was observed in data obtained from glacial ice cores taken in Antarctica. The industrial age just speeded up the release of these gases into the environment. We will study these ice core results in greater detail in Chapter 7.

REDUCING GREENHOUSE GASES

About 25% of the needed greenhouse gas decreases, about 370 billion tons, could be achieved by stopping tropical deforestation, restoring and conserving degraded lands, and improving land productivity worldwide through the use of best practices in agriculture and forestry. This can be done at local and regional levels, giving individuals an important role.

International climate policy usually focuses on lowering emissions by adopting alternative energy options. Using alternative energy, such as solar or wind power, without any accompanying adverse climate and biodiversity effects is crucial to reducing remaining energy-related emissions.

No analysis of climate change is complete without finding ways to lower greenhouse gas emissions from the burning of fossil fuels. The good news is that many environmental leaders and organizations are working with industry to reduce companies' carbon dioxide emissions, as well as enhance their competitive advantage. Industry is looking to energy-efficient methods like solar or wind power to save money that can, in turn, be used for research and development of even better methods, like nanotechnology energy transmission.

These new conservation strategies have led companies to invest in CO_2 reduction while also funding local conservation projects and economies. This augments biodiversity protection, and improves conditions in local communities. Some of the world's largest companies with strong environmental leadership records—like British Petroleum (energy), Intel (semiconductors), and SC Johnson (household products)—have committed large amounts of funding to researching environmentally friendly ways of doing business.

We should strive to create positive CO_2 reduction programs that help people, biodiversity, industry, and the atmosphere. Alone, this strategy will not reverse the global warming trend; industrial countries must still greatly reduce CO_2 emissions. However, it will help stabilize greenhouse gases while offsetting the massive deforestation currently taking place in global hotspots and wilderness areas. Along with atmospheric inventories and research to monitor habitat changes caused by climate change, such Earth-friendly plans will eventually stop and even reverse (think hundreds of years) the Earth's intensifying atmospheric plight.

Quiz

1. Greenhouses work
   (a) when the walls and ceiling are painted green
   (b) by trapping the sun's heat
   (c) best when arctic mosses are grown
   (d) when generators are used to maintain temperature

2. Greenhouse gases include all but which of the following?
   (a) Graphite
   (b) Water vapor
   (c) Nitrous oxide
   (d) Carbon dioxide

3. The burning of fossil fuels (oil, coal, and natural gas) that creates global warming and changes the environment is known as the
   (a) deforestation
   (b) enhanced greenhouse effect
   (c) lithification
   (d) biodiversity

4. Compared to carbon dioxide, how do nitrogen oxides trap heat in the atmosphere?
   (a) They don't trap heat at all
   (b) Much less efficiently
   (c) About 10 percent less efficiently than carbon dioxide
   (d) Much more efficiently

5. Of the 92 percent of Americans who had heard of global warming, what percentage thought the United States should reduce its greenhouse gas emissions?
   (a) 40 percent
   (b) 50 percent
   (c) 75 percent
   (d) 90 percent

6. When the earth's natural atmospheric gases decrease the amount of heat released from the atmosphere, it is known as the
   (a) wind power
   (b) winter solstice
   (c) greenhouse effect
   (d) green thumb effect

7. What percentage of total greenhouse gases comes from carbon dioxide?
   (a) 30 percent
   (b) 50 percent
   (c) 70 percent
   (d) 90 percent

8. The 2004 United Nations Convention on Biological Diversity suggests that global warming is causing
   (a) poor television reception
   (b) species habitat and migration shifts averaging 6.1 km/decade toward the poles
   (c) an increase of mosquitoes in southern climates
   (d) species habitat and migration shifts averaging 6.1 km/decade toward the equator

9. If the amount of human-produced carbon dioxide being absorbed into the oceans gets high enough, the ocean's top layer may become increasingly
   (a) acidic
   (b) basic
   (c) neutral
   (d) murky

10. British Petroleum, Intel, and SC Johnson have all committed to
    (a) supporting breast cancer research
    (b) sponsoring a 5K race on Earth Day
    (c) finding environmentally friendly ways of doing business
    (d) using wind power to power all their processes

Part One Test

1. Which kinds of clouds are often seen around mountain peaks?
   (a) Contrails
   (b) Stratus
   (c) Mammatus
   (d) Orographic

2. The large-scale clearing of all the trees from a land area by disease, burning, flooding, erosion, pollution, or volcanic eruption is called
   (a) sustainable use
   (b) deforestation
   (c) glaciation
   (d) toxicology

3. In early Greek mythology, the Earth goddess *Gaia,* mother of the Titans, was honored as
   (a) a great cook
   (b) the first Nobel prize–winner in earth sciences
   (c) an all-nourishing deity
   (d) an excellent stag hunter

4. In which layer does all the weather we experience take place?
   (a) Outer space
   (b) Troposphere
   (c) Mesosphere
   (d) Stratosphere

5. Weak tornadoes (F0/F1) make up roughly what percentage of all tornadoes?
   (a) 25%
   (b) 50%
   (c) 75%
   (d) 100%

6. Atmospheric pollutants from volcanic eruptions affect
   (a) only insects and plants
   (b) all living organisms worldwide
   (c) only humans living at the base of the mountain
   (d) polar climates only

7. In the United States, what percentage of the species on the endangered animals list live in or rely on wetlands?
   (a) 20%
   (b) 35%
   (c) 50%
   (d) 60%

8. Greenhouse gases are
   (a) hypothetical
   (b) measurable
   (c) mostly made up of argon
   (d) immeasurable

9. The land below the levels of the seas is known as the
   (a) oceanic crust
   (b) equator
   (c) continental crust
   (d) wetlands

10. Wetlands are
    (a) good places to find cactus
    (b) not found in the United States

(c) natural filters through which water trickles down to underground reservoirs
(d) not used as resting spots by migratory birds

11. The greatest impact on the greenhouse effect has come from
    (a) industrialization
    (b) increases in benzene
    (c) cigar smoke
    (d) hunters and gatherers

12. The oceanic crust
    (a) is thinner than the continental crust
    (b) is thicker than the continental crust
    (c) is about the same thickness as the continental crust
    (d) cannot be compared to the continental crust

13. Due to the undeniable evidence of global warming, the National Academy of Sciences has
    (a) decided to ignore greenhouse gases since they produce colorful sunsets
    (b) ordered fans for all its scientists to help them keep cool while they work
    (c) predicted global temperature increases of 3° to 9°F in the next 100 to 200 years
    (d) become discouraged and now refuses to tell anyone anything

14. The Gaia hypothesis which stated that the Earth (Gaia) existed as a single living organism was written by
    (a) Marc Williams
    (b) Richard Gordon
    (c) George Bennett
    (d) James Lovelock

15. What do scientists call the landmass that was in one big chunk or continent before it broke up into separate continents?
    (a) Panacea
    (b) Pangaea
    (c) Pongo Pongo
    (d) Atlantis

16. For hurricanes to begin forming, meteorologists have found that ocean water must be warmer than
    (a) 18.5°C
    (b) 24.5°C
    (c) 26.5°C
    (d) 36.5°C

17. Rain forests and plants are important because they absorb CO_2 from the air through
    (a) osmosis
    (b) butterflies
    (c) the Krebs cycle
    (d) photosynthesis

18. Dinosaurs like *Triceratops* and *Allosaurus* are
    (a) extinct
    (b) about the size of German Shepherd dogs
    (c) usually found in shopping malls
    (d) increasing in number

19. Nitrogen dioxide, which gives the sky that yellow-brown look, is also known as
    (a) the Aurora Borealis
    (b) smog
    (c) sleet
    (d) turbidity

20. The thinner, extended edges of a continental landmass found below sea level is called the
    (a) continental shelf
    (b) transcontinental bridge
    (c) continental bookcase
    (d) intercontinental wedge

21. In the Gaia concept, humans are seen as
    (a) tourists
    (b) the singular most important species
    (c) dumber than dirt
    (d) one species among millions

22. The first phase in the formation of a hurricane is that
    (a) ducks fly south
    (b) temperature rises
    (c) barometric pressure drops
    (d) barometric pressure rises

23. How well do nitrogen oxides trap heat in the atmosphere compared to carbon dioxide?
    (a) Much less efficiently
    (b) Much more efficiently
    (c) About the same
    (d) Nitrogen oxides do not exist in the atmosphere

24. When unique plant or animal species naturally occur in only one area or region, they are said to be
    (a) exotic
    (b) adventitious
    (c) epidemic
    (d) endemic

25. Which of the following is a colorless, odorless, flammable hydrocarbon, released by the breakdown of organic matter and the carbonization of coal?
    (a) Methane
    (b) Silicon
    (c) Graphite
    (d) Sulfur

26. In 2004, scientists reported in *Science* that the ocean had taken up what percentage of the carbon dioxide gas that humans had released into the air since the 1800s?
    (a) 30%
    (b) 40%
    (c) 50%
    (d) 70%

27. The Fujita scale is used to measure
    (a) fish
    (b) wind damage
    (c) total body fat
    (d) rain damage

28. Old-growth forests are made up of trees that are often
    (a) 10 years old
    (b) 50 years old
    (c) 100 years old
    (d) several thousand years old

29. Which cloud type is the most familiar of the basic cloud shapes?
    (a) Nimbostratus
    (b) Billows
    (c) Cumulus
    (d) Lenticular

30. As global warming has increased, many people around the world have
    (a) started building bigger cars
    (b) begun developing "clean" energy alternatives
    (c) started going to the beach more
    (d) stopped buying jackets

31. A complex community of plants, animals, and microorganisms linked by energy and nutrient flows is called an
    (a) aquarium
    (b) equilateral atoll
    (c) ecosystem
    (d) alternate dimension

32. The differences in the climate's average state, or its regular cyclic variability, over an extended period of time is known as
    (a) early frost
    (b) climate regularity
    (c) spring planting
    (d) climate change

33. When an animal or plant has a specific relationship to its habitat or other species, it is said to fill
    (a) an ecological niche
    (b) a crevasse
    (c) a sedimentary layer
    (d) a liturgical niche

34. Air pressure is measured with a
    (a) thermometer
    (b) pH meter

(c) stethoscope
(d) barometer

35. One example of the Earth's overall equilibrium is seen in
    (a) ocean salinity levels
    (b) Hollywood disaster films
    (c) growth rings of trees
    (d) roadside mileage signs

36. What is the primary cause of declining biodiversity?
    (a) Increase of mating pairs
    (b) Habitat loss
    (c) A hole in the ozone
    (d) Pesticides

37. Continents are broken up into how many major land masses?
    (a) 4
    (b) 6
    (c) 8
    (d) 10

38. Which of the following is considered one of the world's "hottest of hot spots"?
    (a) Sicily
    (b) Melbourne, Australia
    (c) The Atlantic Forest
    (d) The Himalaya Mountains

39. The ever-changing total water cycle, part of the closed environment of the Earth, is called the
    (a) magnetosphere
    (b) hydrosphere
    (c) lithosphere
    (d) trophosphere

40. Which North American mammals were killed in huge numbers during and following the construction of the first east-west railroad?
    (a) Horned toads
    (b) Lemmings
    (c) Gophers
    (d) Buffalo

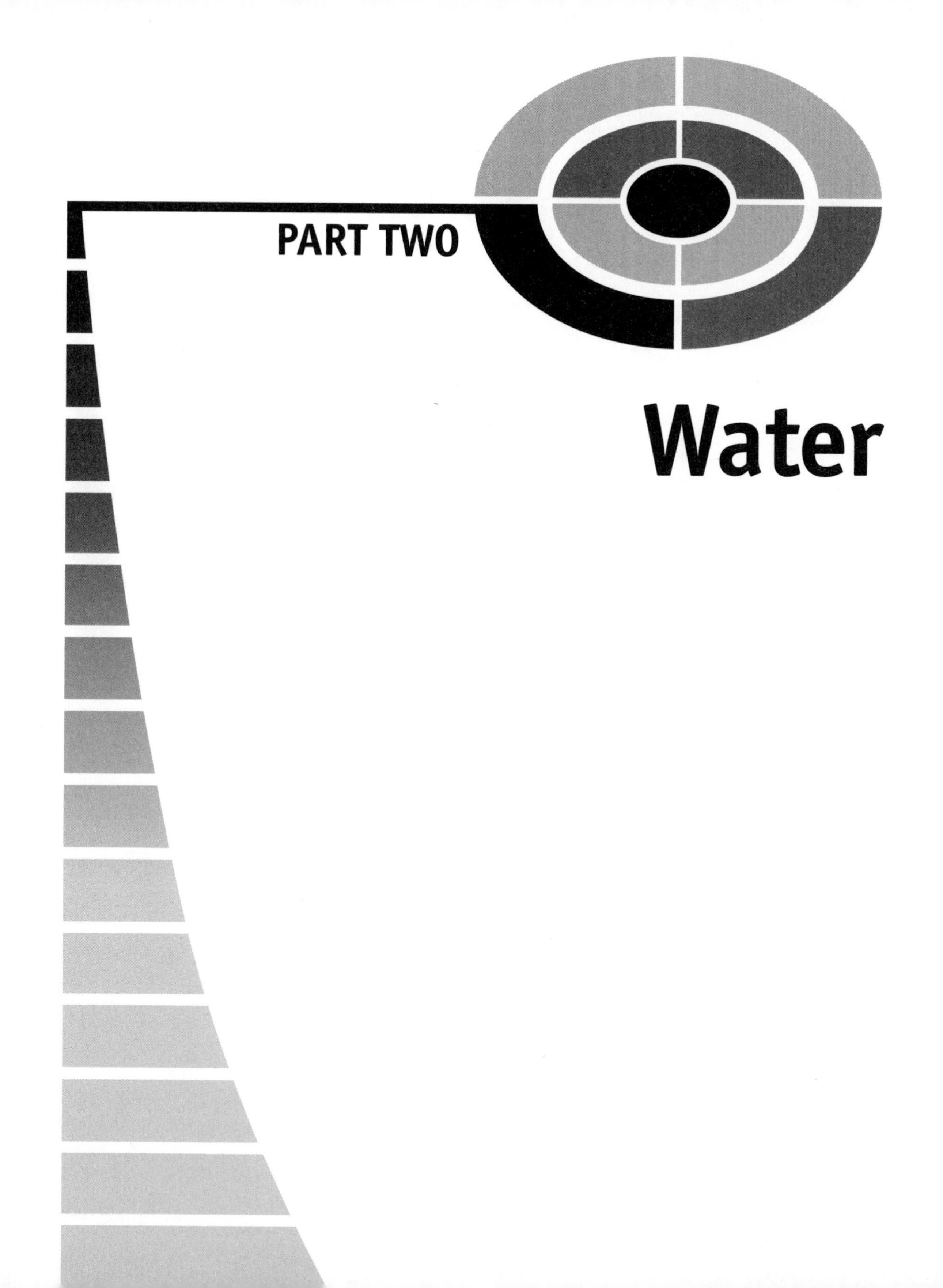
PART TWO
Water

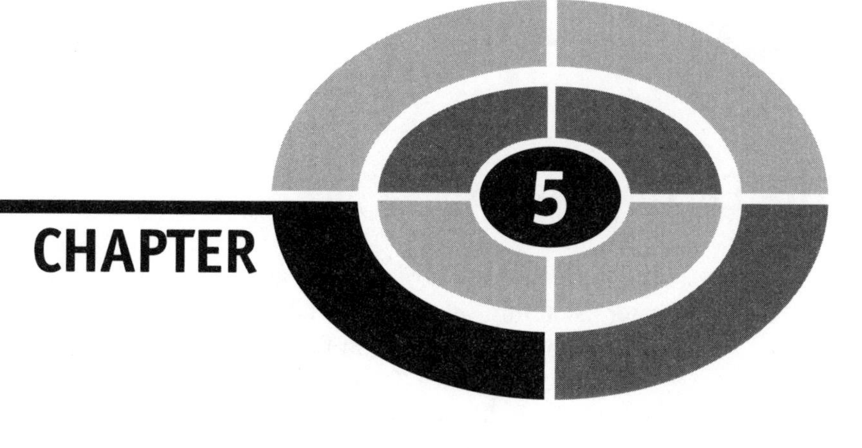

CHAPTER 5

The Hydrologic Cycle

We can't escape our need for water, even if we wanted to. Our bodies are made up of 66% water and require water daily! We developed in amniotic fluid; we can't last more than two or three days without water; we wash ourselves and nearly everything else in water; we get much of our food from water (oceans); and, we travel on water. Water is king!

> **Hydrology** is the study of the occurrence, distribution, and movement of water on, in, and above the earth.

If water is king, then the *hydrologic cycle* is queen. This natural, give-and-take balancing system is the agent of the planet's water supply. It recycles the earth's crucial water supply in many different forms.

Since the beginning of time, the Earth's water has been used over and over. Recycling is not a recent idea. The water you're drinking today may have been, at one time, part of a tropical bay with a basking *Brachiosaurus* family enjoying the beach nearby. Thanks to the hydrologic cycle, that same swallow of water has probably been a liquid, solid, and gas infinitely many times throughout geologic time. Time after time, water plays important, but different, ecological roles as it moves from one form to another.

The Earth's Water

Water covers over 70% of the Earth's surface, but it's hard to picture that much water. Standing on a beach and looking seaward, ocean water stretches to the horizon and seems to go on forever.

The oceans hold 97% of the Earth's water, the land masses hold 3%, and the atmosphere holds less than 0.001%. The water on the land masses is stored as fresh water in glaciers and icecaps, groundwater, lakes, rivers, and soil. The annual precipitation for the Earth has been estimated at more than 30 times the atmosphere's total ability to hold water. This points to the fact that water is quickly recycled between the Earth's surface and the atmosphere. Table 5-1 shows how long water stays in a reservoir before recycling.

Even the world's total amount of rainfall is incredibly large. To give you an idea of the volume of water we are talking about, think of this: If all the rain that falls on the Earth in one year fell on the state of Texas (total area around 692,408 km^2) in one day, the entire state would be covered with approximately 560 meters of water!

The oceans are the source of most of the atmosphere's evaporated moisture. Of this, around 90% is returned to the oceans through precipitation. The remaining 10% is blown across land masses where temperature and pressure changes

Table 5-1 Water is stored for different time periods in different places.

Water source	Time in location
Atmosphere	8 days
Rivers	16 days
Soil moisture	70 days
Snow	140 days
Glaciers	50 years
Lakes	100 years
Shallow groundwater	200 years
Oceans	1,000 years
Aquifers and deep groundwater	10,000 years

lead to rain or snow. Any other water not lost through evaporation and rainfall balances the cycle through runoff and groundwater that flows back to the seas.

Water in the atmosphere is thought to be replaced every eight days. Water found in oceans, lakes, glaciers, and groundwater is recycled more slowly. Water exchange in these reservoirs may take place over hundreds or thousands of years.

A water **reservoir** is a place in the atmosphere, ocean, or underground, where water is stored for some period of time.

Some water resources (like groundwater) are being consumed by humans at rates faster than can be resupplied. When this happens, the water source is said to be *nonrenewable*. Table 5-2 illustrates the amount of water held in different natural reservoirs and the amount of fresh water from different sources.

WATER IS KING

In order to understand how water became "king," let's take a look at some of water's unique properties.

Of all the chemical formulas that are written, the one that most people remember is H_2O, or the formula for water. Water is made up of two hydrogen atoms bonded to an oxygen atom. However, because of other bonding factors, water has a variety of special properties that make it universally important.

Table 5-2 Fresh water is mainly found in glaciers.

Source	Volume of water (km^3)	% of total	% of fresh water
Oceans	1,328,000,000	97.20	0
Groundwater	8,400,000	0.62	22
Glaciers	29,300,000	2.15	72
Fresh and saline lakes, inland seas	230,000	0.02	6
Rivers and atmospheric humidity	14,000	0.001	Trace

Water changes forms very easily. It slips easily between three different forms:

- Solid (ice);
- Liquid (water); and
- Gas (water vapor).

At sea level, pure water boils at 100°C and freezes at 0°C. At high altitudes and low atmospheric pressures, the temperature at which water boils goes down. When salt is dissolved in water, the freezing point of water is lowered. To get rid of ice on streets, trucks sprinkle salt on snowy streets. The salt makes it harder for water to freeze into ice.

Water absorbs or releases more heat for each degree of temperature rise or fall than any other substance. For this reason, it's an excellent cooling liquid and heat transference medium for thermal and chemical experiments involved in industrial processes.

When temperatures between lakes and rivers and the outside air are different, the outside air reacts in different ways. Fog forms if a cold lake drops the temperature of the surrounding air enough to cause saturation. When this happens, tiny water droplets form and float in the air.

As a drop of rain falls through the air, it dissolves atmospheric gases. Then, when it hits the ground, rivers, or oceans, it influences their compositions.

Oceans and large lakes, like the Great Lakes between the United States and Canada, have a big impact on the climate. They serve as the world's heat reservoirs and heat exchangers. Because of this ability, they account for much of the water that falls yearly as rain and snow over nearby continental land.

Water has the highest *surface tension* of any liquid except mercury. This tight bonding of water molecules forms a strong layer, allowing water to support things that are heavier and denser. A pine needle, hair, or insect (water skipper) can float or move around on this strong, thin surface layer.

Surface tension measures the strength of a liquid's thin surface layer.

Surface tension is a big player in the wind's ability to use energy to push water into waves. Waves, in turn, are crucial to the rapid circulation of oxygen in lakes and oceans.

Another kingly aspect of water's personality is its ability to dissolve other compounds. Nearly everything dissolves in water. In fact, water is called the *universal solvent.* Without water, life wouldn't be possible on Earth. It plays a huge part in the transfer of essential nutrients to all life on the planet.

Hydrologic Cycle

Remember from Chapter 1 that the hydrosphere, crust, and atmosphere combined make up the biosphere. The hydrosphere includes all the water in the atmosphere and on the Earth's surface.

When the sun heats the oceans, the cycle starts. Water evaporates and then falls as precipitation in the form of snow, hail, rain, or fog. While it's falling, some of the water evaporates or is sucked up by thirsty plants before soaking into the ground. The sun's heat also keeps the cycle going.

> The **hydrologic cycle** is made up of all water movement and storage throughout the Earth's hydrosphere.

Water is constantly circulating between the atmosphere and the Earth and back to the atmosphere through a cycle involving *condensation, precipitation, evaporation,* and *transpiration.* This is called the *hydrologic cycle.* Fig. 5-1 illustrates the many ways water is transported through the hydrologic cycle.

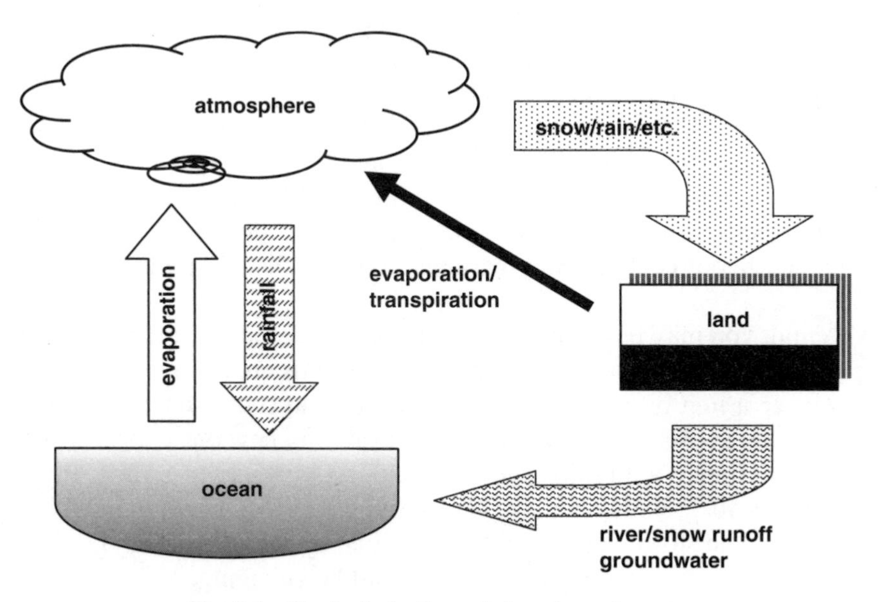

Fig. 5-1 The hydrologlic cycle is a dynamic system.

Water vapor is carried by wind and air currents throughout the atmosphere. When an air mass cools down, its vapor condenses into clouds and eventually falls to the ground as *precipitation* in the form of snow, rain, sleet, or hail.

Water takes one route from the atmosphere to the ground, but can take a variety of paths and time periods to get back up into the atmosphere. These paths include the following:

- Absorption by plants;
- Evaporation from the sun's heating;
- Storage in the upper levels of soil;
- Storage as groundwater deep in the earth;
- Storage in glaciers and polar regions;
- Storage or transport in springs, streams, rivers, lakes; and
- Storage in the oceans.

When water is stored somewhere for any length of time, it is called a *water reservoir*. A reservoir is a holding area. Nature's reservoirs are oceans, glaciers, polar ice, underground storage (*aquifers*), lakes, rivers, streams, the atmosphere, and the biosphere (within living organisms).

Surface water in streams and lakes returns to the atmosphere as a gas (vapor) through the process of *evaporation*.

Water held inside plants returns to the atmosphere as a vapor through a biological process called *transpiration*. When plants pull water up through their roots from the soil, use some of the dissolved minerals to grow, and then release the water back through the leaves, the entire cycle is known as *evapotranspiration*. This happens the most during times of high temperatures, wind, dry air, and sunshine. In temperate climates, this occurs during the summertime.

When air currents rise into the colder atmospheric layers, water vapor condenses and sticks to tiny particles in the air. This is called *condensation*. When a lot of water vapor coats enough particles (dust, pollen, or pollutant), it forms a *cloud*.

As a child, you may have lain on your back and stared up at the clouds while trying to spot different shapes. It's fun because clouds are always changing. Just when you see a lion or armadillo, the cloud drifts into the shape of a dog or a tree. Clouds are constantly reevaporating and forming new ones. Unless you live in a very dry climate with only one small cloud in the sky at a time, the cloud you see now will probably not be the one you see in 15 minutes. Unless, of course, it's a large storm system getting ready to dump a boat-load of water on your picnic! Even then, the cloud shapes would be changing.

As the air gets wetter and wetter (saturated), water droplets accumulating within the cloud get bigger and bigger. When these droplets get too heavy, gravity wins and they fall as *precipitation.*

> **Precipitation** can take the form of rain, snow, sleet, or hail, depending on temperature and other atmospheric conditions.

[*Note:* Throughout this chapter, I often use the words "rain" and "rainfall" interchangeably with "precipitation." Just know that precipitation includes rain, drizzle, mist, fog, sleet, hail, snow, and any other wet stuff falling from the sky.]

After rain hits the ground, it can evaporate quickly, be absorbed (by the land or the sea), or run off into storm sewers, streams, or rivers.

Even though the hydrologic cycle balances what goes up with what comes down, one part of the cycle gets stuck in polar regions during the wintertime. In cold climates, rain is stored as snow or ice on the ground for several months. In glacial areas, the time period can extend from years to thousands of years. Then, as the temperature climbs in the spring, the water is released. When this happens in a very short period of time, flooding occurs.

The hydrologic cycle is an endless loop. With transpiration and evaporation happening constantly, the cycle goes on and on and on. Every time a molecule of H_2O goes through the hydrologic cycle, it is recycled and is ready to begin the adventure all over again.

Let's look at the different parts of the hydrologic cycle more closely. Because we are talking about water, each process has some variations.

EVAPORATION

The sun provides the energy that powers evaporation. When water is heated, its molecules get excited and vibrate so much that they break the bonds holding them together. When this happens, solar energy (light and heat) causes water to *evaporate* from oceans, lakes, rivers, and streams. Rising from the earth's surface, warm air currents sweep water vapor up into the atmosphere.

> When water changes its form from a liquid to a gas (vapor), it is said to **evaporate.**

Because of the huge amount of water in the oceans, it makes sense that roughly 80% of all evaporation comes from the oceans, with the extra 20% coming from inland water and plant transpiration. Wind currents transport water vapor around the world, influencing air moisture worldwide. For example, a humid summer day in Iowa is brought about by warm tropical winds blown northward from the Gulf of Mexico. When water vapor isn't in a cloud, it generally exists as *humidity*. Evaporation takes place during hot summer months and in equatorial regions.

The hydrologic cycle constantly keeps water moving and fresh. Hydrologists (scientists who study the Earth's water cycle) estimate that 100 billion gallons of water a year are cycled through this process.

Without the hydrologic cycle, life on Earth would not have developed. Every description of creation tells how the oceans were formed before the continents and their inhabitants. We use water for everything, both internally and externally. Without water, life would not exist on Earth. It is second in importance only to the air we breathe.

CONDENSATION

This part of the hydrologic cycle begins with *condensation,* when water vapor condenses in the atmosphere to form clouds. Condensation takes place when the air or land temperature changes. Water shifts forms when temperatures rise and fall. You see this in the early morning when dew forms on plants.

As water vapor rises, it gets cooler and eventually *condenses,* sticking to minute particles of dust in the air. Condensation describes water's change from its gaseous form (vapor) into liquid water. Condensation generally takes place in the atmosphere when warm air rises, then cools and loses its ability to cling to water vapor. As a result, extra water vapor condenses to form cloud droplets.

When climatic conditions are right, clouds form, winds blow them around the globe, and water vapor is distributed. When clouds cannot hold any more moisture, they dump it in the form of precipitation, usually rain or snow.

TRANSPORT

Next in the hydrologic cycle, *transport* is the movement of water through the atmosphere. Commonly, this water moves from the oceans to the continents. Some of the earth's moisture transport is visible as clouds, which consist of ice crystals and/or tiny water droplets. Clouds are propelled from one place to another by the jet stream, surface-based circulations (like land and sea breezes), or other mechanisms. However, a typical 1-kilometer-thick cloud contains only enough water for roughly 1 millimeter of rainfall, whereas the amount of moisture in the atmosphere is usually 10 to 50 times greater than that.

Most water is transported in the form of water vapor, which is actually the third-most abundant gas in the atmosphere. Water vapor may be invisible to us, but not to satellites, which are capable of collecting data about the atmosphere's moisture content. The United States Geological Survey and National Weather Service are constantly monitoring atmospheric pressure, temperature, currents, and weather systems to better understand yearly fluctuations as well as predict severe weather.

PRECIPITATION

Precipitation is the main way that water is transported from the atmosphere to the Earth's surface. There are different types of precipitation, but the most common is rain.

Water moisture falls in the form of rain, snow, and hail from clouds circling the globe, propelled by air currents. Clouds are fairly active. For example, when clouds rise over mountain ranges, they cool, becoming so full of moisture that water falls as rain, snow, or hail, depending on local air temperatures.

Rainfall levels vary widely by location. Across the United States, depending on the terrain, rainfall amounts can be very different. For example, some deserts in Nevada average less than 3 centimeters of rainfall annually. The Midwest receives approximately 35 centimeters of rainfall per year, while tropical rain forests in places like Hawaii get more than 150 centimeters of rainfall per year. It rains almost every day!

Rainfall can change from one year to the next. In 1988, Kansas and Nebraska experienced a severe drought, which stressed crops and reduced yields. Then in 1993, the same areas went through heavy flooding, which again damaged crops, but this time from too much water.

Some people believe the Pacific Northwest of the United States gets excessive rain and that the sun rarely shines. Well, like anything else, it depends on the time of year and location. The annual rainfall in some regions of Washington state is more that 450 centimeters per year, while other areas get only 25 centimeters per year or less. This rainfall difference comes from the nature of the land and its topography (shape). High rainfall amounts are mainly found on the western side of the Cascade Mountains, while light rainfall is found on the eastern, or *rain shadow,* side of the mountain.

TRANSPIRATION

Another type of evaporation that adds to the hydrologic cycle is *transpiration.* This is a little more complicated. During transpiration, plants and animals release moisture through their pores. This water rises into the atmosphere as vapor.

Transpiration is most easily seen in the winter when you can see your breath. When exhaling carbon dioxide and used air, you also release water vapor and heat. Your warm, moist exhalation on a frosty winter morning becomes a small cloud of water vapor.

Transpiration from the leaves and stems of plants is crucial to the air-scrubbing capability of the hydrologic cycle. Plants absorb groundwater through their roots deep in the soil. Some plants, like corn, have roots a couple of meters in length, while some desert plants have to stretch roots over twenty meters down into the soil. Plants pull water and nutrients up from the soil into their leaves. It is estimated that a healthy, growing plant transpires 5 to 10 times as much water volume as it can hold at one time. This pulling action is driven by the evaporation of water through small pores in plants' leaves. Transpiration adds approximately 10% of all evaporating water to the hydrologic cycle.

Groundwater

The water found below the Earth's surface is commonly called *groundwater.* This is the water found in subterranean spaces, cracks, and the open pore spaces of minerals. Depending on the geology and topography, groundwater is either held in one area or flows away toward streams. It can also be tapped by wells. Some groundwater is very old and may have been stored for thousands of years.

It is sometimes easy to think of groundwater as all the water that has been under the land's surface since the Earth was formed, but remember that nearly all water is circulating through the hydrologic cycle.

The *residence time,* or length of time that water spends in the groundwater portion of the hydrologic cycle, differs a lot. Water may spend as little as days or weeks underground or as much as 10,000 years. Residence times of tens, hundreds, or even thousands of years are not unknown. Conversely, the average turnover time of river water, or the time it takes a river's water to completely replace itself, is roughly two weeks. Look back to Table 5-2 to see the different times water is stored in various places.

Groundwater represents all the water that has soaked into the ground and is found in one of two soil layers. The layer closest to the soil's surface is called the *zone of aeration,* where spaces between soil particles and types are filled with both air and water.

Under this surface layer lies the *zone of saturation,* where all the open spaces have become filled with water. The depth of these two layers is often dependent on the topography and soil makeup of an area.

> The **water table** is found at the upper edge of the zone of saturation and the bottom edge of the zone of aeration.

The *water table* is the boundary between these two layers. As the amount of groundwater increases or decreases, the water table rises or falls accordingly. When the entire area below the ground is saturated, flooding takes place. Any later rainfall is forced to remain above ground. Fig. 5-2 shows the location of the water table line in relation to other ground layers.

The amount of open space in the soil is called soil *porosity.* A soil that is highly porous has many gaps for water storage. The rate at which water moves through the soil is affected by the *soil permeability.* Different soils hold different amounts of water and absorb water at different rates.

Hydrologists keep track of an area's soil permeability in an effort to predict flooding. As the soil pores fill, more and more water is absorbed and there are fewer places for the extra water to go. When this happens, subsequent rainwater can't be absorbed by otherwise thirsty soil and plants, and is added to flood waters. Flooding happens often in the winter and early spring, since water can't penetrate frozen ground. Rainfall and snow melt have nowhere to go and become runoff.

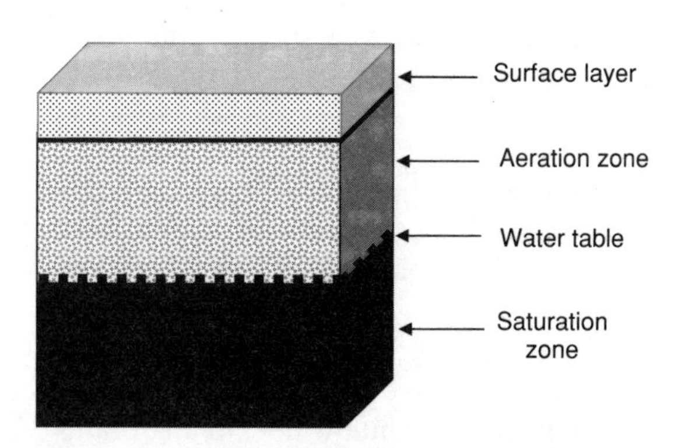

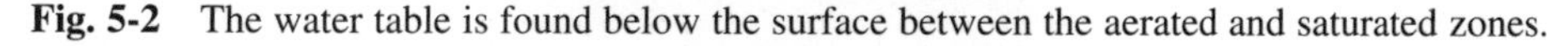
Fig. 5-2 The water table is found below the surface between the aerated and saturated zones.

Watershed

Streams and rivers get their water from the areas around them, including hills, valleys, and mountains. The water runs downhill to existing grooves like stream or river beds. During hot months, these dry out but are refilled by rain or melting snow later in the season.

> The geographical region from which a stream gets water is called a **drainage basin** or **watershed.**

The boundary between two watersheds is called a *divide.* The continental divide in North American is the high line running through the Rocky Mountains. Rainfall and streams on the east side of the Rocky Mountains drain to the Atlantic Ocean or Gulf of Mexico, while flowing water from the western slopes of the Rocky Mountains runs to the Pacific Ocean.

Drainage basins are complex structures that can have as many as thousands of streams and rivers draining them, depending on their geographical size. Streams are described by the number of tributaries that drain into them.

> A **tributary** is a stream that flows into another stream.

For example, a first-order stream is a stream with no tributaries. A second-order stream has only first-order tributaries flowing into it. The best way to remember this is that the higher the stream order, the larger the watershed that drains into it. Fig. 5-3 shows the ordering of streams.

> A **stream gage** refers to a specific site along a stream's length where flow measurements are taken.

In the United States, the United States Geological Survey (USGS) is the main federal agency that keeps records of natural resources. Within the USGS, the Water Resources Division monitors water resources. It has collected and tabulated USGS data for many years, with much of it available on the Internet.

The USGS has counted approximately 1.5 to 2 million streams in the United States. These streams have increasingly larger areas of drainage and flow rates. The Mississippi River, a tenth-order stream, drains 320 million km^2 of land area.

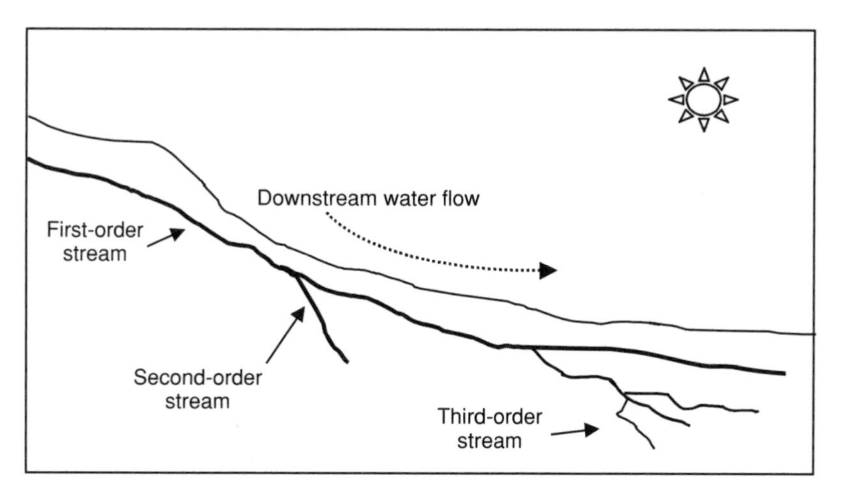

Fig. 5-3 Streams are ordered according to how many other streams empty into them.

Hydrology also includes the study of water motion and water-borne elements carried along as dissolved quantities or in separate phases. A related aspect of hydrology determines flow rates for rivers and streams. This information, important in the design and evaluation of natural and manmade channels, bridge openings, and dams, is a critical factor in understanding drainage effects.

Stream drainage systems are usually in one of the following four channel configurations:

- Dendritic
- Trellis
- Rectangular
- Radial

Most streams follow a branching drainage pattern called *dendritic,* from the Greek word dendros, meaning "tree." Dendritic drainage patterns are irregular with tributaries at various angles from the main stream (tree trunk).

As with most other water patterns, the surrounding landscape has an effect on drainage patterns. In *trellis drainage,* the tributary streams link up with larger streams at sharp angles. Trellis drainage is also found where there are several parallel, hard rock ridges that direct water flow between the ridges.

Rectangular drainage has right angle (90°) bends in all the streams. The subsidiary stream sections are also at right angles and are about the same length.

Radial drainage is found flowing from the top of a mountain peak. There is a high central point from which all streams flow downward in all directions.

Fig. 5-4 illustrates the four types of watershed drainage seen in streams of different orders. Each drainage type is dependent on local landforms and the types of rock and soil present.

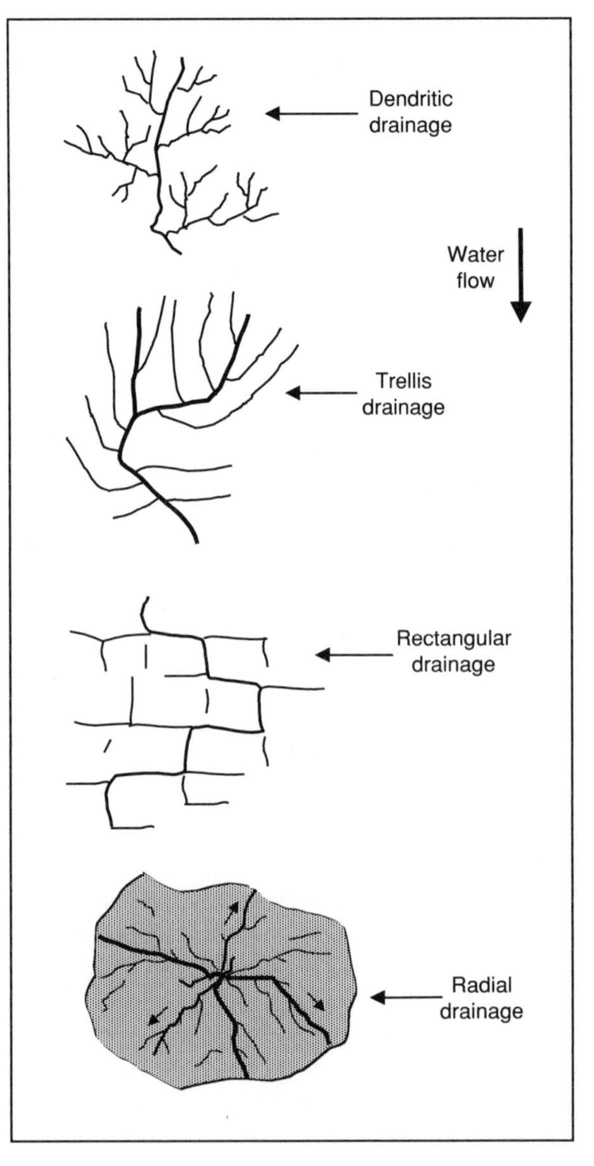

Fig. 5-4 There are four main types of stream drainage patterns.

Aquifers

Groundwater is stored in *aquifers.* Aquifers are large underground water reservoirs. There are two main types of aquifers: *porous media aquifers* and *fractured aquifers.*

Porous media aquifers are made up of combined individual particles such as sand or gravel. Groundwater is stored in and moves through the spaces between the individual grains.

Permeable soil includes lots of interconnected cracks or spaces that are large enough to let water move freely. In some permeable materials, groundwater moves several meters a day, while other regions only allow water to flow a few centimeters in 100 years.

When soil is very porous (grains not touching each other), the soil is called *unconsolidated.* However, when sand grains are cemented and squashed together, such aquifers are known as *consolidated porous aquifers.* Sandstone is an example of a consolidated porous material.

Fractured aquifers, as the name implies, are made up of broken rock layers. In fractured aquifers, groundwater moves through cracks, joints, or fissures in otherwise solid bedrock. Fractured aquifers are often made of granite and basalt.

Limestones can also form fractured aquifers, but often their gaps and splits are expanded when the limestone is dissolved away by flowing mineral-rich water. When this happens, wide underground channels are formed. When there have been a lot of chemical reactions and dissolving of limestone over a long period of time, creating a cave or cavern, the area is called a *karst.*

Porous material like sandstone is so tightly cemented or recrystallized that all the original open space has filled in. The rock has become solid. However, it is still possible for it to crack and serve as a fractured aquifer. We will look at karst development more closely in Chapter 11 when we study the calcium cycle.

We learned that when water soaks into the ground, gravity pulls it down through open spaces, pores, cracks, or fissures until it reaches a depth where all the soil or rock spaces are full. At this point, the soil or rock becomes saturated, and the water table line is drawn. However, the water table is not always at the same depth below the land's surface.

During heavy rainfall periods (months, years), the water table can rise. Conversely, during times of low rainfall and high evapotranspiration, the water table falls. Remember the area above the water table is the unsaturated zone, while the area below the water table is called the saturated zone. Groundwater is found predominantly in the saturated zone.

Rainfall soaks into the soil and moves downward until it hits impenetrable rock, at which time it turns and begins to flow sideways. Aquifers are often found in places where water has been redirected by an obstacle. Hydrologists are concerned about surface pollutant runoff that makes its way down into underground aquifers.

Groundwater in soil or rock aquifers within the saturated zone adds up to large quantities of water. Aquifers forming a water table that separates the unsaturated and saturated zones are called *unconfined aquifers* since they flow right into the saturated zone.

Groundwater doesn't flow well through fairly impermeable matter such as clay and shale. Some aquifers, however, lie beneath layers of impermeable materials like clay. These are known as *confined aquifers.* These aquifers do not have a water table that separates the unsaturated and saturated zones.

Confined aquifers are more complex than unconfined aquifers that flow freely. Water in a confined aquifer is commonly under pressure. This causes the well water level to rise above the level of the water in the aquifer. The water in these reservoirs rises higher than the top of the aquifer because of the confining pressure. When the water level is higher than the ground level, the water flows freely to the surface forming a *flowing artesian well.*

A *perched water table* occurs when water is blocked by a low-permeability material below the aquifer. This disconnects the small, perched aquifer from a larger aquifer below. They are separated by an unsaturated zone and a second, lower water table.

Groundwater goes back to the surface through aquifers that empty into rivers, lakes, and the oceans. Fig. 5-5 shows how water flows across the landscape and aquifer locations. The flow of groundwater is much slower than runoff, with speeds usually measured in centimeters per day, meters per year, or even centimeters per year.

> When water flowing from an aquifer gets back to the surface, it is known as **discharge.**

Groundwater flows into streams, rivers, ponds, lakes, and oceans, or it may discharge in the form of springs, geysers, or flowing artesian wells.

One problem that concerns hydrologists is the depletion of water from aquifers faster than they can naturally be refilled. Humans are the primary cause for aquifer depletion, called *groundwater mining.* Groundwater is pumped out of aquifers for drinking water or to irrigate crops.

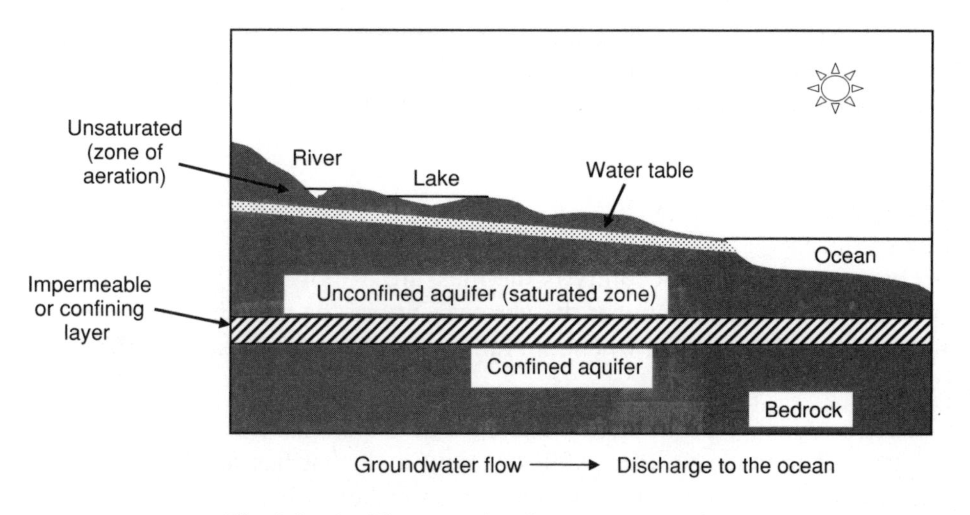

Fig. 5-5 Aquifiers are subsurface water reservoirs.

When water enters an aquifer it is called *recharge*. Groundwater often discharges from aquifers to replenish rivers, lakes, or wetlands. An aquifer may receive recharge from these sources, an overlying aquifer, or from rainfall or snow melt followed by infiltration.

Infiltration takes place when precipitation soaks into the ground. This depends mostly on the soil porosity. Once in the ground, infiltrated water eventually becomes groundwater and is stored in aquifers. However, if it is quickly piped back out to the surface, the levels can't build up.

> The **aquifer recharge zone** is that area, either at the surface or below ground, that supplies water to an aquifer and may include most of the watershed or drainage basin.

Groundwater discharge adds considerably to surface water flow. During dry periods, some streams get nearly all their water from groundwater. Throughout the year, the physical and chemical characteristics of underground formations have a big effect on the volume of surface runoff.

While the rate of aquifer discharge controls the volume of water moving from the saturated zone into streams, the rate of recharge controls the volume of water flowing across the earth's surface. During a rainstorm, the amount of water running into streams and rivers depends on how much rain the underground area can absorb. When there is more water on the surface than can be absorbed by the groundwater zone, it runs off into streams and lakes.

RUNOFF

We have seen that permeability is the measure of how easily something flows through a material. The higher the permeability of the soil, the more rain seeps into the ground. However, when rain falls faster than it can soak into the ground, it becomes *runoff.*

> **Runoff** is made up of rainfall or snow melt that has not had time to evaporate, transpire, or move into groundwater reserves.

Water always takes the path of least resistance, flowing downhill from higher to lower elevations, eventually reaching a river or its tributaries. All of the land that eventually drains to a common lake or river is considered to be in the same watershed or runoff zone. Watershed drainages are defined by topographic division that separate surface flow between two separate water systems.

Water runoff takes place when water flows over land into local streams and lakes; the steeper the land and the less porous the soil, the greater the runoff. Rivers join together and eventually form one major river that carries all of the sub-basins' runoff into the ocean.

Overland runoff is mainly visible in urban areas. Land use activities in a watershed can affect the quality of surface water as contaminants are carried away by runoff and groundwater, especially through *infiltration* of pollutants. Understanding the factors that affect the rate and direction of surface and groundwater flow helps to determine where good water supplies exist and how contaminants migrate.

FLOODING

In July of 1976, a summer storm grew over the Colorado Rockies in an area stretching between Loveland and Estes Park, Colorado. The Big Thompson Canyon River collected the rainfall from the storm, which had little chance of absorption along its steep rocky walls. Rainfall amounts were so heavy that transpiration, groundwater, and runoff couldn't keep up and went into overload. The river became a torrent, quickly doubling and tripling in size, as everything in its path gave way to the floodwaters.

Within a few hours' time, a flash flood was moving boulders 3 meters in diameter down the canyon, trapping people in their cars and destroying homes and businesses. The National Weather Service had measured normal flow of the river at 137 ft^3/sec of water, but during the flood surge, the river's flow rate was expanded to 31,200 ft^3/sec. In all, over 20 centimeters of water fell in less than two hours. Following the devastation, 145 people were dead or missing, 418 houses were destroyed, and 138 others were damaged. Fifty-two businesses were destroyed and over $35 million in damages was chalked up to one bad storm.

During the fall of 1996, an area along the Red River in the Midwest also received record amounts of rain. That winter, extremely cold air froze the excess water before it could run off, and the record rainfall was followed by record snowfall. For several winter months, snow levels grew. Then, when the temperatures finally began to rise, melting became a big problem. Not only did the winter's snow melt, but the frozen rainwater from the previous fall melted, too. The Red River and surrounding drainage basin streams couldn't handle this record excess. So, in the spring of 1997, a large area of the Northern Plains region suffered massive flooding. River levels rose to over 9 meters above flood stage and towns along the Red River like Grand Forks, North Dakota, were shut down because of flooding. Cities were completely incapacitated, with entire downtown sections under meters of water. Large chunks of floating ice blocked the river's normal flow and forced it out of its banks into nearby homes and businesses as well. Table 5-3 lists record amounts of rainfall.

Table 5-3 Rain record amounts can fall in very short time periods.

Rain Duration	Rain amount (cm)	Date	Location
1 minute	3.8	Nov. 26, 1970	Barot, Guadeloupe, West Indies
42 minutes	30.5	June 22, 1947	Holt, Missouri, USA
2 hours, 10 minutes	42.3	July 18, 1889	Rockport, W. Virginia, USA
2 hours, 45 minutes	55.9	May 31, 1935	D'Hanis, Texas, USA
9 hours	108.7	Feb. 28, 1964	Belouve, La Réunion, Indian Ocean
10 hours	140	Aug. 1, 1977	Muduocaidang, Nei Monggol, China
24 hours	109	July 25–26, 1979	Alvin, Texas, USA

Soil and Rock

As we've seen, soil is made up of tightly packed particles of different shapes and sizes. As water hits the ground, it sinks down into pores and between soil particles. A high-porosity soil can hold large amounts of water because of its many pore spaces. If the pores are interconnected and permit water to flow easily, the soil is considered permeable. Sands, gravels, and other rocky soils allow rapid infiltration due to high permeability. However, clay particles are tiny and the pore arrangement between clay particles causes clay soils to be fairly rainproof and resistant to infiltration.

Soil's water retention is a big factor in whether an area will flood. Frequently, soil layers do a good job for a while, but with continued rain, soil reaches saturation and then it's all over—runoff begins.

A soil's initial water content is also important. Commonly, water infiltrates dry soils faster than wet soils. A huge storm that lasts for a week, like a tropical storm, with high amounts of rainfall also affects infiltration. When rain or snow melt reaches the soil surface faster than it can seep through the pores, the water collects at the surface or may travel downhill to the nearest stream channel in a drainage basin. This aspect of soil's absorbent ability is one of the reasons why brief, high-intensity storms, like the Big Thompson Canyon storm, often create more flooding than light rains over a longer period of time.

Whether water is able to move through the ground depends on a region's glacial and bedrock geology. In the northern United States, during the last Ice Age, glaciers covered much of the land surface and left behind *till, outwash,* and lake deposits. *Till,* a rock-and-soil mixture of all sizes, has low permeability when clay is present. *Outwash* is made up of highly permeable sand and gravel open to groundwater flow. Lake deposits can be clay, silt, or sand, with permeability dependent on sediment type.

The type of bedrock structures under glacial deposits also plays a part in groundwater transport. Sandstone channels water when the spaces between grains are connected, creating high permeability. Limestone fractures with many interconnecting splits can also conduct water easily. Finely grained rocks like shale and slate, however, usually have low permeability.

Water Use and Quality

Land use and soil types are linked with human activity. Pesticide and fertilizer use on crops affects water purity if runoff joins with surface water or groundwater. Industrial use and land storage are related to the processing or disposal of hazardous chemicals. This may also present a hazard to streams or underground aquifers. Unfortunately, most water quality problems come from populated areas and improper land use.

A variety of land uses brings pollutants into the overall water cycle. The United States has spent over $300 billion in the last 30 years on controlling pollution. However, there are still a lot of heavily polluted streams, rivers, and lakes.

> Water and air pollutants come from **point source** and **non–point source** locations.

A *point source pollutant* comes from one specific place like a factory or oil refinery. These water pollution sources are easier to find and fix since everyone knows exactly where the problem is.

Non–point source pollution from agriculture, storm runoff, lawn fertilization, construction, and sewer overflows is harder to trace and correct. Hydrologists focus a lot of attention on correcting and regulating non–point sources of pollution. We will look more closely at contaminants in Chapter 8 when we study water pollution.

Land use is also important in other ways. Its impact on the hydrologic cycle and on soil and water quality is direct as well as indirect. Direct impacts increase soil erosion. Indirect effects come from land use and land cover impacts upon the environment. Climate and environmental changes have an indirect impact upon water.

For example, land cover can limit the water available for groundwater recharge, reducing groundwater discharge to the river. The effects of land cover on the hydrologic cycle and water quality also takes place through changes in downstream flows. These direct hydrological surface impacts involve flood, drought, river and groundwater changes, and water quality.

The Future

Water is critical to life. Too much or too little water can have tremendous consequences. In the next 1000 years, conservation and an acute understanding of the way water is transported and stored around the planet will be essential.

The National Weather Service (NWS), the National Oceanic and Atmospheric Association (NOAA), and the United States Geological Survey are all working to expand and improve hydrologic forecast and warning information in order to preserve and protect our water (the king!)

Quiz

1. A pollutant that comes from one specific place like a factory or oil refinery is a(n)
   (a) open source
   (b) point source
   (c) non–point source
   (d) menace to the community

2. Groundwater is stored in
   (a) pitchers
   (b) milk jugs
   (c) moon craters
   (d) aquifers

3. All of water's movement throughout the Earth's hydrosphere is known as the
   (a) hydrologic cycle
   (b) geologic cycle
   (c) atmospheric cycle
   (d) climatic cycle

4. The layer closest to the surface, where spaces between soil particles are filled with both air and water, is called the
   (a) zone of hydration
   (b) zone of aeration
   (c) zone of acclimation
   (d) zone of precipitation

5. When water changes from a liquid to a gas or vapor, it is called
   (a) aeration
   (b) precipitation
   (c) evaporation
   (d) inoculation

6. Surface tension measures the
   (a) depth of a liquid in a lake
   (b) depth of the water table
   (c) configuration of salt crystals
   (d) strength of a liquid's thin surface layer

7. Aquifers that form a water table that separates the unsaturated and saturated zones are called
   (a) confined aquifers
   (b) saturated aquifers
   (c) unconfined aquifers
   (d) dams

8. Outwash is made up of
   (a) impermeable clay and silt
   (b) detergents
   (c) highly permeable sand and gravel open to groundwater flow
   (d) impermeable sand and gravel closed to groundwater flow

9. The time that water spends in the groundwater part of the hydrologic cycle is called
   (a) geologic time
   (b) cosmic time
   (c) aeration time
   (d) residence time

10. Infiltration takes place when
    (a) rainfall soaks into the ground
    (b) deserts get drier and drier
    (c) plants use water during photosynthesis
    (d) food is absorbed by the stomach

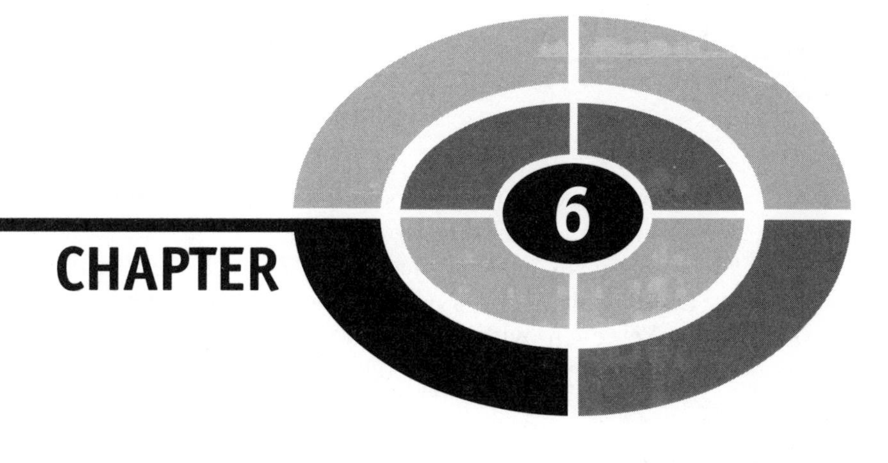

Oceans and Fisheries

Although humans live mostly on land, the Earth is truly a water planet. In our solar system, the "blue planet" is unique, covered with seemingly endless oceans. These deep ocean stretches give us food, trade items, trade routes, recreation, and entertainment. From surfers to developers, people are drawn to ocean shores. Everyone wants to see the latest natural treasure washed up from the depths.

Nearly 72% of the Earth's surface is covered by oceans. The oceans hold 1,300,000,000 km^3 of water. So it should come as no surprise that the interaction between global climate and the oceans has fascinated scientists for years.

The work of French oceanographer and undersea biologist Jacques Cousteau opened up the unseen ocean depths world to everyone. Cousteau invented the equipment that made ocean exploration possible. *SCUBA* (self-contained underwater breathing apparatus) and the iron lung allowed divers to reach depths previously impossible. Today, SCUBA diving is probably at its all-time high as a recreational sport. More and more people are fascinated by the complex and beautiful world beneath the waves.

In 1974, Cousteau started the Cousteau Society (300,000 members) to educate people and to protect and preserve the world's oceans.

Aboard his research ocean vessel, *Calypso*, Cousteau made films like *The Silent World* and *World Without Sun*, which won Academy Awards for best documentary. Cousteau's books include *The Living Sea*, *Dolphins*, and *Jacques Cousteau: The Ocean World*.

Jacques Cousteau died in 1997, but he left behind a legacy of wonder and hope for the world's oceans that he shared with children and adults worldwide.

Yet even with all the sophisticated instrumentation that scientists and oceanographers possess today, the ocean still keeps many of its secrets.

Oceans

It is thought that the first life developed in the oceans as microorganisms. The oceans have served as cradle, restaurant, and recreation. Oceans have kept people apart and brought them together. Even now, when continents have been mapped and their interiors made accessible by road, river, and air, the oceans are largely a mystery. Fig. 6-1 shows the different depths to which the oceans have been explored.

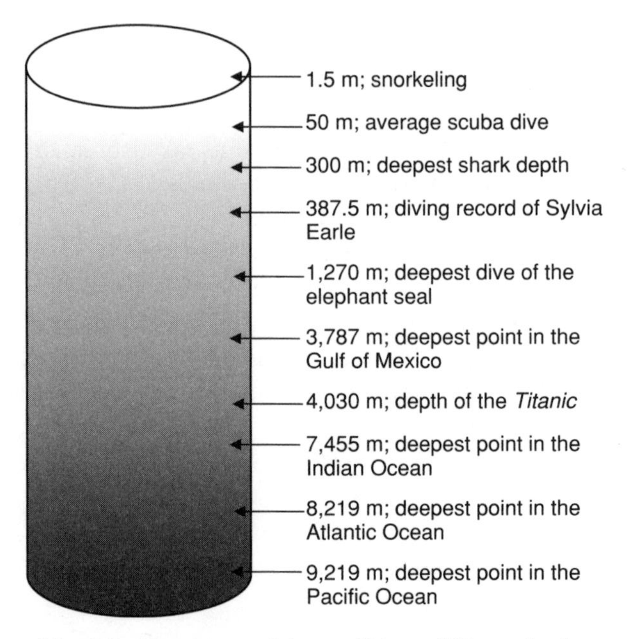

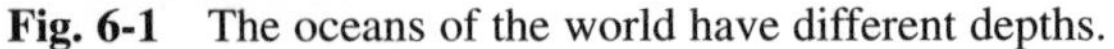

Fig. 6-1 The oceans of the world have different depths.

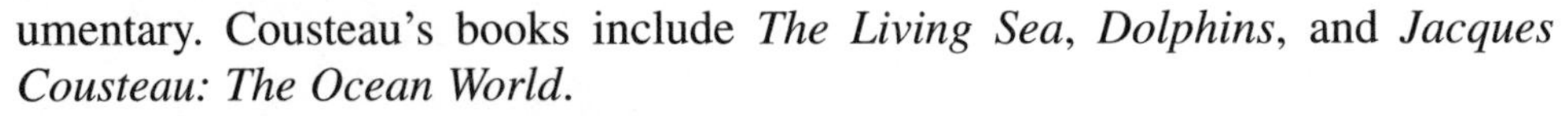

SALINITY

It has been estimated that if all of the oceans' water were poured off, there would be enough salt to cover the continents to a depth of 1.5 meters. That's a lot of salt!

Salinity is the amount of salt found in one kilogram of water. Salinity, or salt content, is written in parts per thousand (ppt) because there are 1000 grams in one kilogram.

The average ocean salinity is 35 ppt. This number does vary as rainfall, evaporation, river runoff, and ice formation change it slightly (32 to 37 ppt). For example, it is said that the Black Sea is so diluted by river runoff, its average salinity is commonly around 16 ppt.

Freshwater salinity is usually less than 0.5 ppt. Water between 0.5 ppt and 17 ppt is called *brackish*. The salt content is too high to be drinkable and too low to be seawater. In areas where fresh river water joins salty ocean water, like estuaries, the water turns brackish.

When salt water gets to the polar regions, it cools and/or freezes, getting saltier and denser. Cold, salty water sinks. The level of ocean salinity increases with depth. Seawater is divided vertically into three layers. The surface layer is mixed depending on rainfall or runoff from the land. The middle layer is called the *halocline* and has a medium range of salinity. The deepest and coldest ocean water has the highest level of salinity.

Fish and animals that live in seawater have worked out ways to survive in a salty environment. Most marine creatures are able to maintain nearly the same concentration of salinity within their bodies as the surrounding environment. If they are moved to an area of much less salinity, they die. You can't put a saltwater fish in a freshwater aquarium!

TEMPERATURE

The ocean has a broad temperature range from warm (38°C), shallow coastal waters near or at the equator to the nearly freezing Arctic waters.

The freezing point of seawater is about –2°C, instead of the 0°C freezing point of ordinary water. Salt lowers the freezing point of seawater. As seawater increases 5 ppt in salinity, the freezing point decreases by –17.5°C.

The ocean is also divided into three vertical temperature zones. The top layer is the *surface layer*, or mixed layer. This warmest layer is affected by wind, rain, and solar heat. Have you ever been swimming in a deep lake? As you get farther away from the surface, which is heated by the sun, it gets colder. Your feet are colder than your upper body.

The second temperature layer is known as the *thermocline* layer. Here the water temperature drops as the depth increases, since the sun's penetration drops, too.

The third layer is the *deep-water layer*. Water temperature in this zone sinks slowly as depth increases. The deepest parts of the ocean are around 2°C in temperature and are home to organisms that either like very cold water or have found specialized environments like volcanic vents that heat the water dramatically.

DENSITY

Temperature, salinity, and pressure come together to influence water *density*, which is the mass of water divided by its volume. Cold seawater is denser than warm coastal water and sinks below the less dense layer.

The ocean waters are similarly divided into three density zones. Less dense waters form a *surface layer.* The temperature and salinity of this layer varies according to its contact with the air. For example, when water evaporates, the salinity goes up. If a cold north wind blows in, the temperature dips and that affects density.

The middle layer is the *pycnocline,* or *transition zone*. The density here does not change very much. This transition zone is a barrier between the surface zone and the bottom layer, allowing little to no water movement between the two zones.

The bottom layer is the *deep zone* where the water stays cold and dense. The polar regions are the only places where deep waters are ever exposed to the atmosphere because the pycnocline is sometimes not present. Fig. 6-2 shows the three transition layers of density (pycnocline), salinity (halocline), and temperature (thermocline) according to depth.

PRESSURE

Although no one really thinks about it, air pushes against us at a constant pressure. At sea level, this pressure is 14.7 pounds per square inch (psi) or 1 kg/cm^2. Our body handles this constant push by pushing back with the same amount of force. On high mountains, the air pressure is less.

Water is a lot heavier than air. The pushing force (pressure) increases when you enter the water. In fact, at 10 meters in depth one atmosphere (14.7 psi) pushes down on you.

It's possible for humans to dive three or four atmospheres with the right scuba equipment, but to go any deeper, tough pressurized vehicles like research vessels and submarines are needed. Whales seem unaffected by pressure shifts.

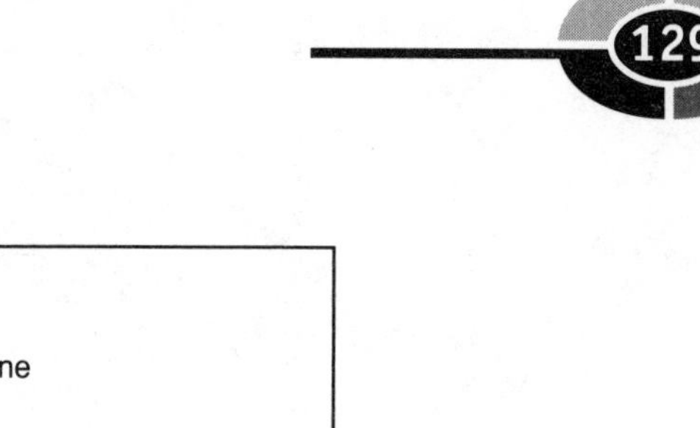

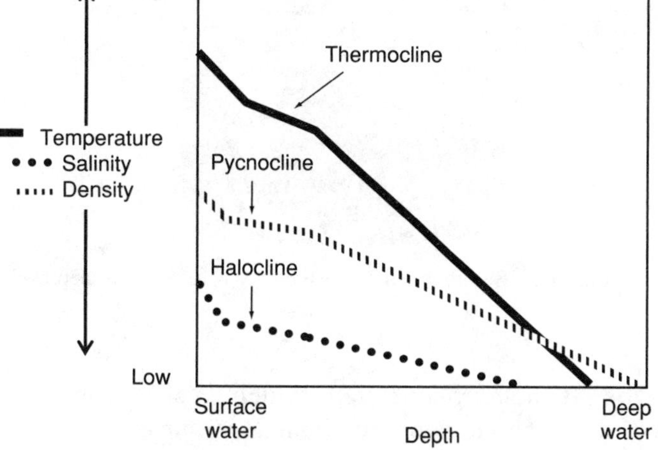

Fig. 6-2 The oceans have temperature, salinity, and density transition zones.

They dive through rapid pressure changes daily. In fact, sperm whales can dive to a depth of 2250 meters and stay down for over an hour.

LIGHT

To thrive, ocean organisms depend on sunlight. Plants and bacteria, such as *kelp*, *sea grass,* and *plankton*, use sunlight to make energy by *photosynthesis*. These organisms provide food for fish and some whales. Fish are eaten by larger fish, sharks, and other predators.

This is the *food web* of the sea. Sunlight is the energy and heat source for the ocean's food web. Surface heat and thermal currents that bring in nutrients make it possible for animals to live in warmed ocean waters.

As sunlight penetrates the ocean, it gets absorbed. Like the other ocean gradients we have seen (temperature, salinity, and density), ocean waters can be divided into three vertical regions based on the amount of sunlight that penetrates.

The first zone, or *euphotic zone*, starts at the water's surface and dips downward to about 50 meters in depth. This depends on the time of year, the time of day, the water's transparency, and whether or not it is a cloudy day. This is the ocean region where there is still enough light to allow plants to carry on photosynthesis. All plankton, kelp forests, and sea grass beds are found in the euphotic zone. Fig. 6-3 shows how these optical regions stack up.

Fig. 6-3 Sunlight penetrates the oceans to different depths.

The next zone is the *dysphotic zone*, which reaches from around 50 meters, or the edge of the euphotic zone, to about 1000 meters in depth. In this zone, there is enough light for an organism to see, but it is too dim for photosynthesis to take place. When divers go deeper into the dysphotic zone, there is less and less available light.

When you get to the *aphotic zone*, there is no light. The aphotic zone extends from about 1000 meters of depth, or the lower edge of the dysphotic zone, to the sea floor. For many years, scientists didn't think there were any animals that lived in this zone, but with deep diving scientific submarines they have found that several specialized species do exist.

Black smokers, the hottest deep-ocean hot springs, have been known to reach temperatures of 380°C. This extremely hot water, mixed with hydrogen sulfide and other leached basaltic trace minerals, is emitted from fractures in the Earth's crust. *White smokers* have a different composition and lower temperatures. In 1977, geologists who had been exploring ocean fractures discovered booming thermal vent communities living without sunlight on the barren sea floor. These big, alien-looking animals used a previously unknown energy process that doesn't include solar heat.

Scientists discovered that the food web depends on sulfur, not sunlight, for energy in vent communities. Deep ocean bacteria transform the chemicals they get from this high-sulfur environment to energy. This energy transformation process is called *chemosynthesis*.

Other dark-living animals eat bacteria, shelter bacteria in their bodies, or consume bacteria-eaters in the web. Vent worms, for example, have no mouth or digestive tract. Instead, they maintain a symbiotic relationship with these chemosynthetic bacteria. The bacteria live in their tissues and provide them with food. Stranger still, scientists found that hemoglobin, which transports hydrogen sulfide to the bacteria, gives vent worms a red color.

BIOLUMINESCENCE

Some animals in the aphotic zone create their own light through a chemical reaction. This is called *bioluminescence.* It is a lot like the reaction that fireflies perform in their warm summer evening dances on land. Microscopic organisms, floating on the surface, produce their own light through bioluminescence. Disturbances by boats, ships, and swimmers can all cause these organisms to glow. This is an amazing sight at night: Because of these glowing party animals of the sea, the wakes of passing ships can be seen for over 10 km!

Ocean Zones

Over 60% of the world's population, roughly 3.6 billion people, lives within 100 km of a coastline. Some countries, like Australia, have all-encompassing coastlines. Canada has the longest coastline of any country, at 90,889 km—or around 15% of the world's 599,538 km of coastlines.

Nearly all continents have shallow land extending off the main continent called the *continental shelf.* This land shelf is fairly shallow and stretches to the *continental slope,* where the land drops away to the deep ocean.

The continental shelf is formed from sedimentary erosion from the land plates, this erosion washes soil into the ocean via rivers and waves. This nutrient-rich sediment provides food for microscopic plants and animals at the beginning of the food web.

The amount of nutrients is so plentiful on the continental shelf that great schools of fish, such as tuna, cod, salmon, and mackerel, thrive in busy communities here. The world's continental shelf regions also contain the highest amount of *benthic life* (plants and animals that live on the ocean floor).

The continental slope connects the continental shelf and the ocean's crust. It begins at the *continental shelf break* where the bottom sharply drops off into a steep slope. It commonly begins at 130 meters' depth and can be up to 20 km wide. Fig. 6-4 illustrates how the continental shelf slopes off to the deeper ocean bottom.

The continental slope, counted as part of the main landmass, together with the continental shelf is called the *continental margin.* Undersea canyons cut through many of the continental margins. Some of these are sliced out by *turbidity currents,* which drive sediments across the bottom.

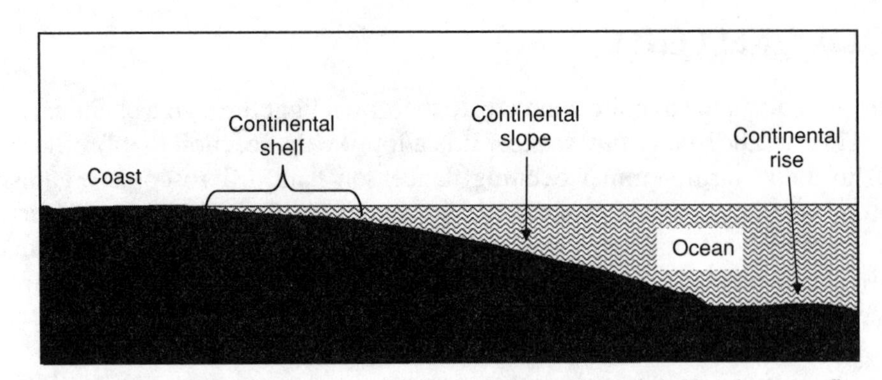

Fig. 6-4 The continental slope stretches beyond the continental shelf to the ocean floor.

Past the continental slope, we find the *continental rise*. As currents flow along the continental shelf and down the continental slope, they pick up, carry, and then drop sediments just below the continental slope. These sediments build up to form the wide, gentle slope of the continental rise.

The deep ocean basin, located at a depth of between 3.7 and 5.6 km, covers 30% of the Earth's surface and has such features as *abyssal plains*, deep-sea *trenches,* and *seamounts* (extinct undersea volcanoes). The abyssal plain is the flat, deep ocean floor. It is almost featureless because a thick layer of sediment covers the hills and valleys of the ocean floor below it. We saw in Chapter 4 how the deep ocean plains serve as a carbon sink for the calcium carbonate remains of phytoplankton. Deep-sea trenches are the deepest parts of the ocean. The deepest one, the Marianas Trench in the South Pacific Ocean, is more than 10.6 km deep.

SHORELINES

Waves, various currents, and tides all intermingle with land, rocks, and plates to give shorelines unique characteristics.

To most of us, a day at the beach is a day of water, sun, surf, and sand. Beaches are composed of mostly sand, pebbles, and rocks depending on the climate and nearby landmass. However, they are constantly changing. Waves and wind are the endless forces that never stop. They build beaches up and wash them away. The main factors that affect the creation and maintenance of shorelines all around the world include:

- Rising of coastal area with associated erosion;
- Sinking of coastal areas with sediment deposition;

- Types of rocks or sediments present;
- Changes in sea level;
- Common and storm wave heights; and
- Heights of tides affecting erosion and sedimentation.

Some beaches go on and on for many kilometers, while others are called pocket beaches with just a nick of sand or smooth stones in a long shoreline. Sometimes beaches are backed by rocky cliffs, while other times they are flanked by belts of sand dunes.

Littoral

The *littoral* zone is a tidal-depth gradient found near the shore. Since there are coastal currents, onshore and offshore winds, reefs, and bays in this area, the shoreline's shape is fairly changeable. All these factors make it harder to forecast water conditions and potential pollution sources accurately.

The littoral area is also where marine organisms, like jellyfish, are found. As many snorkellers know, it's more common to see marine life in the shallower depths near the shore than in the open ocean. The littoral zone reaches from the shoreline to nearly 200 meters out into the ocean.

Currents

Ocean waters are constantly on the move. Currents are affected by wind, salinity and temperature, bottom topography, and the Earth's rotation. In fact, if it were not for currents that transport nutrients as well as pollutants, the world's oceans would be in even more trouble than they are now. Just as trapped water like that found in the Dead Sea (five times the salinity of the open oceans) eventually becomes super-concentrated and unlivable for marine life, noncirculating oceans would be no fun.

Determined by forces such as wind, tides, and gravity, currents keep our oceans stirred up. Currents move water thousands of miles. There are many different currents, but the seven main currents are the *Antarctic Circumpolar current* (also called the *West Wind Drift*), the *East Wind Drift*, the *North* and *South Equatorial currents*, the *Peru current*, the *Kuroshio current,* and the *Gulf Stream*. These currents move in large rotating circles called *gyres* and are driven by atmospheric winds that whip the planet all year round. Look back to Fig. 1-2 to see the general circulation of ocean currents in the Northern and Southern Hemispheres of the globe.

A global circular current takes place when deep water is created in the North Atlantic, sinks, moves south, roams around Antarctica, and then heads northward to the Indian, Pacific, and Atlantic basins. Oceanographers estimate that it takes roughly a thousand years for water from the North Atlantic to wind its way into the North Pacific.

The *Somali current*, off Africa's eastern coast, is a current with a twist. It does the impossible: It reverses direction twice a year. From May through September, the Somali current runs northward. Then from November through March it runs southward. April and October serve as the turnaround months, like the roundhouse for trains. As the Somali current flows northward, upwelling takes place and brings nutrients to marine life. However, when the current turns southward again, the banquet table is put away.

Most people have heard of the Gulf Stream. It has a strong effect on the east coast of the United States. The Gulf Stream surface current is a strong, western boundary current and part of a greater current system that includes the North Atlantic current, Canary current and the North Equatorial current. It is warm, deep, swift, and fairly salty, separating open ocean water from coastal water. The Gulf Stream commonly travels at a speed of nearly 4 knots.

Atlantic Ocean

There are several differences between the Atlantic and Pacific Oceans. The Atlantic Ocean is divided into two sections by the Midatlantic Ridge, the margins of the North American and Eurasian plates. The Atlantic is also a spreading ocean, except for a small subduction zone in the Caribbean Sea. The Atlantic continental margins are considered passive by geologists because volcano and earthquake activity across the plate is rare.

The continental shelf in the Atlantic Ocean is a fairly flat, wide, sandy, area. It stretches gently out from the landmass of North and South America for 50 to 100 meters before dropping down to the continental slope (4° angle) for a distance of around 1 km. The depths are between 2000 to 3000 meters here.

Eventually, the continental slope begins to rise slowly. This large area, hundreds of kilometers wide, is called the *abyssal plain*. It is found at depths of 4000 to 6000 meters with only the occasional *seamount* (extinct volcano) for interest. The plain rises until it comes to a hot, active volcanic gap and ridge area, the Midatlantic Ridge.

The Midatlantic Ridge (nearly 2000 meters) is the margin between crustal plates where new seafloor is created and being pushed up from the magma below.

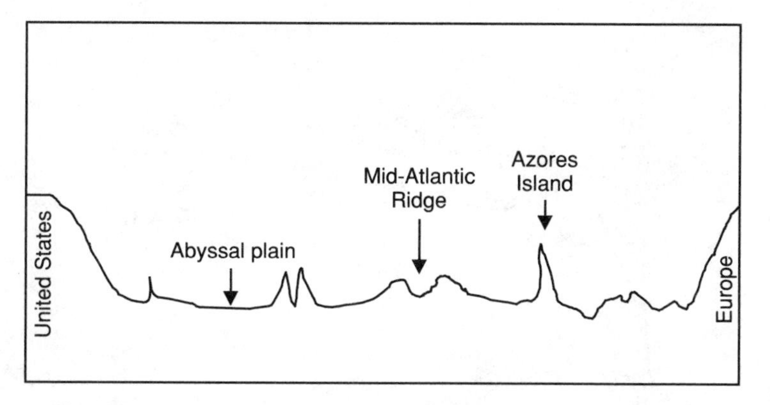

Fig. 6-5 The Atlantic Ocean is spreading at the Midatlantic Ridge.

The other side of the ridge, closest to Europe and Africa, is essentially a mirror image of the first half, revealing similar ocean depths and underwater landscape. Fig. 6-5 provides a rough topographical outline of the Atlantic Ocean from North America across to Europe.

Pacific Ocean

The first thing that most people notice about the Pacific Ocean is its size. It is huge! It's the largest ocean on Earth. But because of its many subduction zones (tectonic plate interactions) along the Ring of Fire, the Pacific Ocean is actually getting smaller. In many millions of years, barring other problems, North America and South America will be a lot closer to Asia and Australia than they are now. By then it may only be a short trip by speed boat!

However, when people think of the Pacific Ocean, they usually think of balmy days, tropical islands, volcanic fireworks, and white beaches. That is fairly accurate, but to geologists, the Pacific Ocean is the hot spot of tectonic plate activity with active continental margins.

Compared to the Atlantic Ocean, the main structural difference of the Pacific Ocean is its depth. In some spots, the Pacific is nearly twice as deep as the Atlantic Ocean. When you start out from the west coast of South America, the continental shelf is short, only 20 to 60 meters long. At its edge, the slope drops off to 8000 meters (like a ball rolling off a table) into the Peru-Chile trench. On the other side of the trench, the depth rises up onto the Nazca plate, around 4000 meters, until it gets to the East Pacific Rise (2200 meters).

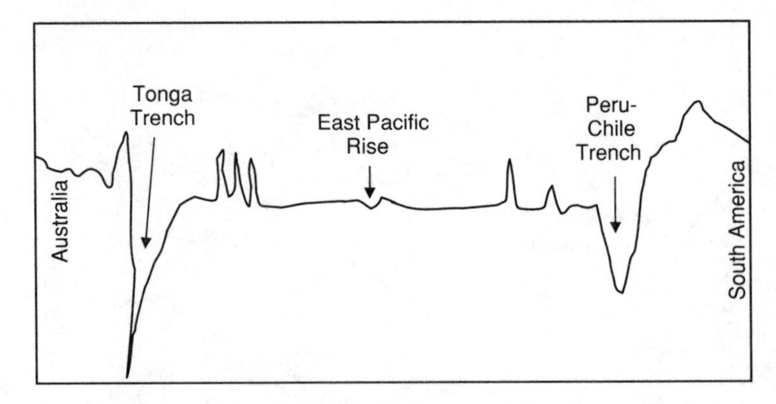

Fig. 6-6 The Pacific Ocean has the deepest trenches on Earth.

On the far side of the East Pacific Rise, the Pacific Ocean floor stays around 4,000 meters in depth with a number of volcanoes, seamounts, and other geology jutting up abruptly from the sea floor. Still traveling east, the Tonga trench drops the ocean depth to nearly 11,000 meters in depth. The ocean floor comes back up from this gash to 4,000 meters until it comes to the short continental shelf of Australia. Fig. 6-6 gives a very general topographical outline of the Pacific Ocean from South America across to Australia.

El Niño

The occasional shifting of warm waters from the western to the eastern Pacific Ocean has a definite effect on the world's climate. This is called the *El Niño* effect.

In the equatorial Pacific region, El Niño causes important changes in the ocean-atmosphere system. These changes (since the Earth is one giant system) impact weather conditions in other areas of the globe, sometimes severely. For this reason, El Niño often gets blamed for all kinds of bad weather.

Some of El Niño's effects include increased rainfall across the southern part of the United States and along the western coast of South America. These can lead to extreme flooding and mud slides. Weather changes can also shift ocean currents that result in drought in the West Pacific and brush fires in Australia. The National Oceanic and Atmospheric Administration (NOAA) operates a network of instrument buoys that measure temperature, currents, and winds in the equatorial area. These buoys send information to scientists and weather forecasters around the world in real time.

Normally, trade winds blow towards the west across the equatorial Pacific. Over distance and time, these winds pile up warm surface water in the west Pacific. This causes the sea surface to be about half a meter higher at Indonesia than at Ecuador.

Ocean temperatures are about 8°C higher in the west, with cool temperatures off South America because of an upwelling of cold water from deep ocean levels. This nutrient-rich cold water increases energy production in the food web and allows a diverse marine ecosystem to flourish.

During El Niño, the trade winds drift across the central and western Pacific, causing a decrease in the thermocline of the eastern Pacific and an increase of the thermocline in the west. When that happens, there is a drop in the water temperature and the thermocline dips to around 150 meters in depth. This stops the upwelling and surface cooling, which disrupts the supply of nutrient-rich thermocline water to the euphotic zone. Consequently, there is a rise in ocean surface temperature and a sharp plunge in the food web's energy production. This has a big impact on the rest of the food web as well as local and commercial fishing interests.

Physical Changes

Human activities can change the seabed directly (by dredging, trawling, and boat groundings) and indirectly (by damming, construction of sea walls, and other things that increase sedimentation and/or cut off the natural flow between land and sea). Coastal landscape changes that replace natural coastal ecosystems—like the building of major urban centers, resorts, hotels, golf courses, ports, and factories—can be huge.

As coastal areas are developed, seabird habitats, marine mammals, and other marine species are decreased or wiped out. Shallow harbors are dredged for the passage of larger ships. Sea walls stop the natural exchange between land and sea. The damming of rivers cuts off important migration routes for salmon and blue crab, which need fresh- and saltwater habitats to survive.

Coastline physical changes can also impact open ocean species and systems, as some species spend part of their life cycle near shore, onshore, or inland.

Another widespread source of physical marine change is caused by *bottom trawling*. Bottom trawlers drag weighted fishing nets across the sea floor, disturbing whatever rocks, coral, and organisms are in their paths. Fish that hide among rocks and coral reefs are left homeless, coral that needs clear water to survive becomes choked with silt, and fishing ecosystems are altered to a large extent.

Bottom trawling is often compared to clear-cutting of forests. However, a forest is clear-cut only once over many years; a specific sea floor area may be bottom-trawled 100 times in a single year. Scientists believe roughly 70 to 95% of the organisms collected by bottom trawlers are not wanted and usually thrown away.

Undersea mining operations, offshore oil exploration, other types of resource extraction activity, ship groundings, and anchorages also disturb coral reefs, ecosystems, and the sea floor.

Dynamite fishing is another method that devastates coral reefs and reef communities. Fishers set off dynamite on coral reefs to kill and stun fish that can then be scooped up with nets. The explosions kill everything near the blast, including corals. Another technique, called *muro ami*, involves pounding reefs with heavy weights to scare fish out of their hiding places, destroying the coral heads in the process.

Fortunately, many of these physical impacts can be prevented or changed through better education on marine environments and newer, less damaging technologies.

Fisheries

The importance of marine resources to humans is obvious. Fish and shellfish give us a valuable source of food protein and a livelihood to many in the seafood industry. Other marine resources meet the needs of the general population and provide jobs. The question is the extent to which the oceans can continue to meet human needs. Scientific studies and marine management based upon current trends are the key to the future.

Although some scientists suggest that there exists a climatic control mechanism that affects many of the largest and most important of the world's fish populations, cyclic variations over 10- to 20-year spans are not well understood. Global marine resource prediction, management, and fish variability need further study.

Fisheries in many countries remain unregulated and inadequately studied, and even where research has been done, political and economic pressures have been known to override scientists' recommendations. In fact, most fisheries, located close to coasts where they are impacted by human activities (pollution), have been overfished. This is the major problem with ocean fisheries worldwide.

The New Zealand orange roughy fishery, which started in the 1970s, declined less than 10 years later. In 1986, New Zealand scientists warned that an 84% reduction in orange roughy fishing was needed or the species would crash. Policy makers did nothing for two years, then set a 20% reduction in 1989. Unfortunately, the orange roughy industry did collapse, and since orange roughy take about 30 years to mature and reproduce, recovery will be slow.

> When nontarget fish species or other marine animals are caught in nets or through other fishing methods, they are called **bycatch**.

Accidental catching and killing of nontarget species, like fish (nonmarketable "trash" fish) and marine mammals (dolphins, sharks, etc.), has also impacted ocean ecosystems. Estimates by the Food and Agriculture Organization (FAO) of the United Nations for 1988–1990 indicate an average of 27 million metric tons of fish per year were discarded as *bycatch* by commercial fishermen. This amount is roughly equal to one-third of the total yearly global catches. This does not include the unintended killing of several hundred thousand sea turtles and marine mammals each year. Fishing methods such as *bottom trawling*, *gill netting*, and *longlining* are especially subject to bycatch. Shrimp trawlers catch and throw away an estimated 9 pounds of nontarget species for every 1 pound of shrimp caught. This is a costly environmental waste.

As people find out about these problems, funding becomes available for research and development of new technologies. For example, when it was learned that thousands of dolphins were caught in tuna nets and drowned before they could be released, new escapable nets were designed. Public demand for the "safer" tuna grew and those companies that didn't change were impacted by sinking sales and income.

Non-native (Alien) Species

When plants and animals found in one part of the world, with their own natural predators, diseases, and ecosystem controls—are transported to a far distant location, they are known as *non-native species*. In their new location, they flourish, die, or do something in between.

Non-native (alien) species are transplanted geographically to a formerly unknown area.

Many times, the new species is not suited to the new environment, and it fails to prosper there. However, in other cases, the new species thrives, pushes out the native species, and, in some instances, wipes out the local ecosystem.

In the ocean, the main way alien species have been introduced is through the ballast water of ships. When ships fill their ballast tanks to adjust their stability in the water, they suck in whatever is in the local water. Later, when the ballast is emptied somewhere else, the stowaway passenger is dumped out. Some scientists think that as many as 3,000 alien species are transported daily in ships around the world.

Non-native species have more of an effect in coastal waters than in the open ocean. In separate coastal zones, where transplanted species are often biologically and ecologically distinct, alien species have a greater impact. Species transport from tropical regions of the Atlantic to the Pacific is frequently more disruptive because there is little natural exchange between these oceans. Transplants between the Indian Ocean and Atlantic are not all that different and do not have the same serious effects since these oceans are connected.

Due to worldwide shipping, the impact of alien species has increased significantly. Global population growth has made it necessary to bring in food and other products from trade partners. Countries who were formerly self-reliant now depend on imported goods.

This shipping increase has caused a huge increase in the introduction of marine alien species into new ecosystems incapable of coping with them. For example, in San Francisco Bay, California, the ecosystem has been overwhelmed by many new alien species. Currently there are over 200 different alien species living in San Francisco Bay, severely impacting native species and the overall ecosystem. Although alien species have not completely wiped out native species, the Bay's natural processes have been changed significantly.

Besides changing native marine ecosystems, alien species can also affect human health. Zebra mussels, an introduced species in the Great Lakes, have done everything from disturbing native species to clogging pipes. However, on the plus side, they increased toxin filtration from the water.

When another alien species, the round goby, arrived, it became a hungry predator on zebra mussels. A problem arose when a popular game fish ate the gobies; these gobies were full of zebra mussels that had filtered concentrated amounts of toxins. When humans ate the game fish (remember the food web?), toxins were passed on to the humans. This is an example of bioaccumulation of toxins in a food web.

Other Ocean Residents

Besides overfishing and introduction of alien marine species, there are environmental concerns for nearly every resident of the world's oceans. Everything from sharks, whales, and dolphins to jellyfish, tube worms, and kelp beds are impacted by the pollution of ocean waters. Since even the smallest members (microorganisms) of the food web are affected by chemicals, turbidity, and temperature increases, any problem takes on a domino effect as larger and larger species are impacted.

> Species in danger of becoming extinct because of overharvesting, habitat/food loss, or threatening pollutants are considered **endangered species**.

Refer to Table 2-2 for the numbers of marine species impacted by environmental issues and for those listed as endangered species in 2004.

The Future

Since over half of the world's population currently lives within 60 km of coastal waters and this population is expected to double within the next three to four decades, the oceans will definitely feel an impact. Six of the world's eight largest cities (population over 10 million) are coastal. Moves to coastal areas are driven by poverty and affluence. Poor people move to cities for jobs, while wealthy people expand shoreline development for resort hotels and seaside homes. The World Resources Institute estimates that about half of the world's coastal ecosystems are threatened by development, with most located in northern temperate and equatorial region, including the coastal zones of Europe, Asia, the United States, and Central America.

Water treatment technology has improved, leading to big improvements in areas where the new technology is used. However, many developing countries cannot afford to implement these new technologies, and population growth in coastal cities has continued to overwhelm existing waste treatment systems.

Population growth affects the world's oceans in complex ways. A lot of interacting factors contribute to the degradation of marine ecosystems and loss of biodiversity. Scientists have a lot of work to do to better understand the underlying causes and effects and what can be done to reverse previous damage. Many scientists are concerned that changes in ocean ecosystems brought about by overexploitation, physical alteration, pollution, introduction of alien species, and global climate change are outpacing efforts to study them, let alone counter their effects.

The scientific community's concern that the oceans are in trouble compelled more than 1100 marine scientists to sign the Troubled Waters Statement in 1998. This statement, a project of the Marine Conservation Biology Institute in Washington, DC, was designed to explain to policy makers and the public how the oceans are in trouble. The statement indicts the massive damage to deep-sea coral reefs from bottom trawling on continental plateaus and slopes, on seamounts, and midocean ridges. These complex ecosystems, developed over thousands of years and destroyed by bottom trawling, support thousands of marine species. The chance of their recovery with continued bottom trawling is slim.

In order to address these crucial concerns, the Troubled Waters Statement asserts:

> To reverse this trend and avoid even more harm to marine species and ecosystems, we urge citizens and governments worldwide to take the following five steps:
>
> 1. Identify and provide effective protection to all populations of marine species that are significantly depleted or declining, take all measures necessary to allow their recovery, minimize bycatch, end all subsidies that encourage overfishing and ensure that use of marine species is sustainable in perpetuity;
> 2. Increase the number and effectiveness of marine protected areas so that 20% of Exclusive Economic Zones and the High Seas are protected from threats by the Year 2020;
> 3. Ameliorate or stop fishing methods that undermine sustainability by harming the habitats of economically valuable marine species and the species they use for food and shelter;
> 4. Stop physical alteration of terrestrial, freshwater, and marine ecosystems that harms the sea, minimize pollution discharged at sea or entering the sea from the land, curtail introduction of alien marine species, and prevent further atmospheric changes that threaten marine species and ecosystems; and
> 5. Provide sufficient resources to encourage natural and social scientists to undertake marine conservation biology research needed to protect, restore, and sustainably use life in the sea.

Although global population also has a big impact on the oceans, it's tough to change in the short term. Slowing or reversing population growth can be augmented in other ways. Actions to reduce the effects of human activities on marine systems include:

- Encouragement of sustainable use of ocean resources;
- Policies and new technologies to limit pollution;
- Creation and management of protected marine regions; and
- Education on ocean preservation the health and its impact on humankind.

The oceans unite people and landmasses of the Earth. In fact, one in every six jobs in the United States is thought to be marine-related and attributed to fishing, transportation, recreation, or other industries in coastal areas. Ocean routes are important to national security and foreign trade. Military and commercial vessels travel the world on the oceans.

To highlight the world's oceans, the United Nations declared 1998 the International Year of the Ocean. This gave organizations and governments a chance to increase public awareness and understanding of marine environments and environmental issues.

Tsunamis and other Hazards

On December 26, 2004, an underground earthquake along the boundary between the Asian continental plate and the Indian Ocean plate took place. Within minutes, tidal shock forces were rushing *tsunami* waves towards both nearby landmasses and those hundreds of miles away. The disastrous impact of these waves on coastal communities in eleven countries was terrible. Entire towns and villages were washed out to sea. At the time of this writing, over 220,000 people have been found dead. Coastal geography was changed dramatically.

> A **tsunami** is a huge, speeding ocean wave created by the force of an earthquake, landslide, or volcanic eruption.

Several days after the Indian Ocean tsunami, scientists from the NOAA Laboratory for Satellite Altimetry in Silver Spring, Maryland, examined data from four spacecraft (*TOPEX/Poseidon* and *Jason*, operated jointly by NASA and the French space agency CNES; the European Space Agency's *Envisat,* and the U.S. Navy's *Geosat Follow-On*). Satellite photos from these Earth-orbiting radar satellites allowed NOAA scientists to measure the height of the devastating tsunami. Some of the waves reported in Sri Lanka, Indonesia, and South India reached heights of 30 meters (100 feet).

Normally, the satellites help scientists chart unexplored ocean basins and forecast the intensification of hurricanes, arrival of an El Niño, and other weather happenings.

Since the photographs from the satellites are seen immediately, the data was not received until several hours after the tsunami had developed, too late to warn people.

In the aftermath of the 2004 tsunami, NOAA researchers plan to use satellite data to improve tsunami forecasts by plotting the ocean bed and determining how tsunami energy is focused or scattered in relation to coastlines. The ability to make depth surveys from space may improve tsunami forecast models. NOAA hopes to fine-tune a tsunami computer model that would help warn at-risk communities and develop a real-time forecasting system for NOAA Tsunami Warning Centers. Satellite data could be used to increase our understanding of how tsunamis roar across the ocean.

As global warming causes ice at the poles to melt, rising seas will flood low-lying towns and cities slowly but surely. Cities like Venice, Italy, and those protected by dikes in the Netherlands will be increasingly inundated along with coastal vacation sites worldwide.

As global coastal regions are affected by rising seas, pollution, overfishing, and catastrophic events like killer hurricanes and tsunamis, our interconnectedness as humans becomes even more important. Policy makers must take the broader view.

Quiz

1. When species are transported geographically to an unknown area, they are called
   (a) the new kids
   (b) jellyfish
   (c) native
   (d) non-native

2. What process causes important changes in the ocean-atmosphere system in the equatorial Pacific region?
   (a) El Corazon
   (b) El Shaddai
   (c) El Niño
   (d) El Abuelo

3. What percentage of light is present in the aphotic zone?
   (a) 0%
   (b) 5%
   (c) 10%
   (d) 20%

4. The Midatlantic Ridge, the margin between crustal plates, is where
   (a) nothing much ever really happens
   (b) new seafloor is created and pushed up from the magma below
   (c) subduction is pushing the continental plate below the Atlantic plate
   (d) lots of ships sink

5. Ocean salinity measures the amount of
   (a) coral reefs formed at an atoll
   (b) salt in one millimeter of mercury
   (c) salt on your skin after swimming in the ocean
   (d) salt found in one kilogram of water

6. Temperature, salinity, and pressure all influence water
   (a) density
   (b) taste
   (c) levels
   (d) fluidity

7. Zebra mussels are
   (a) the largest of the mussels
   (b) an introduced species in the Great Lakes
   (c) actually more spotted than striped
   (d) related to zebra clams

8. SCUBA stands for
   (a) self-counted umbrella banking account
   (b) self-contained underwater belching apparatus
   (c) sea coral growth and barracuda activity
   (d) self-contained underwater breathing apparatus

9. What percentage of the earth's surface is covered by oceans?
   (a) 48%
   (b) 56%
   (c) 66%
   (d) 72%

10. Which current reverses direction twice a year?
    (a) California current
    (b) Gulf Stream
    (c) Somali current
    (d) Australian current

CHAPTER 7

Glaciers

When you think about major catastrophic impacts on the shape of the land, earthquakes, hurricanes, volcanic eruptions, and mudslides come to mind. These sudden, violent upheavals cause chaos when they occur. News images of death and destruction after an earthquake, for example, are definitely to be avoided if you want to have a good day.

But what about glaciers? How does a huge chunk of ice compare to the other tricks up Nature's sleeve? The answer is surprising. Glaciers are just as powerful as their terrifying cousins, but they are stealthy. Glaciers move rock, carve valleys, and deposit soils in new places, but in slow motion. Watching a glacier flow is like watching grass grow—only slower. But for those with patience, glacier-created views are spectacular.

Roughly 75% of the earth's fresh water is stored as ice in glaciers. Water falls on the ice as snow and then leaves it through melting or evaporation. In the Pacific Northwest of the United States, Washington and Oregon get over 700 million gallons of water each summer from melting glaciers.

The amount of water found in glacial ice is about equal to the amount of rainfall falling constantly over the whole planet for half a century. Glaciers cover over 15 million square kilometers. If all the ice on land melted, the ocean level could rise as much as 80 meters globally.

In the past, ancient people thought glaciers were crafted by spirits. When glaciers overran villages, people thought it was a hint from the gods that they had

either built in the wrong place or had done something to anger the Earth powers. Luckily, because of a glacier's slow movement, people usually had plenty of warning. However, when glacier ice moves beyond the edge of a cliff, its massive weight can suddenly fracture underlying rock and cause a rock and ice avalanche. This happens with no warning and is disastrous to plants, animals, humans, and the landscape.

Most ancient people learned to leave the high glaciers to sure-footed goats and polar bears while they wisely chose to live at lower elevations. They didn't go out onto the ice or tundra except in hunting parties. *Tundra* is the barren treeless area found south of an ice zone. The ground is frozen for hundreds of meters in spots, but warms enough in the summer for the thin surface soil to melt and allow tiny plants, like mosses and lichens, to grow. Tundra is also found just below the frozen snow and ice of high mountain peaks. It's also found above the tree line.

Glaciers are old and new. They are formed by snow that fell thousands of years ago, which is then pressed down harder and harder by the weight of later snows. After a while, the extreme weight of the accumulated snow packs down, pushes out most of the air, and recrystallizes into coarser, denser ice crystals. Some glacier ice crystals can grow to be as large as tennis balls.

Have you ever made a snow ball? If the snow is packed loosely, it will fly apart when you throw it. If you mold and pack snow into a tight ice ball, it will fly straight like a rock when you throw it. (Don't try this at home!)

Glacier ice is made from packed snow that becomes large enough to shift and move from its original home. As snow accumulates, it goes through a middle stage, called *firn,* before it turns into solid, rock-like ice. This material is a lot like the packed mounds of ice and rock along snow-plowed roads in the winter and early spring in northern climates. Depending on temperature and snowfall, it can take between 10 to 3500 years for snow to transform completely into glacial ice. When enough ice accumulates on an incline, it begins to slide or flow downhill.

When glacier ice gets very dense, it looks blue like the water in a swimming pool. After years of pressure, the ice is compressed and becomes much denser, pushing out all the tiny air spaces between ice crystals. When it becomes super-dense, the ice absorbs all other colors in the rainbow and reflects blue. If glacier ice looks white, it means that it still holds a lot of tiny air bubbles.

Pleistocene Era

The Pleistocene Era of overall glacier growth began around two million years ago when average temperatures in the Northern Hemisphere lowered to around 2°C. It ended about 10,000 years ago. At that time, huge ice masses covered the land, spreading over millions of square kilometers, advancing and retreating. Glaciologists study this era more than earlier ones since it is more recent and the effects are easier to see.

For glaciers to form, it must be cold enough to keep snow on the ground year round, but there must be a lot of moisture as well. For example, some mountain ranges with an ocean on one side will form glaciers on the slopes facing the sea, while the dry side that gets little snow from year to year has no glaciers.

During the Pleistocene Era, about 30% of the Earth's land was covered by glaciers. Geologists think that glaciers are an exception to common Earth formation. Only about a dozen or so major ice ages are thought to have existed during the Earth's nearly four-billion-year-old cooling and crust-forming period. Each ice age of continent-wide glacier contact is thought to have lasted a few million years. This is fairly short when you think of the Earth's long history. Glaciers, present during all of human history, have existed as only a passing player in the Earth's total land development. Pleistocene ice reached at its farthest distances from the North Pole and South Poles only about 25,000 years ago. At its peak, Pleistocene ice reached as far south as Chicago in North America and Paris in Europe. Canada was completely covered with ice over a kilometer thick in most areas.

The environment got so cold between the years of 1350 and 1650 that scientists called it the Little Ice Age. We are still living in a glacier ice period.

Glaciologists, scientists who study ice movement, have found that today roughly 10% of the earth's land is covered by thick layers of glacier and polar ice. Glaciers continue to carve mountains, smooth highlands, scrape and groove plains, enlarge valleys, and change the face of the land. The effects of ancient and current glaciers can be seen over much of the Northern Hemisphere. In North America, some ancient glaciers (even older than those from the Pleistocene) pushed as far south as New Mexico, while others are thought to have covered the single supercontinent Pangaea entirely at the end of the Paleozoic era, over 200 million years ago. This was before the huge crustal plates broke apart and the land started to shift around the globe.

The *cryosphere,* encompassing all the ice on the planet, has been a part of the Earth in larger or smaller amounts since the planet was formed. In the very beginning, it's likely that the entire planet was a big, round snowball.

> The **cryosphere** encompasses all the ice on the planet. This includes land ice and floating sea ice (*icebergs*).

There are two major types of glaciers, *alpine (valley) glaciers* and *continental glaciers. Alpine glaciers,* the most widely found glacier type, form in the high-elevation mountains, and then move down a valley from the colder mountain peaks. Alpine glaciers are the glacier type seen in Glacier National Park located in the Canadian Rockies. Most of the 70,000 to 200,000 known glaciers in the world are alpine glaciers.

Continental glaciers are the much larger, U-shaped glaciers with the deepest ice found in the curved middle. This heavy center area drags on the surrounding glacial edges causing the ice to flow. The two largest continental glaciers, in Antarctica and Greenland, cover a lot of land. The Antarctic ice sheet is large enough to cover the continental 48 United States to a depth of over 3 km of ice in places! This largest of the continental glaciers, though huge by today's standards, is small compared to ancient ice that covered most of the Earth's landmass.

Glaciers are the farmers of the Earth. Their movement southward and back again to cold arctic regions drags pulverized rock over millions of square kilometers. The scattering of bits of rock by glacial pressure nourishes soil with added minerals and nutrients. Some of the continents' richest farming areas have been directly in the path of advancing and retreating glaciers.

Glacial ice has also captured and spread the products of volcanic eruptions. They do this in two ways: 1) by melting and mixing with lava at the time of an eruption, and 2) by trapping volcanic ash, dust, and particles during and between snowfalls. The carbon found in volcanic ash is an important nutrient on continental plains.

Ice Caps and Sheets

Ice caps are found in polar and subpolar areas of high mountain ranges with ice and snow on the peaks all through the year. Ice caps are smaller versions of ice sheets. They cover an area less than 50,000 square kilometers. Ice caps can be thousands of meters long and hundreds of meters deep. The mountain ranges of Alaska and Norway, as well as polar islands like Iceland, have ice-capped peaks. When an ice cap covers a single peak like Mount Rainier, it is called a *carapace* ice cap.

> An **ice cap** is the area of ice that spreads out equally in all directions from the center of a glacier.

When snow that falls on high mountain peaks and plateaus builds up year after year with little summer melt, an alpine *ice field* is formed. After the snow gets to be about 30 meters deep, the bottom layers become compressed into ice. Snow keeps falling on the mountains and the snow gets deeper. Eventually, the snow overflows nearby valleys and starts flowing downhill. When this happens, you have a glacier!

An **ice sheet** or **ice field** is a thick layer of pressure-packed ice located on a high, nearly level area of land.

Canada has experienced four major Ice Ages. Of these, the Columbia Ice Field remains as part of a huge sheet of ice that originally covered nearly all of the Rocky Mountains. Though currently shrinking from global warming, the Columbia Ice Field's glaciers once extended much farther south during the last Ice Age. This ice field includes nearly all the parts of an active ice field and glacier system.

The Columbia Ice Field, the largest body of ice in the Canadian Rocky Mountains, measures 325 km^2 in area and is estimated to be 365 meters thick. Mount Columbia is the highest point in the ice field at 3745 meters. The average snowfall on the Columbia Ice Field is about 7 meters per year. The Columbia Ice Field spread out during the last ice age and has three main outlets, including the Athabasca Glacier.

When an ice cap carves a valley and emerges at the lower end, it is called an **outlet glacier.**

Ice sheets can be divided into two types: *land-based ice sheets* and *marine-based ice sheets.* The base of a land-based ice sheet lies mostly above sea level. The East Antarctic Ice Sheet and the Greenland Ice Sheet are land-based ice sheets.

Conversely, the greater part of the West Antarctic Ice Sheet lies below sea level. The West Antarctic Ice Sheet is a marine-based ice sheet. Much of the West Antarctic Ice Sheet is grounded well over a kilometer below sea level and more than two kilometers below sea level near its center. Its surface slope is less than that of the East Antarctic Ice Sheet. This sloping profile leads glaciologists to believe that the West Antarctic Ice Sheet is flowing and shrinking faster than its East Antarctic counterpart.

A *continental ice sheet* is the largest of the glacial masses. It builds up in the coldest regions with the most yearly snowfall. As the snow gets deeper and the base ice gets thicker, the ice sheet covers mountains and spreads over and across everything in its path. When a continental ice sheet comes to a valley, it pours into it and fills it. Tall peaks are pushed up against and around shaping ice at the sheets' edge to bend and melt forming streams. The longest glacier in North America is the Bering Glacier in Alaska, extending 204 km.

Greenland's continental ice sheet, with ice over 125,000 years old, covers an area of over 1 million km^2. This sheet is the largest glacier in the Northern Hemisphere. It reaches a maximum thickness of nearly 3000 meters near the center of the island where it lies on the land's surface. Near the coast, large masses of the glacier break off to form icebergs. Ice covers everything except coastal areas and the tops of some mountain peaks. These summits, sticking though the ice, are called *nunataks.* Narrow outlet glaciers drifting down from the northern continental ice sheet in Greenland drop over 10,000 icebergs into the Atlantic Ocean each year, some the size of office buildings.

Pack ice describes the first ice formed on the surface of the sea. Exposed to different climatic conditions, it becomes broken, piled up, and packed together. Pack ice is constantly on the move, driven by the winds and ocean currents. Stress from the wind forms cracks in thinner ice, which the wind then widens. A large crack in the ice is called a *lead.*

During frigid winter months, the water in a lead will freeze again quickly. If the water is calm, it forms *nilas.* Nilas is a gray, greasy-looking ice that freezes into a thin, transparent ice layer filling the lead. Since it is so thin, the nilas is easily cracked again by the wind. New thin ice is also easily crushed by thicker floes to form a *pressure ridge.*

Pack ice presents a challenge to navigation. However, there are powerful ice-breaking ships that can slice a path through pack ice. Breaking ice is not just a matter of forcing ice out of the way, as you might think, but happens when a ship rides up onto and over the ice in front of it. Icebreakers are specially designed for this and have sloping bows, heavy displacement (weight) for their size, and lots of power. They are also specifically reinforced to handle the force of the ship hitting the ice at speed. Then, once the ship is on top of the ice, its huge weight (around 13,000 to 15,000 tons) breaks and crushes the ice beneath it. Countries and companies who trade in and around the northeast and northwest passages of the Arctic use icebreakers.

GLACIER TEMPERATURES

Glaciers are classified into two basic types according to their temperature.

The first type is *temperate* or *warm glacier* where ice temperature is at its melting point. This type has a thin frozen surface with winter temperatures measuring below 0° Celsius. These glaciers are found in mountain regions outside the Arctic and Antarctica.

The second type of glacier is the *cold glacier.* Cold glaciers are found in areas where the yearly temperatures average considerably below the melting point of ice. In these areas, such as the polar regions of the Arctic and Antarctica, the bulk of the ice is permanently frozen. Although there are a few water pockets located thousands of meters beneath the ice, meltwater streams are only found on the surface or close to the glacier edges during short summer months, before they quickly refreeze.

MELTWATER STREAMS

In the summer, temperate glaciers feed meltwater streams that flow even a small amount in the winter. The drainage of meltwater from temperate glaciers is very different than that of cold glaciers' drainage systems. In temperate glaciers, meltwater flows down and inside to the glacier bed before reaching the leading edge, sometimes called the *snout.* The meltwater stream flows out from under the glacier and makes an opening, or portal, through the leading edge of the glacier.

Cold polar glaciers are frozen so solidly that melting ice only flows at the glacial margins. Glaciologists have the most trouble with this type of glacier, since the meltwater forms wide icy streams at the edges that can be tricky to cross. Marginal meltwater streams flow around the snout's leading edge and sometimes join in front of the frozen leading edge to run on down the valley.

The thermal characteristics of glaciers strongly influence the way they affect the land beneath them. In countries once covered by glacier ice and meltwater streams, widening valleys and deposited rock provide a history about passing glaciers. Temperate glaciers erode rock much more completely as they grind along. Colder glaciers, frozen to the bedrock below them, are less abrasive. In fact, they protect the land from weathering and erosion by wind and high meltwater flow.

Antarctica

Antarctica is a continent of extremes. The continent of Antarctica, double the size of Australia, contains the world's largest ice sheet. It is the coldest place on Earth, with a record low temperature of –89.2°C. It is also the windiest place on Earth, with winds of over 150 km/hr lasting for a week or more at a time.

The Antarctica ice sheet is 4200 meters thick and has ice over 200,000 years old. Ice covers all of Antarctica except for a very few mountain peaks in the

Transantarctic Mountains. Antarctica is the highest continent. Its ice dome extends 2700 meters at its highest point. The height of the ice sheet from bedrock to the top is roughly 2000 meters above sea level.

Antarctica's ice sheet extends over the ocean as a continental *ice shelf* between 500 and 1000 meters thick and has existed for about 40 million years. Antarctic ice shelves have been known to create smooth, flat-topped icebergs, sometimes hundreds of kilometers across. Ice shelves can be stable over hundreds of years, but, depending on climate and temperature, some break up quickly. The Wordie Ice Shelf in the Antarctic Peninsula has almost completely disappeared in less than 40 years.

Antarctica's Ross Ice Shelf, the world's largest, measures approximately 800 by 850 km and covers over half a million square kilometers of the Ross Sea.

Except for the much smaller Ward Hunt Ice Shelf on Ellesmere Island in the Canadian Arctic, ice shelves are only found in Antarctica.

Antarctica's harsh environment prevents most land animals from making a home there. There are no large land animals, no trees, and no grass. Although most people consider the Arctic and Antarctica to be about the same, the Arctic is much more habitable.

Only on *subantarctic islands* surrounding Antarctica do four species of seals and twelve species of birds find places to live and enjoy the scenery year around. These remote islands also offer nesting sites for albatrosses and snow petrels.

Thousands of seals mob the rocky shorelines during the spring breeding season. Weddell seals, the most common species, live in coastal waters all year long feeding on squid and fish. They dive to depths of more than 600 meters and can stay underwater for over an hour when hunting for food.

The most common Antarctica citizens are *penguins.* Five of the sixteen known penguin species are only found in Antarctica. Of these, the *Adélie penguin* is the most common and southernmost located. Although they come ashore in the spring to breed in huge colonies on rocky islands, they head back out to sea the rest of the year.

The largest of the penguins is the *Emperor penguin.* The Emperor stands 1.2 meters high, about the size of a large dog. Emperor penguins breed at the end of the summer.

The Emperor female lays one egg and then goes back to the sea to shop or attend her aerobics class. The Emperor male becomes Mr. Mom. He tends a single egg the entire winter. It rests on the tops of his feet, just below his belly. This warm protected spot and an occasional turning protects the egg until it is ready to hatch in the spring when ocean food supplies are most abundant. Emperor babies then have all summer to grow, mature, and play before harsh winter conditions set in.

Arctic

The Arctic Ocean is the smallest of the world's oceans. It lies north of most of Asia, Europe, and North America. Some scientists consider the Arctic Ocean to be part of the Atlantic rather than a separate ocean. They call it the Arctic Sea.

The northern region found within the Arctic Circle (66°30′ North) is commonly called the *Arctic* or *North Pole.* The North Pole is near the center of the Arctic Ocean and ice covers much of the ocean the year around. A *polar ice mass* forms a somewhat jagged circle with the North Pole as its center and covers about 70% of the Arctic Ocean. It rotates in a clockwise direction, driven by the polar easterly wind and ocean currents. The permanent polar ice mass is as thick as 50 meters in some spots.

Arctic climate is less harsh and dry than its cousin to the south, Antarctica. In fact, some Arctic areas were ice-free even during various ice ages. Its friendlier climate, with fewer glaciers and more ice-free areas, allows the Artic to contain a large diversity of plants and animals able to live in its more habitable environment.

There are a total of 48 species of animals currently found in the Arctic, including marmot, lemming, hare, fox, wolf, bear, reindeer, and caribou. Most of these animals migrate south to avoid the cold, harsh winters.

The plants and organisms that grow in thin Arctic soils are mostly mosses, algae, and lichen. Over time, the buildup of organic debris, clay, and sand allows for the growth of larger species such as grasses, flowers, and trees in some areas.

Glacier Zones

Glaciers have two main zones, or sections. The first zone is known as the *accumulation zone.* This is just what it sounds like. It is the part of the glacier that is building up or accumulating, the area that gets bigger in size and depth. The second zone is called the *ablation zone.* This is the part of the glacier that is shrinking.

Accumulation zone = more snow and ice
Ablation zone = less snow and ice

Glaciers increase in colder periods when lots of snow falls. That's easy to remember! Glaciers decrease by 1) *melting,* 2) *sublimation* (evaporation of the

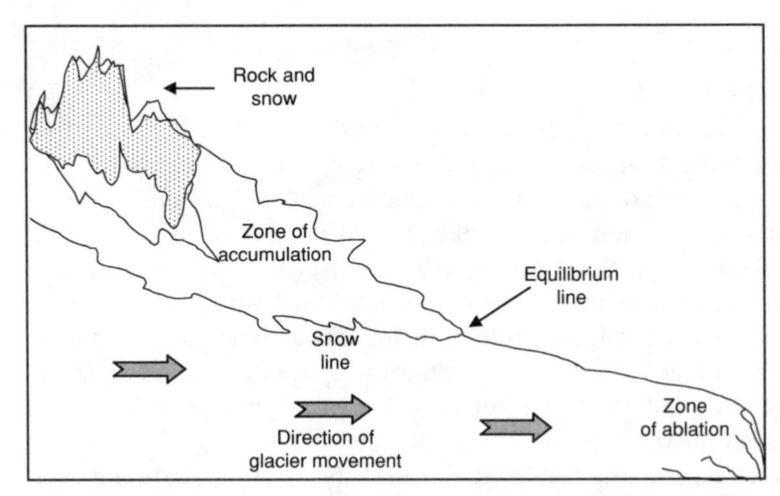

Fig. 7-1 Glaciers grow and retreat in accumulation and ablation zones.

water in ice that goes directly to a gas and doesn't become liquid in between), and 3) *calving* (small glacier pieces breaking off). Actually, glaciers contain sections that grow and shrink within the same glacier. Fig. 7-1 shows the different glacier zones and their location within a glacier.

The line between accumulation and ablation zones is called the *equilibrium line.* This is the boundary where the new glacial material added and old glacial material lost is balanced. This equilibrium line changes with the seasons. When more snow and ice are added in the winter, accumulation is at its highest. The reverse is true for ablation. The greatest amount of ablation takes place when temperatures rise in the hot summer months.

Icebergs

When large masses of ice break off from the lower ends of a glacier and slowly make their way into the sea, *icebergs* (from the Danish *berg,* meaning "mountain") are formed. This sudden or gradual splitting off is called *calving.* Calving happens when a chunk of ice breaks off from a larger piece of land or sea ice.

Icebergs are large chunks of ice broken off from the lower ends of massive glaciers.

Shiploads of tourists visit Alaska's coast and Norway's fjords yearly to experience calving and hear the loud crack of splitting ice just before a huge cliff of ice shears away and falls into the sea with a crash. When a large chunk of ice breaks off into the water and rises to the surface, it becomes a floating ship-killer. No wonder the *Titanic* had a problem!

Icebergs are tricky. It is important to give them lots of space. Because of their tremendous weight and the way trapped air makes them float, icebergs bob along with only 10% of their total size showing. The other 90% is below the water line. In other words, only a bit of the iceberg's total volume shows, with the rest remaining unseen. See Fig. 7-2 to get an idea of an iceberg's visible size compared to what is hidden from the eye.

Floating ice includes sea ice and frozen water in lakes and rivers. Sea ice typically covers around 15 million km^2 of the Arctic Ocean and about 19 million km^2 of the southern ocean around Antarctica during their winter seasons. Icebergs that shear off from an ice shelf stretching out over the water can float free in the sea or become trapped in shallower arctic waterways. This causes problems for ships. The seasonal cycle of extended sea ice influences both human activities

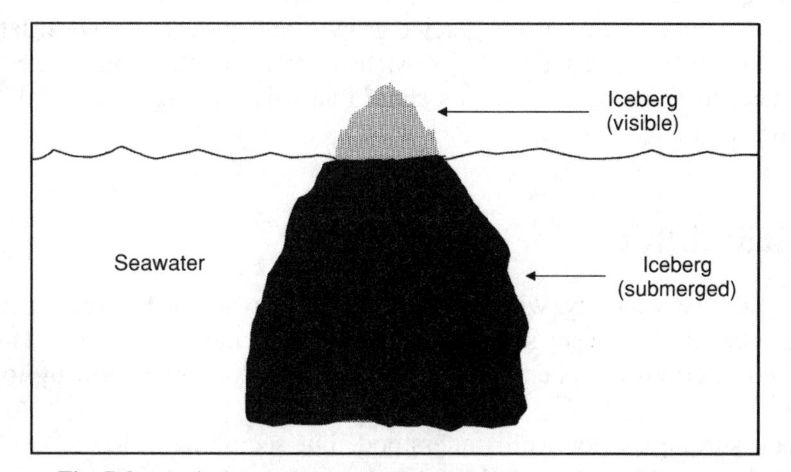

Fig. 7-2 An iceberg's greatest bulk is below the surface of the water.

and biological habitats. Many polar mammals, such as penguins, polar bears, and seals, depend entirely on sea ice for habitat. It provides landing sites for migrating birds and relaxing seals.

Glacier Speed and Movement

In 1863, Swiss geologist Louis Agassiz, a founding member of the National Academy of Sciences and later regent of the Smithsonian Institute, decided to measure the speed of glacier flow. He and his students hammered stakes into the ice of a glacier in a straight line. Over the next twelve months, the stakes moved with the glacier's movement. The stakes in the middle of the flow moved the farthest—about 75 meters! Long vertical tubes pounded deeply into the ice showed that the ice at the bottom of the glacier moved faster than upper layers.

Originally, observing the glaciers in native Switzerland, Agassiz found several telltale clues of the past presence of glaciers. They included: 1) glacier marks on the earth, 2) deep-cut valleys, 3) large boulders carried long distances, 4) scratches and rock smoothing, and 5) mounds of debris pushed up by glacial advances. He also found that in several places glacial signs could be seen where no glaciers were currently found. Agassiz gathered all this information together into a theory that a great ice age had once covered the Earth. He published his theory in *Étude sur les Glaciers* in 1840. As a result of his studies, he became known as the Father of Glaciology.

Today glaciologists measure glacier growth and movement with data from satellites and airborne radar readings. Measurements of the Antarctic continental ice sheet by modern instruments show that it is moving northward about 9 meters per year.

GLACIER SURGE

Glaciers flow very slowly, with different layers flowing at different speeds. On average, glacial movement slips along at about 20 meters per year. However, during short periods of speedy movement or *surge,* they may flow up to 10 km per day.

Glacial surging is not well understood, but seems to follow a cycle. The Variegated Glacier in southern Alaska is known to surge every 16 to 21 years. The Medvezhiy Glacier in the Soviet Pamirs surges approximately every 10 years. These glaciers are easier to study than most other glaciers that have longer irregular surge cycles. Predicting surge cycles in glaciers that have been inactive or very slow for decades is not easy. No one really knows why glaciers are asleep one day and wake up and take off the next.

Some glaciologists think earthquake-caused avalanches may set off a surge. Others think surges depend on glacier size, shape, position, gradient, temperature, climate, and bedrock type.

It has been noted that before a surge, the upper part of a glacier thickens, while the lower part thins. As the top gets denser, the glacier gets steeper and the stresses in the base increase. The meltwater streams are squashed and pushed together until they form a thin sheet of water under the entire glacier. When this happens, friction between the glacier and the underlying bedrock goes way down and the glacier slides easily. It is as if the sheeting water serves as rollers under the heavy glacier.

In 1936, the Black Rapids Glacier surged, forming an icy, crevassed cliff over 100 meters high at its normally flat-front edge. During that surge, the glacier cruised along at speeds up to 66 meters a day, threatening a tourist lodge, a highway, and damming a major river. Luckily, it slowed and stopped short of these targets. However, in the past ten years, glaciologists have been watching the Black Rapids Glacier closely for surges, since another major surge would take it across the Alaska Pipeline and a major highway.

The Kutiah Glacier in Pakistan holds the glacier speed record. In 1953, it zoomed more than 12 km in three months, or nearly 112 meters per day, making it the fastest recorded glacial surge that has ever been recorded.

Glacier flow happens much the same way as flow in frozen rivers and lakes. Ice layers at the base of a glacier are under extreme pressure and become elastic, or flowing. These moldable glaciers are able to flow over jagged bedrock without breaking or cracking. Commonly, flow at the top and center of a glacier is faster since it doesn't have ground friction to slow it down.

When a glacier moves downhill, it drags along a massive amount of rock and debris. Glacier sides undercut nearby cliffs, pulling rock down. Extreme temperatures do the same thing by loosening rock along the mountain sides. This rock falls down and is carried on the top of the glacier, while the base ice scrapes and grinds the underlying bedrock. Ice pressure against the ground and bedrock also causes the glacier to melt slightly and flow on the *meltwater.* The water between the glacier's base and the bedrock makes movement possible.

Think of it in terms of ice skating. When friction melts ice beneath the blade of an ice skater, she zooms around the ice doing figure-eights and dramatic spins. Water molecules between the skate blade and the ice act like tiny ball bearings, allowing the skater to cross the ice with ease.

Glaciers carry large and small eroded bits of rock downstream, eventually dropping them where the ice melts. Rocks that were picked up by glacier movement and left off somewhere along the way were called *drift* by early scientists who couldn't figure out where odd, displaced rocks came from.

All deposited material of glacial origin found on the land or at sea is called **drift.**

Drift that has been grabbed, eroded, sorted, and spread out by glacier meltwater streams is called *outwash.*

Crevasses

Thin upper layers of glaciers, not weighed down by tons of compacted snow, have very little pressure on them. The upper layers of a glacier are stiff and rock-hard. They flow very little, if at all. Instead, these top ice layers, pulled from below and stressed by the more fluid flowing ice, crack open to form *crevasses.*

Crevasses are the deep, vertical fractures in active glaciers caused by structural stresses. Glaciers with the most movement have hundreds of crevasses.

Some crevasses, or *fissures,* form cracks that can extend 10 to 30 meters. These cracks break up surface ice into lots of different size blocks. They are most often found where ice pulls against bedrock walls, where a valley turns, and along steep slopes.

There are several types of crevasses. They are named according to their character and structure: *longitudinal, marginal, transverse, splaying,* or *echelon.* Fig. 7-3 shows several different forms of crevasses. When many crevasses intersect, they create a wildly disrupted surface with ice towers called *séracs.* Unstable in the extreme, séracs collapse easily.

When a glacier flows over a drop or cliff, it first breaks up into transverse crevasses, then cracks apart fully to form chaotic landscapes know as *icefalls.* When the pressure upon upper layers gets too great, massive icefalls shear away huge curtains of ice from the glacier's face. These are the most dangerous features of glacier ice to skiers and glaciologists. Sometimes hidden by recent snowfalls, the unwary person must always be on the lookout for these unseen traps.

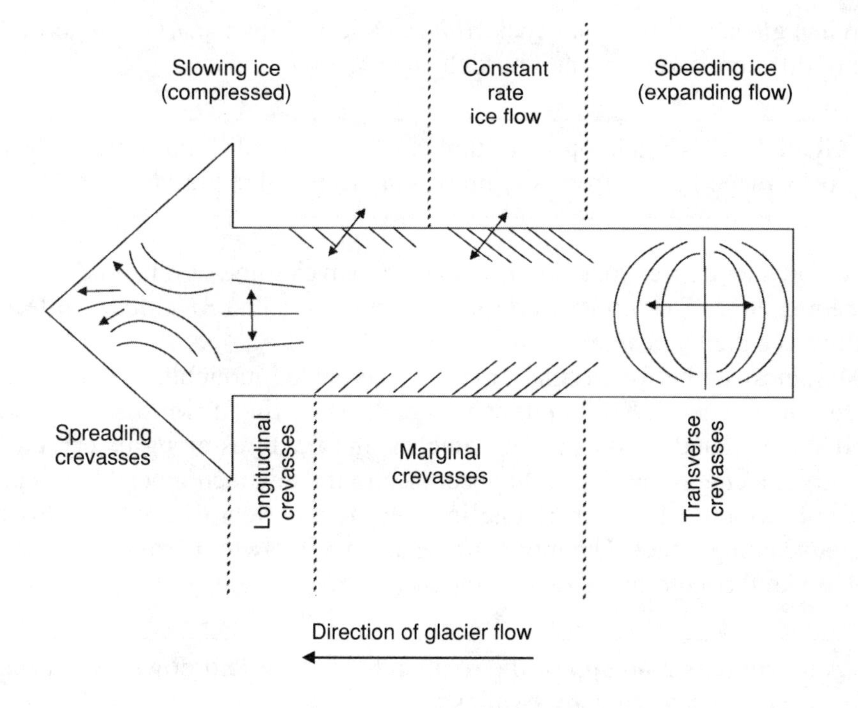

Fig. 7-3 Glacial crevasses open at different rates and directions.

A *bergschrund* is a unique type of crevasse that is found at the head of a glacier where the gradient is very steep and the main part of ice pulls away from the more stable ice clinging to the steep mountainside. They are jagged and can span hundreds of meters. These types of crevasses are very difficult to cross.

Glacial Till and Moraines

Rock pulverized by shifting ice is gathered and carried downhill as the glacier moves. As temperatures increase and/or a great distance is traveled, the glacial ice begins to melt. When this happens, rock is deposited as a sediment called *glacial till.* Large boulders picked up in one place and dropped in another by a glacier are called *erratics* since their composition doesn't usually match surrounding rock where they are found.

When glacial till becomes rock (*lithifies*), it is known as *tillite,* a porous mixture of different-sized bits and particles.

Glacial till is made up of a jumble of rocks of different sizes. These rocks range in size from a grain of sand to massive boulders.

As a glacier carries rock down a valley, it leaves some of it behind at the sides or leading edge. This ground-up rock, sometimes called *rock flour,* collects into landforms called *moraines.*

Moraines can be found lining most of the jagged mountain scenery found in and around ice fields. Rock under pressure doesn't grind down smoothly like polished stones found in streams; it crunches and fractures along its crystal faces. This gives a craggy outline and appearance to the untamed glacial landscape.

The Sawtooth Mountain Range in Idaho is a perfect example of this brutal glacier grinding affect. The mountain peaks in the Sawtooth range are sharp and jagged like the teeth of a saw.

A **moraine** is made up of the rock and sediment laid down by passing glaciers as they cut across valleys.

There are three types of moraines: *lateral*, *terminal*, and *ground moraines*. Table 7-1 compares these different types.

A *lateral moraine* is found along the edges of a glacial valley. It is made up of long, knife-edged ridges of broken rock piled along the steep sides of an alpine glacier as it cuts through a valley. Because of their weight and larger size, primary glaciers often cut deeper, larger valleys than smaller, less powerful glaciers. These mammoth glaciers muscle past and block smaller glaciers with the vertical buildup of lateral moraines along their sides.

A *terminal moraine* is created at the leading edge of a moving glacier. Long and horizontally curved across a glacier's head, a terminal moraine indicates the farthest limit that a valley glacier or continental glacier has moved. Lakes are often formed behind terminal moraines, like the Great Lakes of Michigan and Zurichsee in Switzerland.

Ground moraines are found as a layer of glacier drift laid down under the ice. Ground moraines range from thin layers with rocks sticking through, to thick layers covering the bedrock completely. The sharp, fine features of the parallel moraines of the Vadret da Tschierva glacier in the Bernina Mountains of Switzerland show the steep buildup of till on either side of the advancing glacier.

Table 7-1 Glacier deposition.

Glacier deposition (moraine)	Description
Drumlin	Old moraines elongated in the direction of new glacier flow
End	Moraine buildup at advancing/retreating glacier edge
Ground	Broad cover of deposited moraine from base-scraping glacier ice
Lateral	Rock piled up along edges of a valley glacier
Medial	Lateral moraine joins with and moves alongside main glacier moraine
Recessional	Moraine left after a glacier retreats
Terminal	Farthermost edge of several advancing glaciers

Geologists study moraines in areas where ancient glaciers have plowed past. Rock deposited along a glacier's path gives clues to its lifespan, direction, and endpoint.

GLACIAL EROSION

As glaciers move, rocks fall down onto the ice from the mountains and are carried along with the glacier as it moves. Over time, these rocks and boulders are covered with new snow and get locked into the glacier's base and sides. Ice, too soft to grind away rock by itself, uses these trapped rocks. The captive rocks act as an abrasive, like rough sandpaper, that scrapes across the land as the glacier moves.

Glacier abrasion is caused by the rocks in the base and sides of a glacier grinding and polishing ground over which the glacier travels.

When glaciers wear away rock it is called *glacial erosion.* The spine-tingling result is that as glaciers flow they shape the land into all kinds of different formations depending on what was there first. Table 7-1 lists the different landscape forms created by traveling glaciers.

Table 7-2 Glacier shapes.

Glacier impact	Description
Arête	Sharp, steep ridge of rock formed between two parallel glaciers
Cirque	Bowl-like dip
Corrie	Widened hollow
Esker	Layers of sediment in long ridges, mounds, and humps
Hanging valley	Short glacier valleys sealed off at one end by passage of main glacier valley
Horn	Pointed peak formed from glaciers flowing down different sides of the same mountain top
Kettle	Hole in the ground left by large chunk of stranded ice that has melted
Moraine	Rock and sediment piled along glacier sides by melting ice
Striation	Parallel grooves in softer rocks
Pyramidal peak	Sharp mountain peak with concave face
Truncated spurs	Projections of rock sheared off by ice
Tarn	Small, rounded lake made from a cirque
Trough	U-shaped, flat bottomed valley

Permafrost

In the frozen areas of the North and South Poles, in arctic and alpine areas, and at high elevations, there are areas where the winter freeze is so deep that the soil is forever frozen and doesn't ever get warm enough to thaw out. This frozen soil is called *permafrost.* Fig. 7-4 shows the permafrost layer. Some geologists think that as much as 25% of the land's surface is frozen as per-

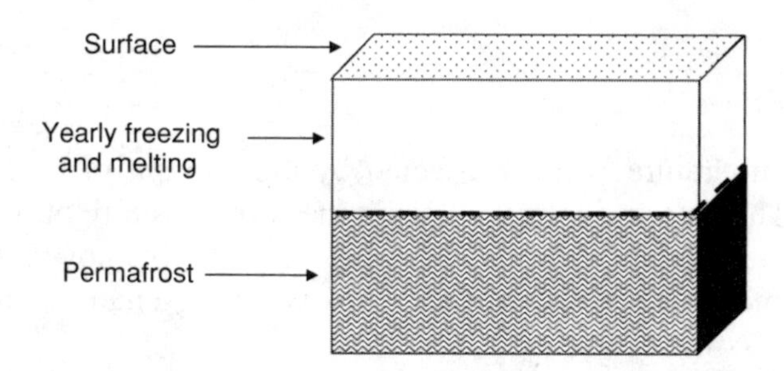

Fig. 7-4 Permafrost is the ground layer that never thaws, even in the summer.

mafrost. The permafrost in Alaska and northern Canada is estimated to be 300 to 500 meters thick.

> **Permafrost** is the layer of ground that remains frozen solid year-round. The upper section of this layer is called the **permafrost table,** growing and shrinking with seasonal temperature changes.

Much of the Northern Hemisphere's exposed land contains permafrost. Permafrost exists in tundra regions and Siberian spruce, fir, and pine forests. Large regions go through seasonal freezing and thawing. Because the melting of frozen ground produces unstable surfaces, understanding permafrost is important to civil engineering and architecture in cold regions.

Frozen permafrost is nearly as hard as rock. It can be built upon with confidence. However, if warmed, it becomes soupy with a lot of melted water and is not safe to use as a construction base.

When the Trans-Alaska oil pipeline was built, the underground sections had to be refrigerated in order to avoid thawing the permafrost. Pipeline supports in the elevated sections are made so that heat is released and doesn't affect the permafrost.

Soil located below the permafrost layer is not frozen, but heated from below by the geothermal energy of the Earth's mantle and core.

Periglacial areas are those bits of land that border a glacier area but are not covered by ice. These are cold, mountainous places, where soil and rock exist as permafrost and only the top few centimeters warm enough to thaw during the summer months. Plant life has a hard time gaining much of a toehold in this hostile climate. It is made of mostly of mosses and grasses.

Global Warming

The Earth's temperature is most affected by the amount of energy it receives from the sun. This energy, coming to us in the form of sunlight, warms the land and oceans. The amount of total warming depends on how much heat gets through the atmosphere to the surface. This warming changes from decade to decade and across the centuries.

There are a lot of different factors that can affect the amount of sunlight that gets through the atmosphere. Heavy cloud cover, atmospheric gases, volcanic dust, and ash all block sunlight, at least for a while.

In order to decide when the next "deep freeze" may strike, glaciologists are trying to figure out the main reasons for past long-term cooldowns. Environmental scientists are also trying to explain why some Ice Ages lasted longer than others.

There are lots of theories. Some scientists think the sun's output was changed for a long period of time. Sunspots that seem to occur in 11-year cycles and affect solar radiation coming to the Earth, however, don't seem to explain the longer cooldowns.

Other scientists think that the Earth's tilt explains the amount of sunlight reaching the poles. While this does explain changes in the amount of sun that reaches the planet's surface, the Earth would have to have been tilted away for a very long time to allow enough cooling to form the massive continental ice sheets.

Still others think a huge meteor hit the Earth and caused a planet-wide winter cold, killing the dinosaurs and encircling the earth in dust like a halo.

We may never know, but one thing's for sure: Glaciers come and glaciers go.

The earth has experienced several ice ages. Even as recently as the mid-14th to mid-19th centuries, a time of human industrialization, the world was in a mini ice age. Temperatures got cold enough for significant glacial growth. Then, with a rise in global temperature, glaciers started melting and disappearing like snowmen on a warm day. They left behind large areas of pulverized rock on ridges, hills, and plains. They left dents and dams that became glacier lakes and valleys.

As we've seen, the greenhouse effect, warming caused by the buildup of carbon dioxide in the earth's atmosphere, is also under careful study. It is thought that the warming in polar regions, especially Antarctica, causes huge amounts of ice to melt. The thinking is that if the Antarctic Ice Sheet were to melt, releasing about 30 million cubic meters of water to the oceans, sea level would rise about 80 meters globally. This increase in ocean levels would wipe out many low-lying and coastal cities such as Buenos Aires, Calcutta, Honolulu, London, Los Angeles, New York, San Diego, Shanghai, Sidney, and Tokyo, to name a few.

Possible evidence of global warming occurred on September 23, 2003, when the largest glacier in the Arctic shelf split into two pieces. Scientists had watched a widening crack form over several months, but did not realize the depth and extent of the growing fissure until it broke.

However, equally possible is the theory that global warming could lead to increased polar snowfall and the growth of many ice sheets. So it's really anybody's guess.

Glacial lifetimes and Ice Ages have only been tracked closely in the past century. Research continues and glaciologists continue to watch the growth and shrinkage of glaciers.

VOSTOK ICE-CORE

Glacial growth is not all black and white. Because of the shape and tilt of the Earth, there are always different temperature zones, even during an Ice Age. Dry deserts, humid jungles, deciduous forests, and frozen tundra all exist at the same time somewhere on this terrific place we call home, the third rock from the Sun.

Geologists discovered fossils that show the average temperature of the Earth in the past 600 million years has been around 22°C. This is not the wintry cold we think of when we think of glaciers. It doesn't describe glacier formation except on high mountain ranges or on polar ice caps.

In the past 20 years, ice cores were drilled in Vostok, Antarctica, near the South Pole. These cores have given scientists much older data on atmospheric carbon dioxide (CO_2) and methane (CH_4) levels. Ice builds up extremely slowly in this area at a rate of 2.5 cm per year. By studying a Vostok ice core section (2.5 km deep from an ice section 3.7 km thick) that represents temperature changes of the past 240,000 years, scientists found greenhouse gas levels from before the last Ice Age. From this information and earlier data, it became obvious that the earth cycles between full glacial and interglacial periods of around 130,000 years with minor ups and downs in between. Comparing current CO_2 and CH_4 levels to those in the Vostok ice core, the Earth seems to be in an 11,000-year interglacial period that began at the end of the last Ice Age. These glacial and interglacial times correspond to the advance and melting of glaciers at the North and South poles, affecting climate, species biodiversity, and habitat.

Future of Glaciers

With the Earth's population expected to double in the next 50 years at current rates and humankind's impact on the global ecosystem growing larger, glaciers have become increasingly important.

We learned that glaciers hold 75% of the world's water. When glaciers advance more than retreat, the impact on our available fresh water is significant. Will there be enough fresh water for everyone on the Earth in the future?

Since our bodies are made up of over 66% water, we don't last much beyond 3 days without replacing the water we lose through breathing, sweating, and urination. Protecting fresh water for drinking is critical to life, not to mention making snow cones and popsicles!

The importance of protecting our natural reservoirs from overuse and contamination is important to our existence as a species. Geologists, geographers, physicists, chemists, mathematicians, and meteorologists in many countries are studying the world's glaciers and building glacier inventories. These scientists are putting together a World Glacier Inventory (WGI) to better understand climate changes and measure total freshwater levels over the Earth's surface.

The World Glacier Inventory contains information on over 67,000 glaciers worldwide. Glacial inventory factors include geographic location, area, length, orientation, elevation, and classification of morphological type and moraines. This information, based upon a single observation in time, should be considered a "snapshot" of any specific glacier. The majority of the information comes from the World Glacier Monitoring Service in Zurich.

Using this and other information, scientists can implement conservation, environmental, and resource decisions that best address the needs of future generations.

Quiz

1. The total amount of ice on a planet is called the
   (a) lithosphere
   (b) mantle
   (c) hydrosphere
   (d) cryosphere

2. The Little Ice Age took place between
   (a) 1000–1400 A. D.
   (b) 1350–1650 A.D.

(c) 1600–1850 A. D.
(d) 1700–1900 A.D.

3. What percentage of the earth's fresh water do glaciers hold?
   (a) 25%
   (b) 40%
   (c) 75%
   (d) 90%

4. When soil is frozen and never gets warm enough to thaw out, it is known as
   (a) rock
   (b) icy mounds
   (c) loess
   (d) permafrost

5. The pieces of land that border a glacier area but are not covered by ice are called
   (a) tectonic margins
   (b) islands
   (c) periglacial areas
   (d) equatorial

6. What percentage of an iceberg is unseen below the surface of the water?
   (a) 30%
   (b) 50%
   (c) 75%
   (d) 90%

7. When glaciers wear away rock it is called
   (a) sanding
   (b) pyroclastic erosion
   (c) glacial erosion
   (d) grooving

8. Icebergs are
   (a) always cold
   (b) large chunks of ice broken off from the lower ends of massive glaciers
   (c) known to be ship killers
   (d) all of the above

9. Who is known as the Father of Glaciology?
   (a) Jack Showers
   (b) Louis Agassiz
   (c) Douglas Williams
   (d) Alfred Wegener

10. The rock and sediment landform deposited by passing glaciers as they cut through valleys is
    (a) moraine
    (b) loess
    (c) glacial dunes
    (d) hardpack

CHAPTER 8

Water Pollution and Treatment

What is water pollution anyway? Is it the gross dumping of industrial waste into lakes and streams? Is it the sometimes massive oil slicks of unfortunate tanker accidents? Is it the runoff of agricultural chemicals into streams and groundwater? Or is it the simple muddying of a forest stream as you slosh through in hiking boots on your way to a scenic peak?

> **Water pollution** comes from the loss of any real or potential water uses caused by a change in its composition due to human activity.

Water is used for everything from drinking and household uses to watering livestock and the irrigation of crops. Fisheries, industry, food production, bathing, recreation, and other services all use water to a large extent. When water becomes unusable for any of these purposes, it is polluted to a greater or lesser degree depending on the extent of the damage caused.

Natural Water

Pure water (H_2O) is completely free from any substances dissolved in it. It is found only in the laboratory. Natural water contains dissolved gases and salts. These substances often make the water suitable for particular uses. For example, water must contain enough dissolved oxygen for fish to survive or they die. Drinking water without dissolved oxygen and with only very low levels of dissolved mineral salts tastes pretty bad. Mineral salts give water its taste.

Water Pollution

Water pollution is caused by the sudden or ongoing, accidental or deliberate, discharge of a polluting material. In Chapter 6, we saw how increasing human populations put pressure on the oceans and marine environment. More and more people on the planet lead to more of the following:

- Sewage produced;
- Fertilizers, herbicides, and pesticides used for crops, lawns, golf courses, and parks;
- Fossil fuels extracted and burned;
- Oil leaked and spilled;
- Land deforested and developed; and
- Various byproducts of manufacturing and shipping generated.

Cultural, political, and economic forces affect the types, amounts, and management of waste produced. Increasing population is just one contributor to increasing pollution. As with everything in the environment, the causes and effects are complex.

Groundwater has been contaminated by leaking underground storage tanks, fertilizers and pesticides, unregulated hazardous waste sites, septic tanks, drainage wells, and other sources. This contamination threatens the drinking water of 50% of the United States' population.

The three major sources of water pollution are *municipal, industrial,* and *agricultural.* Municipal water pollution comes from residential and commercial waste water. In the past, the main way to treat municipal wastewater was to reduce suspended solids, oxygen-demanding materials, dissolved inorganic compounds, and harmful bacteria. Today, the focus is on the improvement of solid residue disposal from municipal treatment processes.

The agricultural midwest of the United States has developed water pollution problems. For example, in Iowa where chemical fertilizers are used over 60% of the state, some private and public drinking water wells have gone over safety standards for nitrates. A substantial number of towns in Nebraska have also shown high nitrate levels and require monthly well testing.

Pollution of marine ecosystems includes runoff from land, rivers, and streams, direct sewage discharge, air pollution, and discharge from manufacturing, oil operations, shipping, and mining.

Although coastal cities have the greatest impact on ocean ecosystems, pollution from runoff is not limited to coastal regions. Runoff from over 90% of the Earth's land surface (inland and coastal) eventually drains into the sea, carrying sewage, fertilizers, and toxic chemicals. Similarly, air pollution from inland as well as coastal cities, including byproducts of fossil fuel consumption, polychlorinated byphenyls (PCBs), metals, pesticides, and dioxins, eventually finds its way into the oceans after rain or snow.

Increased oil demand has increased offshore oil drilling operations and oil transport. These activities have resulted in many oil spills. The number of oil spills worldwide of between 7 and 700 tons and greater than 700 tons has varied in the past 30 years, with some years being better than others. Table 8-1 lists the top fifteen oil spills (excluding acts of war) recorded by the International Oil Tanker Owners' Pollution Federation (IOTPF). The IOTPF measures all oil lost to the environment, including that burned and released into the atmosphere or still held in sunken ships.

For comparison, the well-known *Valdez* spill of 1989, thought by many to be a particularly bad spill, was not the worst ever seen. That spill (rated 35th in largest spills), off the coast of Alaska, was not as extensive as many others, but had a huge impact on the delicate and pristine polar environment.

Spills account for only 10% of marine oil pollution. About 50% of oil pollution in marine waters comes from ongoing low-level sources such as marine terminal leaks, offshore dumping, oil drilling mud, land runoff, and atmospheric pollution from incompletely burned fuels. We will learn more about the various polluting effects of fossil fuels in Chapter 13.

Cumulative pollution effects on ocean ecosystems are very serious. For example, in the Gulf of Mexico, scientists have identified "dead zones" in once highly productive waters. These zones have been traced to excessive nutrients from farms, lawns, and inadequately treated sewage. This stimulates rapid plankton growth that in turn leads to oxygen depletion in water.

Table 8-1 Oil spill impacts are as related to spill location as total volume lost.

Rank	Ship	Year	Location	Volume (tons)
1	Atlantic Empress	1979	Off Tobago, West Indies	287,000
2	ABT Summer	1991	700 nautical miles off Angola	260,000
3	Castillo de Bellver	1983	Off Saldanha Bay, South Africa	252,000
4	Amoco Cadiz	1978	Off Brittany, France	223,000
5	Haven	1991	Genoa, Italy	144,000
6	Odyssey	1988	700 nautical miles off Nova Scotia, Canada	132,000
7	Torrey Canyon	1967	Scilly Isles, UK	119,000
8	Sea Star	1972	Gulf of Oman	115,000
9	Irenes Serenade	1980	Navarino Bay, Greece	100,000
10	Urquiola	1976	La Coruna, Spain	100,000
11	Hawaiian Patriot	1977	300 nautical miles off Honolulu	95,000
12	Independenta	1979	Bosphorus, Turkey	95,000
13	Jakob Maersk	1975	Oporto, Portugal	88,000
14	Braer	1993	Shetland Islands, UK	85,000
15	Khark 5	1989	120 nautical miles off Atlantic coast of Morocco	80,000
35	Exxon Valdez	1989	Prince William Sound, Alaska, USA	37,000

Blooms of toxic *phytoplanktons* and red tides have increased in frequency over the last two decades, and may be linked to coastal pollution. For example, storm water runoff carries with it suspended particulates, nutrients, heavy metals, and toxins. The effects of the storm water runoff often cause *dinoflagellate* (red tide) blooms following storms. These events cause mass mortality among some fish species, resulting in marine mammal deaths, and can be a serious threat to human health.

pH

The measurement of *pH* is important into wastewater treatment before its release into natural water ecosystems.

> The measurement of the number of hydrogen ions in water, on a scale from 0 to 14, is called the water's **pH.**

A solution with a pH value of 7 is neutral, while a solution with a pH value <7 is acidic and a solution with a pH value >7 is basic. Natural waters usually have a pH between 6 and 9. The scale is negatively logarithmic, so each whole number (reading downward) is ten times the preceding one (for example, pH 5.5 is 100 times as acidic as pH 7.5).

$$pH = -\log [H+] = \text{hydrogen ion concentration}$$

The pH of natural waters becomes acidic or basic as a result of human activities, such as acid mine drainage, emissions from coal-burning power plants, and heavy automobile traffic.

DISSOLVED OXYGEN

Oxygen enters the water by direct atmospheric absorption or by aquatic plant and algal photosynthesis. Oxygen is removed from water by respiration and the decomposition of organic material.

> **Dissolved oxygen** is the amount of oxygen measured in a stream, river, or lake.

Dissolved oxygen is also an important marker of a river or lake's ability to support aquatic life. Fish need oxygen to survive and absorb dissolved oxygen through their gills.

The actual level of dissolved oxygen present in even the cleanest water is extremely small. The amount of dissolved oxygen in water depends on several factors, including temperature (the colder the water, the more oxygen that can be dissolved), volume and velocity of water flow, and number of organisms using oxygen for respiration.

Oxygen solubility in water at a temperature of 20° Celsius is 9.2 milligrams oxygen per liter of water (about 9 parts per million).

The amount of oxygen dissolved in water is expressed as a concentration, in milligrams per liter (mg/l) of water. Metropolitan activities that affect dissolved oxygen levels include the removal of native vegetation, runoff, and sewage discharge. Many of these pollutants are nontoxic, so how do they cause pollution problems?

The answer comes back to the oxygen levels. A rapidly flowing stream reaches saturation of around 9 ppm (100% saturation). This permits healthy growth of natural flora and fauna (animals and bacteria). The bacteria are mostly aerobic (they require oxygen) and their numbers are controlled by the availability of food (digestible organic matter).

When there is a big discharge of organic materials, like sewage, milk, or agricultural waste, there is an immediate abundance of food. Bacterial growth is greatly stimulated and populations increase rapidly, consuming and depleting available oxygen levels in the water.

The amount of oxygen depletion over time depends on the speed with which the stream takes up oxygen from the atmosphere (*reaeration capacity*). Fast-flowing, turbulent streams are getting new oxygen quickly, while deep, slow-flowing rivers take up oxygen much more slowly. Lakes reaerate even more slowly.

Oxygen loss may be counteracted by the photosynthesis of green plants, which produce oxygen during daylight. In heavily polluted areas, however, plant processes just can't keep up and oxygen levels drop drastically. As dissolved oxygen levels decrease, fish die. Plants also respire 24 hrs/day, seven days a week. In the worst cases, anaerobic conditions (complete lack of free oxygen) result and fish die.

TURBIDITY

Turbidity is a measure of water's cloudiness. The cloudier the water, the higher the turbidity. Water turbidity is caused by suspended matter such as clay, silt, and organic matter. It can also result from increases in plankton and other microscopic organisms that interfere with the passage of light through the water. Turbidity itself is not a major health concern, but high turbidity can interfere with disinfection and provide a suspended medium for microbial growth. High turbidity also points to the presence of microorganisms. High turbidity can be caused by soil erosion, urban runoff, and high flow rates.

Water Treatment

Whenever water is used for humans, it must be treated from two different angles. First, any surface water from rivers that is used in cities is treated for drinking, usually by chlorination. After water is used for drinking, washing, lawns, toilets, and so on, it has to be treated at a wastewater treatment plant before it can be released back into the environment. Fig. 8-1 illustrates the path that water takes from initial *water treatment* (chlorination) to urban use, and then *wastewater treatment* before its release back into the environment.

SURFACE WATER TREATMENT

Most municipal water purification systems use several steps to treat water, from physical removal of surface impurities to chemical treatment.

Before raw water is treated, it passes through large screens to remove sticks, leaves and other large objects like plastic bottles. Sand and grit settle out or fall to the bottom of a tank during this stage. During *coagulation,* a chemical such as *alum* (aluminum sulfate) is added to the raw water. The water is mixed with

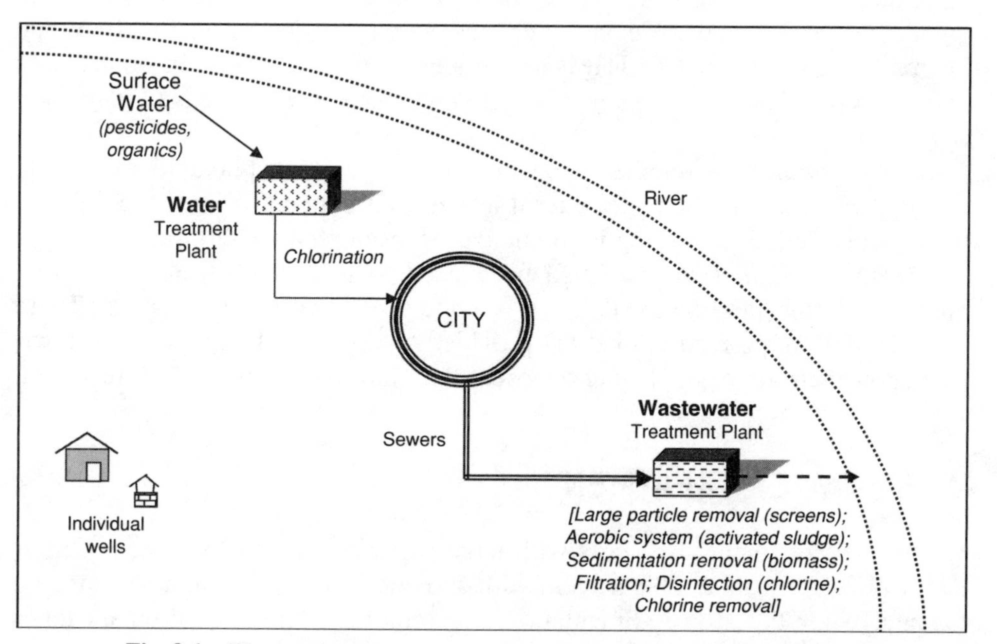

Fig. 8-1 Water and waste treatment plants use various methods to purify water.

the alum, forming sticky blobs that snag small particles of bacteria, silt, and other impurities.

Flocculation is typically the removal of impurities (or desired materials depending on the process) by a froth or bubbles. A *floc* is commonly taken from the top of the tank rather than the bottom.

The water is then pumped very slowly through a long basin called a settling basin. This is done to remove much of the remaining solid material. It collects at the bottom of the basin, in a process called *sedimentation* or *clarification.*

The next step includes the removal of microorganisms like viruses, bacteria and protozoans, as well as any remaining small particles. This is done by *filtration* of the water through layers of sand, coal, and sometimes other granular materials. This purification step copies water's natural filtration method as it moves through the soil. After the water is filtered, it is treated with chemical disinfectants to kill any organisms not collected in the filtration step.

Chlorine is a good disinfectant. *Chlorination* is used in many water treatment plants. Chlorine is commonly added as a chloroamine to help prevent *trihalomethane* (THM) formation. It is added as ammonia (NH_3) and chlorine (Cl) at the head of the plant, before alum. Chloroamine or chlorine is added again after filtration when the organic load is small to provide residual chlorination in the distribution lines.

In many cities, chlorine is the only chemical disinfection method used on surface water. Unfortunately, it is not without problems. When chlorine mixes with organic material, like dead leaves, it creates potentially dangerous THMs. While big treatment plants remove THMs to maintain a safe level, those in small towns often do not. THM warnings have been announced in parts of Newfoundland and Nova Scotia.

Ozone oxidation is another good disinfectant method, but unlike chlorine, ozone doesn't stay in the water after it leaves the treatment plant. It doesn't kill bacteria that might be lurking in municipal or residential water pipes.

Ultraviolet light (UV) has also been used to treat wastewater by killing microorganisms, but like oxidation, it is a one-time treatment at the plant. There is no continuing protection like that provided by chlorine. Even with this drawback, however, some people that oppose water chlorination prefer UV treatment.

WASTEWATER TREATMENT

Wastewater treatment also starts with large particle removal by screen. This is then followed by an aerobic system with activated sludge that removes organic loading. After this, the sedimentation and removal of organic biomass takes place, often recycling more than once.

Biomass sludge is removed from wastewater before the filtration step. Wastewater filtration, followed by disinfection with chlorine and its removal, occurs before the clean water is finally discharged into the environment.

CONTAMINANTS

Water pollution is bad news. It poisons drinking water, poisons food animals (through buildup of environmental toxins in animal tissues), upsets the biological diversity of river and lake ecosystems, causes acid rain deforestation, and lots of other problems. Most of these problems are contaminant specific.

In general, four main contaminants types exist: *organic, inorganic, radioactive,* and *acid/base.* These are released into the environment in a variety of different ways, but as we learned in Chapter 5, most pollutants enter the hydrologic cycle two ways, as direct (point source) and indirect (non–point source) contamination.

Point sources include effluents of various qualities from factories, refineries, and waste treatment plants that are released directly into urban water supplies. In the United States and elsewhere, these releases are monitored, but some pollutants are still found in these waters.

Non–point sources include contaminants that come into the water supply from soil/groundwater systems runoff and from the atmosphere through rainfall. Soils and groundwater contain fertilizer and pesticide residues and industrial wastes. Atmospheric contaminants also come from gaseous emissions from automobiles, factories, and even restaurants.

In 1987, the United States Environmental Protection Agency (EPA) issued the Clean Water Act, Section 319, which calls for federal cooperation, leadership, and funding to help state and local efforts in tackling non–point source pollution.

Sometimes, just when we think we are helping the environment, we find we've created another problem. The case of *methyl tertiary-butyl ether* (MTBE), an oxygenate from natural gas that increases octane levels, is a good example. In 1990, the Clean Air Act mandated that MTBE be added to the gasoline in parts of the country that had ozone problems. So for the next few years, MTBE was added to gasoline. Marketing campaigns even convinced us that higher octane gasoline was better for your car. It was. Gasoline burns cleaner and with fewer emissions when MTBE is added. Unfortunately, it was not the perfect additive everyone thought it was.

It turns out that MTBE is a serious groundwater pollutant. MTBE is less toxic than benzene, the most hazardous gasoline-related groundwater pollutant in the United States. But like benzene, MTBE can change the color of groundwater and cause it to smell and taste like turpentine in the smallest quantities.

Because of this, California and 11 other states are phasing out MTBE and again looking for alternatives to help reduce atmospheric pollution and fossil fuel emissions.

CHEMICALS

Many pollution sources come from sewage and fertilizers containing nutrients like nitrates and phosphates. In high levels, nutrients overstimulate the growth of water plants and algae. Uncontrolled growth of these organisms clogs waterways that use up dissolved oxygen and as they decompose, keeps sunlight from penetrating into deeper water. When this happens, the photosynthetic cycle of good water plants and organisms is affected. This hurts fish and shellfish that live in the affected water.

Nitrogen

Nitrogen is needed by all organisms to form proteins, grow, and reproduce. Nitrogen is very common and found in many forms in the environment. Inorganic forms include *nitrate* (NO_3), *nitrite* (NO_2), *ammonia* (NH_3), and *nitrogen gas* (N_2). Organic nitrogen is found in the cells of all living things and is a part of proteins, peptides, and amino acids. High levels of nitrate, along with phosphate, can overstimulate the growth of aquatic plants and algae, causing high dissolved oxygen consumption, killing fish and other aquatic organisms. This process is called *eutrophication.* Nitrate, nitrite, and ammonia enter waterways from lawn fertilizer runoff, leaking septic tanks, animal wastes, industrial wastewaters, sanitary landfills, and discharges from car exhausts.

Phosphorus

Phosphorus is also a nutrient needed by all organisms for basic biological processes. Phosphorus is an element found in rocks, soils, and organic material. Its concentration in fresh water is usually very low. However, phosphorus is used widely in fertilizer and other chemicals, so it is often found in higher concentrations in populated areas. Phosphorus is commonly found as phosphate (PO_4^{-3}). High levels of phosphate and nitrate cause eutrophication. The main sources of phosphates in surface waters are detergents, fertilizers, and natural mineral deposits.

ORGANIC MATTER

Pollution also takes place when silt and other suspended solids like soil run off from plowed fields, construction and logging sites, urban areas, and eroded river banks after a rain. This can also happen with snow melt.

Lakes, slowly moving rivers, and other areas of water go through eutrophication (in the presence of excess nitrogen and phosphorus), which gradually fills a lake with sediment and organic matter. When these sediments dump into a lake, for example, fish respiration is impacted, plant growth and water depth are limited, and aquatic organisms asphyxiate. Removal of nitrogen and phosphorus help to prevent eutrophication.

Organic pollution enters waterways as sewage, leaves, and grass clippings, or as runoff from livestock feedlots and pastures. When bacteria break down this organic material (measured as *biochemical oxygen demand,* BOD), they use the oxygen dissolved in the water. Many fish and other water inhabitants cannot live when dissolved oxygen levels drop below 2 to 5 parts per million. When this happens, it kills water organisms in huge numbers and the food web is hit hard. Most of the swimming organisms leave the affected zone if possible. Downstream in rivers, the water usually contains higher oxygen levels as long as additional organic loading doesn't take place.

The pollution of rivers and streams with chemical contaminants has become one of the most critical environmental problems of the past 100 years. Chemical contaminants flowing into rivers and streams cause a domino effect of destruction.

PATHOGENS

It's important to include *pathogens* as a type of pollution with wide-reaching health risks. Pathogens (disease-causing microorganisms) cause everything from typhoid fever and dysentery to respiratory and skin diseases. They include such organisms as bacteria, viruses, and protozoa. These living nemeses enter waterways through untreated sewage, storm drains, septic tanks, farm runoff, and bilge water. Although microscopic, biological pathogens have a monstrous effect in their ability to bring about sickness.

Fecal coliform bacteria, present in the feces and intestinal tracts of humans and other warm-blooded animals, can enter rivers and lakes from human and animal waste. If any fecal coliform bacteria are present, it is an indication of the possible presence of pathogenic microorganisms. Pathogens are commonly found in such small amounts that it is not practical to monitor them directly.

However, high concentrations of bacteria in water are caused by septic tank failure, poor pasture and animal keeping methods, pet waste, and urban runoff.

The largest concern in water treatment is assuring human health. Removal of potential pathogens, turbidity, hazardous chemicals, and nitrates are all important factors in keeping water safe.

Most underdeveloped and war-torn countries around the world have little or no clean drinking water. Disease in many of these countries is directly related to the numbers of people who are forced to use polluted water supplies. They have no other choice.

Acid Rain

Rain, naturally acidic because of the carbon dioxide in the earth's atmosphere, reacts with water to form carbonic acid. While regular rain has a pH of 5.6 to 5.7, most pH readings depend upon the type and amount of other gases like sulfur and nitrogen oxides present in the air.

We learned in Chapter 4 that the most important natural acid is carbonic acid, formed when carbon dioxide dissolves in water ($CO_2 + H_2O \rightarrow H_2CO_3$). Rocks, like limestone and marble, are particularly sensitive to this type of chemical (acid) weathering.

> **Acid precipitation** describes wet acid pollution forms found in rain, sleet, snow, fog, and cloud vapor.

Precipitation, normally between 5.0 and 5.6 in pH, is normally slightly acidic. Some sites on the eastern North American coast have precipitation pH levels near 2.3 or about 1000 times more acidic than pure water. Vinegar, which is very acidic, has a pH of 3.0.

Even in the 17th century, industry and acidic pollution was known to affect plants and people. In 1872, Angus Smith published a book called *Acid Rain.* However, it wasn't until the past forty years that problems connected to acid deposition became an international and recognized problem when fishermen noticed fish number and diversity declines in lakes throughout North America and Europe.

> **Acid rain** refers to all types of precipitation (rain, snow, sleet, hail, fog) that is acidic (pH lower than the 5.6 average of rainwater) in nature.

Acid rain is formed when chemicals in the atmosphere react with water and return to the earth in an acidic form in raindrops. When acid rain falls on limestone statues, monuments, and gravestones, it can dissolve, discolor, and/or disfigure the surface by reaction with the rock's elements. The process is known as *dissolution.* Historical treasures such as statutes and buildings, hundreds to thousands of years old, suffer from this kind of weathering. Acid rain is a byproduct of the industrial revolution: Industry releases the chemicals that form the acid in the atmosphere.

The sulfur and nitrogen oxides that form acid rain are released mostly from industrial smokestacks and automobile, truck, and bus exhausts, but they can also come from burning wood. When they reach the atmosphere, they mix with moisture in the clouds and change to sulfuric and nitric acid. Then, rain and snow wash these acids from the air.

Acid rain has been measured in the United States, Germany, Czechoslovakia, Yugoslavia, the Netherlands, Switzerland, Australia, and elsewhere. It is also becoming a big problem in Japan, China, and Southeast Asia.

Acid rain affects lakes, streams, rivers, bays, ponds, and other bodies of water by increasing their acidity until fish and other aquatic creatures can no longer live there. Aquatic plants grow best between pH 7.0 and 9.0. As acidity increases, submerged aquatic plants die and waterfowl lose a basic food source. At pH 6, freshwater shrimp cannot survive. At pH 5.5, bottom-dwelling bacterial decomposers die. When they are gone, leaf litter and other organic debris do not get broken down and subsequently build up on the bottom. When this eliminates plankton's food, they die off, too. Then, below 4.5 pH, all fish die.

Acid rain also harms vegetation. The forests of Germany and other nations in Western Europe are thought to be dying from acid rain. Scientists think that acid rain has damaged the protective waxy coating on leaves, which then allows acids to penetrate. This interrupts the water evaporation and gas exchange so that the plant no longer can breathe. The plant's nutrient conversion and water uptake is also damaged.

The most significant effect of acid rain on forests results in nutrient leaching and concentration of toxic metals. Nutrient leaching takes place as acid rain adds hydrogen ions to the soil, which then react with existing minerals. This strips calcium, magnesium, and potassium from soil and robs trees of nutrients.

Toxic metals such as lead, zinc, copper, chromium, and aluminum are deposited in the forest by the atmosphere. When acid rain interacts with these metals, they then stunt tree and plant growth, along with mosses, algae, nitrogen-fixing bacteria, and fungi.

Economic losses from acid rain in the United States are estimated to be over $15 billion annually in the eastern part of the nation and could cause nearly $2 million yearly in forest damage.

Acid Deposition

Acid deposition forms in two ways. It is belched out as hydrochloric acid directly into the atmosphere or from *secondary pollutants* that form during the oxidation of *nitrogen oxides* or *sulphur dioxide* gases released into the atmosphere. These pollutants can travel hundreds of kilometers from their original source.

> Acidic pollutants, dropped from the atmosphere to the Earth in wet and dry forms, act together in **acid deposition.**

Acid precipitation also takes place when nitrogen oxides and sulphur dioxide settle on the land and interact with dew or frost. Roughly 95% of the elevated levels of nitrogen oxides and sulfur dioxides in the atmosphere come from human actions. Only 5% comes from natural processes. The five main nitrogen oxide sources include:

1) Burning of oil, coal, and gas;
2) Volcanic action;
3) Forest fires;
4) Decay of soil bacteria; and
5) Lightning.

Nitrogen oxide and sulfur dioxide concentrations are much lower than atmospheric carbon dioxide, primarily responsible for the natural acidity of rainwater. These gases are much more soluble than carbon dioxide, however, and have a greater impact on the pH of precipitation.

ACID RAIN EFFECTS

In the mountains, many acidified aquatic ecosystems go through *acid shock.* This happens when acidic deposits buildup in the snow pack during the winter. With the spring melt, the stored snow melts and acids are released suddenly at concentrations 5 to 10 times more acidic than precipitation. Although most adult fish survive this acid shock, eggs and fry of many spring-spawning species are hit hard by this acidification.

Acid precipitation can also damage plant leaves, particularly in the form of fog or clouds when their water content can be up to ten times more acidic than regular rainfall.

These acid pollutants disrupt the ability of leaves to retain water in dry or drought conditions. Acidic deposition leaches nutrients from the plant tissues and weakens their structure. Altogether, these impacts decrease plant growth rates, flowering ability, and crop yields. It also makes plants more susceptible to insects, disease, drought, and frost.

The effects of acidic deposition on human health are surprising as well. Toxic metals, like mercury and aluminum, are released into the environment through the acidification of soils and make their way into ground water.

High metal levels are toxic to humans. Aluminum has been linked to *Alzheimer's disease.* High concentrations of sulfur dioxide and nitrogen oxides have been correlated to increased respiratory illness in children and the elderly.

Acid deposition also affects jobs. Acidic lakes and streams cause fish declines. Lower fish numbers hit commercial fishermen and industries that depend on sport fishing revenues. Forestry and agriculture are affected by crop damage.

Lastly, acid deposition affects stone, pitting and wearing it away. Limestone buildings, statues, and head stones are easily attacked by acids, as well as iron or steel structures. Automobile paint fades when attacked by acid deposition. I guess that's a good reason to wash your car a lot!

TREATMENT

There are a number of things that can be done in order to alleviate the problems of acid deposition. For example, liming is done to normalize the pH of lakes that have been acidified. This involves adding large amounts of hydrated lime, quick lime, or soda ash to lake waters to increase the alkalinity (make the water more basic). This fix has some drawbacks, as some lakes are unreachable, too big and therefore too costly to treat, or have a high flow rate and quickly become acidic again after liming.

The best way to slow or stop acid deposition is to limit the chemical emissions at their source. In several environmentally conscious countries, regulations now limit the amount of sulfur and nitrogen oxide emissions that can enter the atmosphere from industrial sources.

Industrial sources limit acidic pollutants three ways: 1) by switching to fuels that have no or a low sulfur content, 2) by using smokestack scrubbers to reduce the amount of sulfur dioxide being released, and 3) by requiring the use of specially designed catalytic converters on vehicles.

In 1991, Canada and the United States established the Air Quality Accord that controls air pollution that crosses across international boundaries. This environmental agreement established a permanent limit on acid deposition emissions in both countries (13.3 million tons of sulfur in the United States and 3.2 million tons of sulfur in Canada).

Oil, Radioactivity, and Thermal Problems

OIL SLICKS

Fossil fuels cause atmospheric pollution from combustion, but they affect water through spills.

Petroleum pollutes rivers, lakes, seas, and oceans in the form of oil spills. The *Exxon Valdez* catastrophe is an example of oil spill water pollution. Large accidental releases of oil are a big cause of pollution along shorelines. Besides supertankers like the *Valdez,* offshore drilling operations add their share of oil pollution. One estimate suggests that one ton of oil is spilled for every million tons of oil transported. It may not seem like a lot, but it adds up.

Oil pollution is increasing and devastating to coastal wildlife. Since oil and water don't mix, even small oil amounts spread quickly across long distances to form deadly oil slicks. Oil tanker spills are a growing environmental problem because once oil has spilled, it is almost impossible to remove or contain completely. Oil floats on the water for long periods of time and then washes up along miles of shoreline. When efforts to chemically treat or sink the oil are made, area marine and beach ecosystems are often disrupted even more.

RADIOACTIVITY

Nuclear medicine is the specific focus of radiology that uses minute amounts of radioactive materials, or *radiopharmaceuticals,* to study organ function and structure. Nuclear medicine imaging is a combination of many different sciences, including chemistry, physics, mathematics, computer technology, and medicine.

Because x-rays travel through soft tissue, such as blood vessels, heart muscle, and intestines, contrast tracers are used in nuclear imaging. Nuclear imaging examines organ function and structure, whereas diagnostic radiology is based on anatomy, bones, and hard tissues.

A relatively new use for radioactive isotopes (radiopharmaceuticals) is in the detection and treatment of cancer. This branch of radiology is often used to help diagnose and treat abnormalities very early in the stages of a disease, such as thyroid cancer. Since rapidly dividing cells like those present in cancer are more vulnerable to radiation than slow-growing normal cells, treatment using medical isotopes works well.

Radon-222, used initially, and now cobalt-60, are used as implants near a cancer, shot as a narrow beam to an inoperable brain cancer or used before surgery to shrink a tumor in lung or breast cancers.

A *radioactive tracer* is a very small amount of a radioactive isotope added to a chemical, biological, or physical system in order to study the system. For example, radioactive barium (^{37}Ba) is used to diagnose unusual abdominal pain, gastroesophageal reflux, gastric or duodenal ulcers, or cancer. Thallium (^{201}Tl) is a radioactive tracer used to detect heart disease. This isotope binds easily to heart muscle that is well oxygenated. When a patient with heart trouble is tested, a scintillation counter detects the levels of radioactive thallium that have bonded to oxygen. When areas of the heart are not receiving oxygen, there is very little thallium binding and they are seen as dark areas.

Radioactively tagged enzymes, or proteins containing radioactive components, are used in a medical specialty called *radioimmunology.* Radioimmunology measures the levels of biological factors (proteins, enzymes) known to be changed by different diseases. Some of these medical isotopes used in medicine include phosphorus (^{32}P), iron (^{59}Fe), and iodine (^{131}I).

Radioactive elements are an important area of chemical study. However, unless these biohazards are disposed of properly, they can find their way into sewage.

Radioactive waste materials are also created by nuclear power plants, industry, and mining of radioactive minerals. Dust and waste rock from uranium and thorium mining and refining processes contain minute levels of nuclear contaminants. When rainfall washes these away from the mining sites, they eventually make their way to waterways. Over time, levels increase and there is a problem.

THERMAL POLLUTION

The final form of water pollution we will study is *thermal pollution.* At first glance, it seems like a fairly harmless form of water pollution, but it can have far-reaching and damaging effects on an ecosystem. Heat pollutes water through its impact on aquatic organisms and animal populations.

> The release of a substance, liquid or air, which increases heat in the surrounding area is known as **thermal pollution.**

Water temperature is important to aquatic life. It controls metabolic and reproductive activities. Most aquatic organisms are cold-blooded. Since they can't control their own body temperatures, their body temperatures are regulated by the water temperature around them. Cold-blooded organisms are adapted to specific temperature ranges. If water temperatures change too much, metabolic activities break down. As we saw earlier, temperature also affects water concentrations of dissolved oxygen and bacteria.

Unlike humans who can adapt to a wide variety of temperature ranges, most organisms live in set temperature niches. When the narrow temperature band of their niche is changed, many aquatic organisms die.

Industries are the main culprit of thermal pollution. Not thinking of thermal pollution as a problem, industries release high-temperature wastewater directly into rivers and lakes without a second thought. The immediate and long-term effects are felt by complex ecosystems at the discharge site and downstream. Although the aquatic environment is less impacted when heated wastewater cools, upstream species near the release location take a thermal hit that affects their population numbers in ecosystems further downstream.

Looking to the Future

Science gives us many practical ways to reduce and treat wastewater before it enters the environment. Through study and modern techniques, the cleanup of past problems is taking place.

However, it is important to always think about how our activities affect the environment. There is a lot we can do to minimize pollution.

Proper disposal of household chemicals (don't pour them down the drain) helps as well. When rains wash away oil, gasoline, and other polluting chemicals, they also pick up excess fertilizers on their way to streams and lakes. There they cause algae and weeds to overgrow in the water.

There are ways to limit polluting runoff. By growing native plants that are resistant to pests, less agricultural treatment is needed. Similarly, if people test their soil for what is actually needed, chemical use is reduced. Putting fertilizers away and not using them when big rainstorms are predicted also helps.

A new way to raise public awareness of runoff problems is through storm drain stenciling. A stencil is made on the top of a storm drain inlet with the name of the stream or water body that it runs into. It's a great way to remind people that whatever goes into the drain will eventually wind up in their favorite trout stream or swimming hole. For more information on drain stenciling, see the following Web site: *http://www.earthwater-stencils.com.*

The Federal Centers for Disease Control estimate that 82% of all Americans have the widely used insecticide Dursban (now banned) in their bodies. In order to limit the use of this chemical and others, control pests, and protect waterways from pollution, people can: 1) use pesticides sparingly, if at all; 2) focus on early identification of pests; 3) use natural controls like predators (think ladybugs that eat aphids); and 4) plant naturally resistant native plants.

There are a lot of other things we can do to keep our air and water clean. Encouraging the food industry to use recycled packaging and natural dyes where possible helps keep toxic dyes out of landfills and the groundwater.

Walking or biking instead of driving also cuts down on acid, hydrocarbon, and nitrogen oxide emissions to the atmosphere and therefore to worldwide freshwater supplies.

With new technology available, water treatment technology has improved significantly in recent years. However, many developing countries cannot afford these new technologies. Population growth in coastal cities has hit existing waste treatment systems hard.

Much more study is needed on the effects of water pollution to help our understanding of its complex problems, different causes, and ways in which damaged ecosystems can be restored.

When everything is said and done, there are lots of personal and societal choices that affect pollution levels locally, statewide, or in the country where we live. Unfortunately, our standard of living and industrial way of life is dependent on progress that is dirtier than that of our distant ancestors. For one thing, there were fewer of them!

Without going backwards in time, the answer seems to point to a combination of small daily changes like paying more for organic or nonpolluting goods and services. These changes may encourage manufacturers to develop cleaner devices (like hybrid cars) and help us all do our part in keeping this terrific place we call home wholesome and beautiful for a long time to come.

Quiz

1. One of the most newsworthy chemical weathering types is known as
   (a) rock grinding
   (b) tornadoes

(c) acid rain
(d) x-rays

2. Radiopharmaceuticals are used
   (a) as fertilizers in farming
   (b) to lose weight without exercising
   (c) to study organ function and structure
   (d) to control insects

3. Storm drain stenciling is a
   (a) new art form
   (b) lot like wall stenciling
   (c) good way to seal open manhole covers
   (d) way to remind people about the final destination of storm runoff

4. Thermal pollution
   (a) can kill off entire marine populations
   (b) is harmless
   (c) is caused by oil spills
   (d) can be prevented by outlawing the use of electric blankets

5. The best way to slow or stop acid deposition is to
   (a) keep people out of chemical labs
   (b) add carbon dioxide
   (c) build taller smokestacks
   (d) limit chemical emissions at their source

6. Acid rain has been measured in
   (a) Germany
   (b) Yugoslavia
   (c) Australia
   (d) all of the above

7. Which of the following is an important marker of a river's ability to support aquatic life?
   (a) Carbon tetraflouride
   (b) Dissolved oxygen
   (c) Free nitrogen
   (d) Not many fisherman

8. What chemical added to wastewater during coagulation forms sticky blobs that snag bacteria, silt, and other impurities?
   (a) Alum
   (b) Carbon
   (c) Sodium
   (d) Polonium

9. What is the reaction called when the surface of limestone statues are discolored and disfigured by rain?
   (a) Glaciation
   (b) Dissolution
   (c) Desertification
   (d) Aging

10. Ultraviolet light, used to treat wastewater by killing microorganisms, is
    (a) a hugely popular water treatment method
    (b) not as effective as ultrapurple light treatment
    (c) a one-time treatment at the plant
    (d) protects water from pollution forever

Part Two Test

1. With more and more people on the planet, a positive environmental step would include more
   (a) sewage produced
   (b) fossil fuels burned
   (c) trees planted
   (d) fertilizers used to increase crop yields

2. Pounding reefs with heavy weights to scare fish out of their hiding places is called
   (a) muro smashi
   (b) muro ami
   (c) muro dumbi
   (d) muro destructi

3. What percentage of runoff from the land surface eventually drains into the sea?
   (a) 25%
   (b) 40%
   (c) 75%
   (d) 90%

4. Continental glaciers are large, U-shaped glaciers with the deepest ice found
   (a) in the curved middle
   (b) at the edges
   (c) in blue patches
   (d) only on separated icebergs

5. The study of the occurrence, distribution, and movement of water on, in, and above the earth is called
   (a) geology
   (b) hydrology
   (c) neurology
   (d) pathology

6. All of the following are stream drainage configurations except
   (a) dendritic
   (b) rectangular
   (c) trellis
   (d) radical

7. In order for fish to survive, water must contain enough
   (a) golf balls
   (b) dissolved oxygen
   (c) water lilies
   (d) dissolved carbon

8. Nonnative species have more of an effect in
   (a) seamounts
   (b) the bathtub
   (c) glacial meltwater
   (d) coastal waters

9. Increased oil demand has increased all of the following except
   (a) oil spills
   (b) oil operations and transport
   (c) sea turtle populations
   (d) offshore oil drilling

10. Sudden dense movements of water that slice deep canyons along the ocean floor are called
    (a) whale tail wake
    (b) turbidity currents

(c) geological uniformity
(d) rapidity currents

11. What percentage of total water is stored as fresh water in glaciers and icecaps, groundwater, lakes, rivers, and soil
    (a) 1%
    (b) 3%
    (c) 5%
    (d) 10%

12. IOTPF is an acronym that stands for
    (a) Indoor Outdoor Tennis Players Federation
    (b) International Octopus, Turtle, and Piranha Foundation
    (c) International Oil Tanker Owners' Pollution Federation
    (d) Indigenous Owls of Temperate Pine Forests

13. Measurement of the number of hydrogen ions in water, on a scale from 0 to 14, is known as the
    (a) mph
    (b) pH
    (c) PhD
    (d) mg/l

14. Forest fires, lightning, bacterial decay, and the burning of oil, coal, and gas are main sources of
    (a) nitrogen oxide
    (b) glucosamine
    (c) titanium dioxide
    (d) bismuth

15. Water in the atmosphere is thought to be replaced every
    (a) 8 days
    (b) 15 days
    (c) 32 days
    (d) 45 days

16. Municipalities, industries, and agriculture are all sources of
    (a) cranberries
    (b) water pollution
    (c) landscaping materials
    (d) graphite

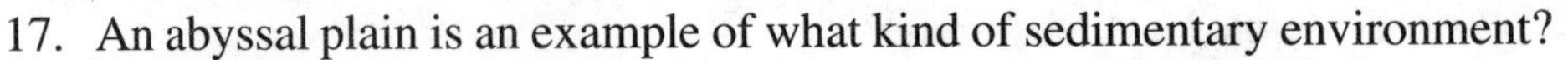

17. An abyssal plain is an example of what kind of sedimentary environment?
    (a) Marine
    (b) Fissure
    (c) Science fiction
    (d) Mountainous

18. The cloudier the water, the higher the
    (a) acidity
    (b) pH
    (c) temperature
    (d) turbidity

19. When a very small amount of an isotope is added to a chemical, biological, or physical system for study, it is called a(n)
    (a) isotopic solution
    (b) placebo
    (c) radioactive tracer
    (d) geographic tracer

20. Natural waters usually have a pH between
    (a) 1 and 2
    (b) 3 and 5
    (c) 6 and 9
    (d) 10 and 12

21. The water table is found at the boundary between the
    (a) littoral zone and deep ocean
    (b) tree line and ridge line
    (c) zone of saturation and the zone of aeration
    (d) mantle and the crust

22. When an ice cap covers a single peak like Mount Rainier, it is called a(n)
    (a) top cap
    (b) carapace ice cap
    (c) cold winter cap
    (d) ice beak

23. Spills account for what percentage of marine oil pollution?
    (a) 5%
    (b) 10%
    (c) 15%
    (d) 20%

24. Fish and shellfish give us
    (a) acne
    (b) a source of whopper stories
    (c) hives
    (d) a valuable source of food protein

25. What measurement calculates the strength of a liquid's thin surface layer?
    (a) Salinity
    (b) Richter scale
    (c) Surface tension
    (d) Geometric pressure

26. Stormwater runoff often causes
    (a) a decrease in street flooding
    (b) red tide blooms
    (c) poor golf course conditions
    (d) orchid blooms

27. Water pollution is caused by all of the following except
    (a) agricultural runoff
    (b) land deforested and developed
    (c) natural, extremely deep underground aquifers
    (d) sudden or ongoing, accidental or deliberate, discharge of a polluting material

28. A stream with no tributaries flowing into it is called a
    (a) first-order stream
    (b) second-order stream
    (c) third-order stream
    (d) fourth-order stream

29. Greenland contains a continental ice sheet with ice over
    (a) 25,000 years old
    (b) 50,000 years old
    (c) 80,000 years old
    (d) 125,000 years old

30. All of the following are intended to reduce the effects of human activities on marine systems except
    (a) encouragement of sustainable use of ocean resources
    (b) increased bottom trawling
    (c) policies and new technologies to limit pollution
    (d) creation and management of protected marine regions

31. When rainfall or snow melt doesn't have time to evaporate, transpire, or move into groundwater reserves, it becomes
    (a) spinoff
    (b) sleet
    (c) runoff
    (d) slush

32. By measuring the number of hydrogen ions in a water sample, scientists can find its
    (a) pH
    (b) dissolved oxygen content
    (c) color
    (d) triple point

33. When plants and animals are transported to a far distant location, they are known as
    (a) native species
    (b) colorful characters
    (c) nonnative species
    (d) contraband

34. When a lot of water vapor coats enough dust, pollen, or pollutant particles, it forms a
    (a) volcano
    (b) cloud
    (c) aqua duct
    (d) cistern

35. What chemical element is most frequently used to purify water?
    (a) nitrogen
    (b) fluorine
    (c) calcium
    (d) chlorine

36. To acknowledge the importance of the world's oceans, the United Nations declared 1998 the
    (a) worst hurricane year in decades
    (b) International Year of the Ocean
    (c) Bermuda Triangle a no-fish zone
    (d) International Year of the Barnacle

37. Scientists first identified "dead zones" in what body of water?
    (a) Black Sea
    (b) Arctic Sea
    (c) Persian Gulf
    (d) Gulf of Mexico

38. The deep ocean basin, located at a depth of between 3.7 and 5.6 km, covers what percentage of the Earth's surface?
    (a) 20%
    (b) 30%
    (c) 40%
    (d) 50%

39. The largest body of ice in the Canadian Rocky Mountains is the
    (a) Columbia Ice Field
    (b) Washington Ice Sheet
    (c) Antarctic Ice Field
    (d) Ross Ice Shelf

40. When a water source is used faster than it can be replaced, it is said to be
    (a) sustainable
    (b) immortal
    (c) nonrenewable
    (d) pristine

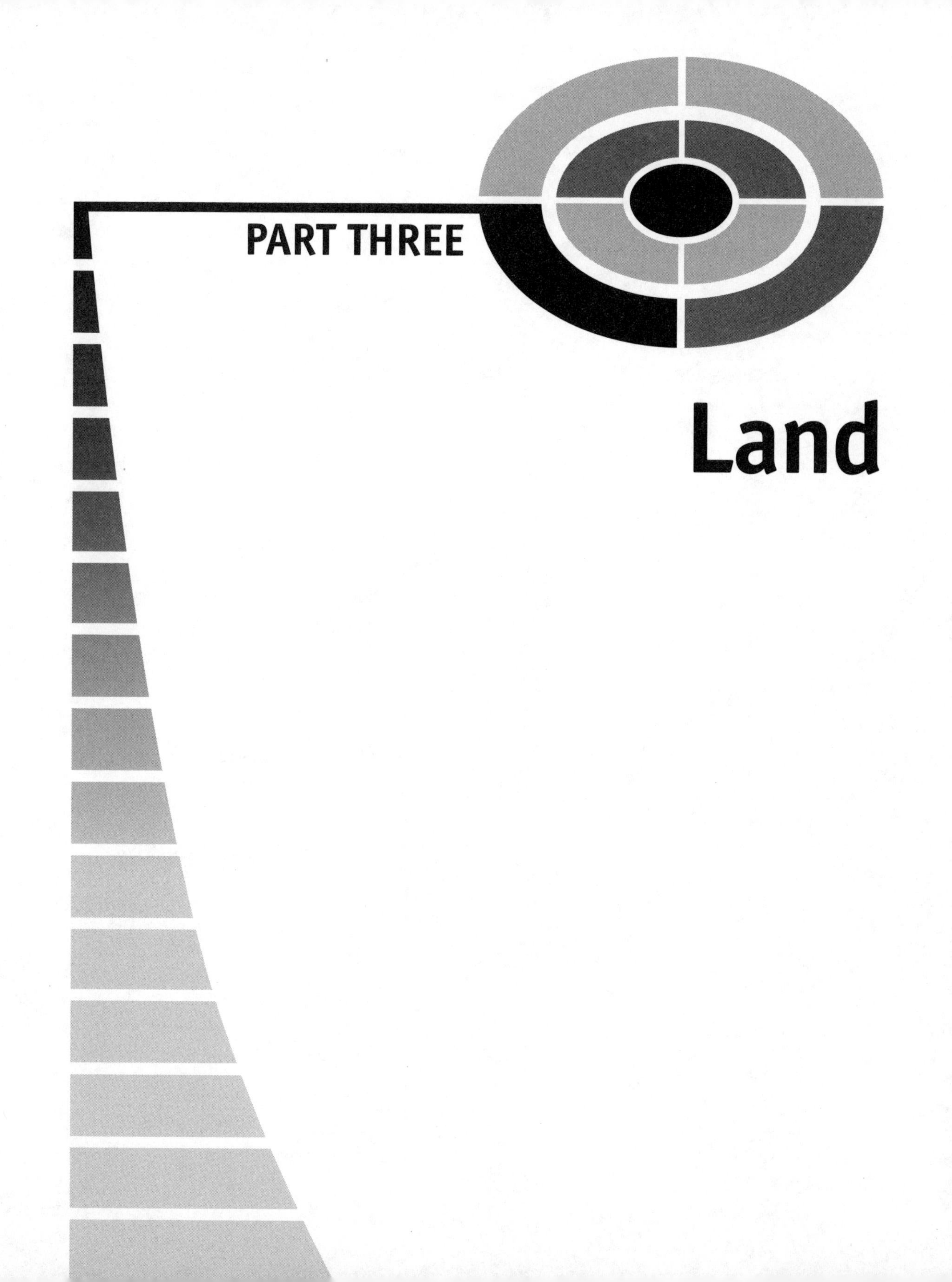

PART THREE

Land

CHAPTER 9

Weathering and Erosion

Rain forest destruction and loss of biological diversity are appearing more and more often in news headlines. To add insult to injury, the clearing of plants and trees in many areas of the world has caused additional environmental problems. In the past 150 years, agricultural growth and changes have caused the widespread destruction of native vegetation around the planet. This has increased *soil exposure* and *erosion.*

Erosion in all forms includes the movement and removal of rock and soil particles from one spot, followed by the slow buildup in a different spot. This is the source of landscape and soil development. Susceptibility to erosion and the speed with which it happens depend on geology, land use, terrain, climate, soil texture and structure, and local vegetation.

Raindrops start water erosion. A raindrop (about 5 mm in diameter) falls through still air and hits the soil at about 32 km/hour. Wind causes drops to slam to the ground harder. Drops act like tiny bombs when falling on bare soil, knocking soil particles aside and destroying soil structure. As rain keeps falling, dislodged grains slip along the surface, filling in larger soil pores and limiting water seepage into the soil. Continued rainfall forms puddles, and if the water isn't absorbed, it keeps flowing downhill as runoff.

Soils erode even more easily when soil grain size decreases. Fine organic particles, silt, and clay are crucial to soil makeup. They contain most of the nitrogen, potash, and phosphoric acid required for life or plant growth. Many forms of life depend on the earth and soil for food, minerals, and habitat. When topsoil is eroded, it takes a very long time to get it back. Some species are lost before this happens.

Denudation

Most people use the word *weathering* as an overall term for erosion and all gradual land wasting, but it actually comes under a larger category called *denudation.* Denudation takes place when surface layers are removed from underlying rock. They are laid bare. When the wind blows constantly on a high mountain slope, there is hardly any soil left on the bare surfaces except for grains that fall into protected cracks and fissures. This is especially obvious in rock outcrops where very little soil gathers.

Denudation takes place when rock disintegrates and is removed from the surface of continents.

Denudation is an umbrella word covering three main types of rock change and removal. These are *weathering, mass wasting,* and *erosion.* All three have the same end product—movement of rock—but they happen in different ways. These processes are compared below.

- *Weathering* takes places when rocks are broken down and transformed at or near the surface by atmospheric and biological agents. Weathering wears away and chemically changes rocks, but there is hardly any rock movement.
- *Mass wasting* is more active than weathering. It happens when there is a shift of broken rock material down a slope due to gravity's pull. Mass wasting, or *rock shift,* can also cause loose rock to suddenly move over short distances, like during a rock slide.
- *Erosion* is a bit like mass wasting, but covers rock/soil transport over much greater distances and is helped along by wind and/or water. Erosion causes fine *grains* (very small rocks, like sand) to travel great distances, like from the mountains across the desserts.

Often, denudation takes place in a step-wise manner. First, denudation starts with weathering. Loosened rock material is then affected by gravity, so we see the following process that ends with mass wasting.

$$\text{Weathering} + \text{Gravity} \rightarrow \text{Mass wasting}$$

If a free-flowing stream or constant wind is added to mass wasting, then erosion occurs.

$$\text{Weathering} + \text{Gravity} + \text{Moving fluid} \rightarrow \text{Erosion}$$

Mass wasting and erosion sometimes overlap. They can also include some type of movement, like a mudflow.

Weathering and Erosion Factors

Rocks are formed at high temperatures and pressures deep within the Earth. Through tectonic activity, they can become exposed to surface conditions like low temperatures and pressures as well as air and water. When rock is exposed to these new conditions, it reacts and begins to change. When this happens, it is known as *weathering.*

There are a few rules of thumb to remember when trying to figure out the weathering pattern in a certain area. These include rock type and structure, texture, existing soil and weathering, slope, climate, and time. These factors, alone or combined, can make a difference in how an area of rock weathers.

Different minerals react differently to weathering action. Quartz is very resistant to chemical weathering, therefore, so are rocks containing any quartz. That is why mountains made up of rock with high amounts of quartz are still standing in areas where their sedimentary neighbors were long ago flattened.

Texture is also an important player in weathering. A rock that is almost totally quartz, like quartzite, will be fairly open to erosion if it has a lot of joints and spaces that allow water to seep in.

The presence of soil, a breakdown product of rock, adds to weathering. When more and more surface area is exposed, either through the washing away of soil from bare rock or as rock gets increasingly jointed, then weathering can reach into more places.

Depending on composition and texture, weathering can proceed at different rates in the same geographical region. This is known as *differential weathering.* It includes variable weathering of rocks with different compositions and structures, as well as a change in intensity. That is why some cliffs with alternating hard and soft rock have curves and ridges where the harder rock has resisted weathering and the softer rock has given way.

Weathering is the breakdown and disintegration of rock into smaller pieces of sediment and dissolved minerals.

An area's slope can speed or slow weathering. When a slope is steep, loosened mineral grains are washed down to the bottom with the help of gravity. They are often carried far away. Additionally, a steep, constantly exposed rock face will lose minerals as weathered grains are removed and new ones exposed. This type of slope, depending on composition of the exposed rock, may weather much faster than a less inclined slope. Gentle slopes also experience less vertical gravity pull. Loosened grains usually collect in piles, many meters thick, along the length of the slope.

Climate can also contribute to weathering changes. Heat and humidity speed up chemical and biological weathering and assist water's penetration of rock. Geologists usually find much greater weathering of limestone and marble in hot, wet climates because of the dissolving effect on calcite. Cold, dry climates have more frost-driven fracturing and weathering. We will look at these weathering factors in more detail.

Time is the biggest weathering factor. Some hard rocks like granite are changed by different weathering factors only over thousands and millions of years. Knowing a rock's composition and its vulnerability to weathering gives geologists important information about a region's geological history. Table 9-1 shows the different factors that affect the rate of speed of rock weathering.

Table 9-1 Rock weathering speed.

Rock characteristics	**Increasing weathering rate →**		
Solubility	*Low*	*Medium*	*High*
Structure	Immense	Has weak points	Highly fractured
Rainfall	Low	Medium	High
Temperature	Cold	Moderate	Hot
Soil layer	Thick	Medium to thin	No soil (bare rock)
Organic activity	Negligible	Moderate	Plentiful
Exposure time	Brief	Moderate	Lengthy

Weathering takes place in two ways, *physical weathering* and *chemical weathering,* which take place at the same time. This is like a one-two punch. When a rock gets broken into pieces, more of its surface is exposed to the air. This causes more of its surface area to be uncovered for chemical weathering, which in turn breaks it down into smaller bits of rock.

Environmental factors can also cause weathering to take place at different speeds. Mechanical methods like rock smashing and cracking take place in cold and/or arid climates where water is meager. Chemical methods take place in warm and/or moist climates where there is lots of water. Biological methods take place in many different environments across the planet.

PHYSICAL WEATHERING

The activities of physical and mechanical weathering create cracks in rock that act as channels for air and water to get deeper into a rock's interior. During weathering, rock is constantly being broken into smaller pieces and the surface area that is exposed to air and water is getting larger. Both of these actions add to the overall chemical weathering.

> **Physical weathering** happens when rock gets broken (cracked, crumbled, or smashed) into smaller pieces without any change to its chemical composition.

Physical or *mechanical weathering* is the breakdown of large rocks into smaller bits that have the same chemical and mineralogical makeup as the original rock. The breakdown of rock into smaller and smaller pieces occurs as follows:

boulders → pebbles → sand → silt → dust

This size-graded breakdown takes place in different ways.

Joints

Many rocks are not solid all the way through. They have many different-sized cracks and fractures caused by stress called *joints.*

Along with joints, rocks can also fracture between individual crystals or grains. Erosion factors, like water, introduce tiny gaps between the grains, leading to grain-by-grain disintegration of rocks and distribution of soil particles.

Frost

Frost wedging is a form of *physical weathering*. Frost wedging is caused by the repeated freeze-thaw cycle of water in extreme climates. Freeze and thaw weathering is common when temperatures drop below freezing at night and rise during the day.

When it rains, rainwater collects in the joints of rocks. As the air cools and temperatures drop at night below freezing, the water inside the joints freezes. As water freezes into ice, it expands.

You have probably seen this for yourself. Have you ever put a drink in the freezer to cool it quickly and forgotten about it? When you come back a few hours or days later, the container is stretched and deformed from the pressure of the expanding liquid.

Within rock fissures, expanding ice puts pressure on the joints in the rock. When the pressure gets too great, the joint expands and soil can sift in, allowing plants to eventually grow. After repeated freezes and thaws, the joints can't take it anymore and finally the rock is cracked.

During warmer summer months, ice melts back to free-flowing water. This seeps into joints and is frozen back into ice at night. This further wedges joints apart, exposing the soil that has collected in the joint to increase erosion.

> **Frost wedging** happens when rock is pushed apart by the alternate freezing and thawing of water in cracks.

Frost action is best seen in wet climates with many freezing and thawing cycles (arctic tundra, mountain peaks). Frost splits rocks into blocks by *joint block separation* and also wears away the edges of blocks grain by grain, rounding the surfaces.

BIOLOGICAL WEATHERING

Biological weathering is a blend of both physical and chemical weathering. Tree and plant roots grow into rock fissures to reach collected soil and moisture. They draw needed minerals from the surrounding rock and collected soil.

You've probably seen weeds growing in tiny cracks in a sidewalk or other paved areas. As the plant grows larger, its roots get thicker and push deeper into the crack. Eventually, this constant pressure cracks the rock farther and pushes it apart. The more the rock is fractured, the easier it is to crack. Both root growth and burrowing in rock gaps cause rock fractures to widen, reaching deeper into the rock and exposing fine soil to additional weathering and loss.

CHEMICAL WEATHERING

As mechanical weathering breaks rock apart, the total exposed rock surface increases and *chemical weathering* increases. Chemical weathering breaks down rock through reactions with a rock's elements and mineral combinations. These change the rock's chemical structure and form. It is much more subtle than physical weathering.

Chemical weathering has been going on for millions of years, but with the addition of manmade industrial pollutants into the Earth's air and water, some forms of chemical weathering have increased.

> When rock and its component minerals are broken down or altered by chemical change, it is known as **chemical weathering.**

The most important natural acid is *carbonic acid,* which is formed when carbon dioxide dissolves in water ($CO_2 + H_2O \leftrightarrow H_2CO_3$). Carbonate sedimentary rocks, like limestone and marble, are especially sensitive to this type of chemical weathering. Gouges and grooves that are often seen on carbonate rock outcrops are examples of chemical weathering.

Chemical weathering replaces original rock minerals with new minerals. These replaced minerals may have very different mechanical characteristics, such as strength and malleability. For example, if clays are formed, they may not be as rigid as the existing rocks might have been, but are much more pliable. Chemical weathering constantly weakens rocks by increasing the chances of *mass wasting.*

- *Oxidation* takes place when oxygen anions react with mineral cations to break down and form oxides, such as iron oxide (Fe_2O_3), softening the original element.
- *Solubility* describes a mineral's ability to dissolve in water. Some minerals dissolve easily in pure water. Others are even more soluble in acidic water. Rainwater that combines with carbon dioxide to form carbonic acid ($H_2O + CO_2 \rightarrow H_2CO_3$) becomes naturally acidic.

- *Hydrolysis* takes place when a water molecule and a mineral react together to create a new mineral. The transformation of the feldspar mineral orthoclase into clay is an example of hydrolysis.
- *Dissolution* happens when environmental acids like carbonic acid (water), humic acid (soil) and sulfuric acid (acid rain) react with and dissolve mineral anions and cations.

In chemical weathering, water nearly always has a part. Carbon dioxide dissolves in rainwater to form carbonic acid, which dissolves limestone rock and carries it away in solution as calcium carbonate. When limestone rock is dissolved away over long periods of time by underground streams, intricate caves and channels are formed.

Soil Types

Just as there are three different soil layers, there are also several factors that determine which type of soil will form. These include structure, rainfall (lots or little), solubility, temperature (hot or cold), slope (gentle or steep), vegetation (types and amount), and weathering time (short or long.) A key factor in naming major soil types is rainfall. Everyone from toddlers making mud pies to petroleum geologists looking for oil can tell whether a soil is wet or dry, hard or soft.

Geologists have named three basic soil types based primarily on water content. These are the *pedocal, pedalfer,* and *laterite.*

The *pedocal* is found in dry or semi-arid climates and contains little organic matter, experiences little to no leaching of minerals, and is high in lime. Most nutrient ions are still present. In places where water evaporates and calcite precipitates in the "B" horizon, a hard layer called the *caliche* or *hardpan,* is formed. Pedocal soil also collects in areas of low temperature and rainfall and supports mostly prairie plant growth.

Pedalfer soil is found in wetter environments and contains greater amounts of organic matter and experiences more leaching. Aluminum and iron are retained after many other soluble nutrients are leached out. This type of soil is found in areas of high temperatures and humid climates with a lot of forest cover.

Laterite, the soggiest soil type, found in tropical and subtropical climates, is high in organic matter. Because of high equatorial rainfall in very wet climates, there is widespread leaching of silica and all soluble nutrients. Iron and aluminum hydroxides are left behind and cause well-drained laterite soils to be red in color. Besides iron and aluminum ores, laterites can also form manganese or nickel ores.

Soil Erosion

We've learned that soil is blown or washed away by wind and/or water action. Soil erosion has been happening for around 450 million years, since the first soil was formed. But rapid soil erosion has been going much more recently. It is a result of human activities like overgrazing or poor agricultural practices, which leave the land exposed during times of erosive rainfall and windstorms.

Accelerated soil erosion impacts agriculture and the natural environment. It is one of the most significant of today's environmental problems and is studied by geomorphologists, agricultural engineers, soil scientists, hydrologists, and others. Local and national policy-makers, farmers, environmentalists, and many organizations have become increasingly concerned.

Erosion converts soil into sediment. Chemical weathering produces clays on which vegetation can grow. A mixture of dead vegetation and clay creates soil that contains minerals necessary for plant growth.

> **Soil** exists as a layer of broken, unconsolidated rock fragments created over hard, bedrock surfaces by weathering action.

Most geologists talk about soil as being part of three layers called *soil horizons* or *soil zones.* These three soil horizons are commonly recognized as "A," "B," and "C," but it is important to remember that not all three horizons are found in all soils. Fig. 9-1 illustrates the way soil horizons are stacked on top of each other.

Soil horizons are described from the top soil layer down to the lowest soil and bedrock level and are as follows:

- **"A" horizon** includes the surface horizon, a zone of leaching and oxidation, where penetrating rainwater dissolves minerals and carries the ions to deeper horizons. It also holds the greatest amount of organic matter.

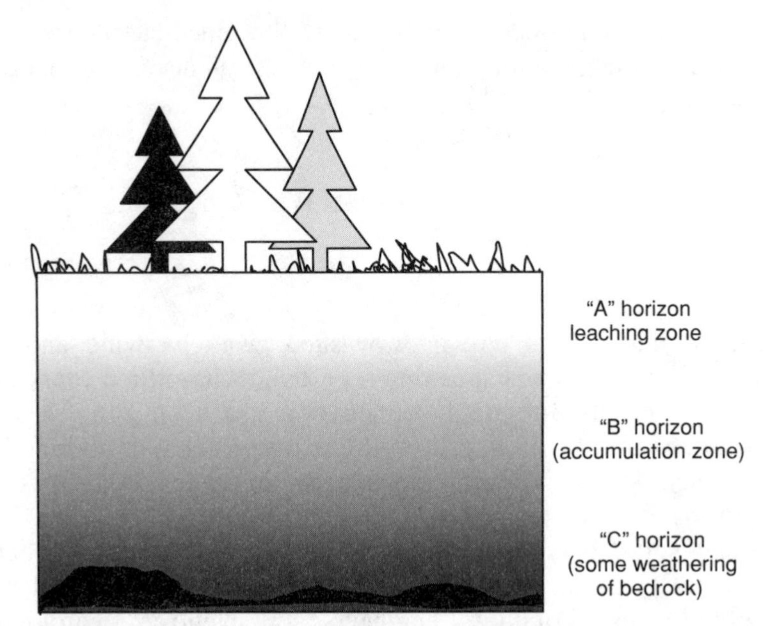

Fig. 9-1 Rock can be divided into a gradient of soil horizons with bedrock at the bottom.

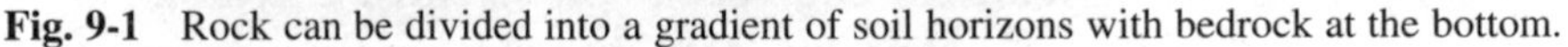

- **"B" horizon** describes the middle horizon, a zone of accumulation, where ions carried down by infiltrating rainwater are reacted to create new minerals. Blocky in texture, it's made up of weathered rock mixed with clay, iron, and/or aluminum.
- **"C" horizon** is the bottom horizon, a zone of unconsolidated, weathered original bedrock.

MASS MOVEMENT

In slides, slips, flows, and landslides, gravity is the main force acting upon soil and rock. When slope stability is changed, a variety of complex sliding movements takes place. Sudden movements of soil or rock that occur when the upper layers separate from the underlying rock and involve one distinct sliding surface are called *landslides.* There is little loose flow.

The rock moves in a solid sheet downward. Slides take place because of a buildup of: 1) internal stress along fractures; 2) undercutting of clay layers and slopes by water (rivers and glaciers); and 3) earthquakes.

Slower long-term soil or rock movement with a series of sliding surfaces and plasticity is known as *creep*. These kinds of movement are often the final event in a series of processes involving slope, geology, soil type, vegetation type, water, external loads, and lateral support.

Plant removal and changes in water use may increase soil water levels and/or soil porosity. Greater water absorption may weaken bonds between soil particles, lowering friction and reducing soil strength and slope stability.

> **Mass movement** is the slow or abrupt movement of rock downslope as a result of gravity.

Mass wasting has several factors that affect it. These include gravity, types of soil and rock, physical properties, types of motion, amount of water involved, and the speed of movement.

Gravity is the main influence on mass wasting. It is always pulling things down. When rocks are piled on a steep mountain slope, there is a high amount of friction that holds them to the slope. However, gravity is pulling the rock downward. The downslope pull of gravity that causes mass wasting is known as *shearing stress*. The steeper the slope, the greater the shearing stress. Movement happens when the weight (shear stress) of the slope material is more than the rock's restraining (shear strength) ability. Fig. 9-2 illustrates the difference of slope angle on shearing stress.

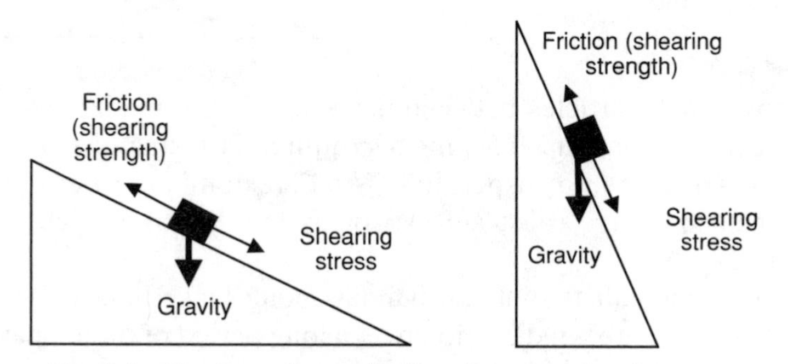

Fig. 9-2 The down slope pull of gravity causes shearing stress.

Shearing stress is linked to the mass of the rock being pulled downward by gravity and to the slope's angle.

There are several different factors that influence and increase shear stress. These include:

- Erosion or excavation undermining the foot of a slope;
- Buildings or embankment loads; and
- Plant removal and loss of stabilizing roots.

The counteracting force that works against shearing stress is friction; with a large body of rock it's called *shear strength.* When the amount of shear stress is higher than the shear strength, something has got to give. A quick movement, like an earthquake, acts as a trigger. It provides just enough energy to overcome the last bit of friction and allow gravity to pull everything down.

Wind Erosion

Wind erosion happens when soils that have been cleared of plants are exposed to high-velocity wind. When wind speed is greater than the soil's gravitational and cohesive forces, it will shift soil and carry it away in suspension.

Wind erosion is caused by the movement and deposition of soil particles by the wind.

Wind moves soil particles between 0.1 and 0.5 mm in size by a bouncing movement and greater than 0.5 mm by rolling. The smallest particles (less than 0.1 mm) separate into suspension. Wind erosion is most easily seen during the suspension stage (dust storms) or as buildup along fence lines and across roads.

When environmentalists spot soil buildup along fence lines or against trees, it's a good sign of severe wind erosion over a long period of time. They also look for damage to young crops from sandblasting.

When fine soil grains are carried as dust during windy conditions, it's known as *smoking paddocks*. After wind erosion has gone on for a while, soil color lightens as organic matter, clays, and iron oxides are blown away. The soil surface becomes smooth with little texture or evidence of previous crops. In sandy soils regions, particles are sorted, leaving the surface covered with a coarse sand layer. When strong winds follow droughts, they stir up huge dust storms that blow fine soil particles over distances of hundreds of kilometers.

Some of the things that can be done to slow erosion include:

- Avoid letting plant cover drop below 30% on sandy soils;
- Control wind access to the soil (leaving stubble or mulch);
- Direct wind speed over a property by protected planting;
- Rotate crop types and frequency of field use;
- Limit tillage and direct-drilling practices; and
- Reduce and/or rotate livestock grazing.

Because soil forms slowly, it is essentially a finite resource. For a long time, many people didn't recognize this. The severity of global erosion is only now becoming widely known.

Rockfalls and Slides

Erosion can also take place on a larger scale. Loose rock can slide and fall without a lot of flowing mud and pebbles. Giant boulders or sections of earth shift position because of shear stress and move as a whole down a slope.

> A **rockfall** takes place when large amounts of rock freefall or shift downward from very steep areas of a mountain slope.

Generally, a *rockfall* happens along the sides of a steep mountain with little plant life, but can also involve cliffs, caves, or arches. Small rockfalls are fairly common, but huge rockfalls are rare because of the amount of force needed to move tons of rock. Small rockfalls take place mostly because of weathering. Earthquakes can produce enough energy to cause large and small rockfalls.

DUNE EROSION

There are two types of *dune erosion,* wind and wave. *Wind erosion* moves sand grains in a series of skipping movements (*saltation*) or by tumbling them along the dune surface (*creep*). Higher and heavier grain size prevents the long-term suspension seen with finer soil particles. In this way, sand from dunes is set in motion to form dunes that creep inland, covering roads, plants, and structures.

> **Dune erosion** happens when plants are damaged or removed, exposing sand dunes to high winds and wave action.

A planted and established dune front serves as a buffer, providing a volume of sand that circulates between the front dune, beach, surf zone, and seabed, depending on sea and wind conditions. Loss of protective vegetation through overgrazing, foot and automotive traffic, grass fires, and building construction exposes sand to high-speed coastal winds and waves.

Australia's coastal zone, an area roughly $^1/_{13}$ of the earth's total land mass, supports the majority of its population. Any erosion across this area impacts millions of people economically, socially, and recreationally. Unfortunately, the increase of human activity in Australia combined with ongoing wind and water influences has added to coastal dune erosion.

Sometimes, *dune blowouts* take place. These happen after the plants on the leading or primary dune have been disturbed or destroyed. After a blowout, a dune aligns with the prevailing wind and forms a U-shape in the dune's lower section. This funnels the wind, raising its velocity and increasing sand loss. We will see a variety of dune shapes and their causes in Chapter 10 when we study desert formation.

Transgressive (creeping) *dunes* cause big problems for landowners inland from the original dune system when sand covers roads, property, or farmland. The construction of protective walls and breakwaters built to protect property on frontal dunes can impact natural beach processes and increase erosion. This makes the artificial replacement of sand necessary. In dangerous situations, beach/cliff homes and businesses are undermined by wave activity, causing structural damage and losts.

Wave energy also affects beach formation by depositing and removing beach materials over time. Constant wave movement carries sand and debris onto the beach, while backwash carries it away. During calm weather, sand buildup forms a beach. During storms, beaches erode away from the violent wave backwash. Buildings and road construction on frontal dunes impact a dune's buffering function in the wider beach zone. When plants are removed for construction, soil becomes unstable, and the dune's ability to trap windblown sand is limited.

There are steps that can be taken to limit dune erosion. These include reestablishing protective vegetation (by controlling or restricting foot traffic, vehicles, and fires) and constructing sand traps or wind barriers. The best way to avoid problems is to protect dune and coastal systems by permitting building development and commercialization behind dune systems.

Water Erosion

The power of water is amazing. Water is heavy and when it's moving quickly, it can easily knock you off your feet. This is why people are warned against trying to cross a raging stream during a heavy rain or flash flood. Entire cars, trucks, and homes can be washed away as easily as a leaf into a sewer drain.

Water impacts soil in a variety of ways. Most of them are good, like watering plants and filling lakes and rivers, but some are destructive.

Rills are formed when water first washes soil grains away through the same downhill grooves over and over again. From the air, they look like fingers extending down a slope from higher elevation.

When water erosion takes place in the same areas over time, rills get big enough to impact roads. When this happens, they become *gullies.*

Large volumes of high-velocity runoff in large rills wash away huge amounts of soil. This cuts deep gullies along depressions and drainage ditches. It can also be a problem when a road is cut along one side of a steep mountain. When there is enough rock and soil washed out from under the roadbed, it collapses. Although it can be fixed, ongoing erosion will probably create the same problem again.

When topsoil is washed away by swift surface water, it creates deep and wide gullies of two kinds: *scour* and *headward* erosion gullies.

In *scour gullies,* rill runoff water takes away soil particles through *sluicing.* Sluicing is the process used by Gold Rush miners of the mid-1800s. They built long, thin, wooden boxes through which running water poured. River current was often directed through sluices so that gold-laden gravel could be sieved faster than using the time-consuming panning method. The moving water washed lighter soil particles away from the heavier gold nuggets (with luck!). The eroded soil was usually the size of fine-to-medium sand. Scour gullies are often found in low, rolling hills.

In *headward erosion* the gully extends upstream as a result of waterfall undercutting and gravitational pull at the gully head. This is often seen in steeper areas. Gullies get bigger through sideways erosion, while undercutting water causes the gully's sides to fall in.

The problem with either of these types of gully erosion is that greater amounts of soil are lost in a shorter time period than with wind. Wide, steep gullies, often nearly 30 meters deep, severely limit the use of the land. Bigger gullies also disrupt farm work, causing access problems for vehicles and livestock. In already ravaged regions of the world this could mean the difference between planting crops or going hungry.

Runoff water flowing over the sides of gullies as well as that pouring down the gully head can cause lateral gullies, also known as *branching gullies.* Buildup of eroded rock and soil at the bottom of slopes or along fencelines is a common sign of branching erosion. The gully keeps getting bigger as additional water rushes down its slopes. When environmentalists are rating gully erosion, small unconnected gullies are not a big deal (minor erosion). Continuous gullies need to be watched (moderate erosion). However, expanding and deeply branching gullies are a sign of severe erosion.

Some of the things to watch for when checking erosion is the formation of *nick points* (rabbit holes, empty root holes, and livestock and/or vehicle ruts) on slopes or drainage areas. These seemingly unimportant grooves in the land can allow water erosion to begin. High water runoff in existing rills, sloped agricultural land, or hillsides can allow water erosion to intensify.

To head off erosion, farmers can reduce or divert some water from reaching the gully. This can be done by increasing regional water use for crops when the rain falls, which reduces runoff. Proper crop maintenance creates a natural way to hold soil against water erosion. Runoff water may also be controlled through creative containment in local holding ponds and regional dams.

The good side of erosion issues is that many solutions can be found by individuals at local levels. People don't have to wait for national mandates and expensive environmental testing equipment to see a problem and take steps to fix it.

Quiz

1. Mass wasting is a combination of
   (a) weathering plus centrifugal force
   (b) gravity plus magma

(c) weathering plus gravity
(d) erosion plus pyroclastic flow

2. Frost wedging is caused by
   (a) Jack Frost
   (b) the repeated freeze-thaw cycle of water in extreme climates
   (c) mudflows
   (d) lava backup

3. Wide destruction of native vegetation across the planet has increased
   (a) soil exposure and erosion
   (b) the number of scenic vacation spots
   (c) oxygen levels in homes
   (d) volcanic eruptions

4. The soil found in dry or semi-arid climates with little organic matter and little to no leaching of minerals is called
   (a) pedocal
   (b) pedalfer
   (c) laterite
   (d) muck

5. Wind erosion blows sand grains in a series of skipping movements called
   (a) tai chi
   (b) vegetation
   (c) blasting
   (d) saltation

6. When rock disintegrates and is removed from the surface of continents, it is called
   (a) an eruption
   (b) denudation
   (c) shedding
   (d) lithification

7. The process that Gold Rush miners of the mid-1800s used to separate rock samples is called
   (a) dipping
   (b) panning
   (c) rolling
   (d) sluicing

8. Chemical weathering happens in all but one of the following ways
   (a) acid action
   (b) hydrolysis
   (c) intoxication
   (d) oxidation

9. When plant roots grow into rock fissures to reach collected soil and moisture, it is known as
   (a) recreational weathering
   (b) biological weathering
   (c) zoological weathering
   (d) desertification

10. To slow or stop erosion, farmers use all of the following methods except
    (a) water diversion and control
    (b) proper crop maintenance and rotation
    (c) overgrazing and frequent tilling
    (d) local holding ponds and dams

CHAPTER 10

Deserts

Most people think of deserts as uninhabitable and devoid of life. These barren stretches are probably the last place on Earth that most people would choose to visit or live. Baking daytime temperatures up to 57° Celsius and cold nights (5–10° Celsius) take people by surprise. Extreme heat and dryness interrupted by unpredictable flash floods, give places like Death Valley and "The Desert of No Return" their names.

Deserts are dry. Growing up in northern Nevada in the United States, I never thought it strange that we only got around 10 to 15 centimeters of rainfall a year. Even the biggest storms only lasted a short time and dropped scant rain or snow. Normally, the wettest deserts get less than 50 centimeters of rainfall a year. You can imagine my surprise when I moved to the Texas Gulf Coast. It gets roughly 130 cm of rainfall a year in massive deluges that flood highways to 4.5 meters and more in places!

In most climates around the globe, rain falls all through the year. The desert is a different story. In many deserts there may be only a few instances of rain each year, with months or years between rains.

In fact, the Antara desert in Chile has spots that only receive rain a few times a century, and some areas have never seen rain.

Another reason deserts are hot and dry is the lack of humidity. Most biomes are protected by their humidity. Temperate deciduous forests, for example, can have more than 80% humidity during the day. Many parts of the humid Gulf Coast of the United States, with large amounts of plant life, have humidity levels hovering around 97 to 99%, with matching temperatures during the summer months. This water vapor (humidity) acts as a temperature regulator or modulator. It reflects and absorbs sunlight and its energy during the day, keeping the temperatures lower. Then at night the water serves as a cover, trapping heat inside the forest and keeping the ecosystem or biome at more moderate temperatures.

Since deserts have between 10 to 20% humidity (on a sweltering day), there isn't much chance of regulating temperatures. With hardly any trees or other plant life to retain heat, deserts cool quickly when the sun sets and heat swiftly again when it rises.

Deserts cover roughly 30% of the Earth's surface. Most deserts, like the Sahara in North Africa and the deserts of the southwestern United States, Mexico, and Australia, are found at low latitudes. Cold deserts are mostly found in the basin and range area of Utah and Nevada and in sections of western Asia. Table 10-1 lists the locations and areas of several of the world's largest deserts.

Table 10-1 The world's major deserts cover a lot of area around the globe.

Desert	Location	Approximate Area (km^2)
Sahara	North Africa	9,065,000
Arabia	Saudi Arabia	2,240,000
Gobi	China	1,295,000
Kalahari	Southern Africa	582,000
Chihuahuan	North Central Mexico, Southwestern US (Arizona, New Mexico, Texas)	455,000
Great Basin	Idaho, Oregon, Nevada, Utah, U.S.	411,000
Great Victoria	Australia	338,500
Patagonia	Peru, Chile	150,000

The common misconception that the desert is a lifeless landscape is really an illusion. Deserts have abundant and unique plant life, as well as unique animals. Soils often contain many nutrients just waiting for water, so that plants can spring to life when it rains. Fires provide a carbon source (ash), and flash flooding mixes everything up to mineralize desert soils.

The dual problems of storing water and finding shelter from the sun's blistering heat and the night's cold make deserts an environment with few large animals. The dominant creatures of warm deserts are reptiles, like snakes and lizards. Only small mammals, like the kangaroo rat of North America, are found in any great number.

Types of Deserts

In 1953, Peveril Meigs divided desert regions on the Earth into three categories (*extremely arid, arid,* and *semiarid*) according to their rainfall per year. In the Meigs system, extremely dry lands can go over a year without rainfall. Arid lands have less than ¼ cm of rainfall yearly and semiarid lands have a mean rainfall of between ¼ and ½ cm. Arid and extremely arid lands are called *deserts,* while semiarid grasslands are known as *steppes.*

Dry areas, formed from global circulation patterns, include most of the Earth's deserts. Deserts are not limited by latitude, longitude, or elevation. They are found everywhere from the poles to the Equator. The People's Republic of China has the highest desert, the Qaidam Depression (2600 meters above sea level), and one of the lowest deserts, the Turpan Depression (150 meters below sea level).

While deserts can be classified in different ways depending on who you talk to, most deserts are distinguished by the following factors:

- Total rainfall
- Number of days of rainfall
- Temperature
- Humidity
- Location
- Wind

The 4 major types of deserts are listed below and include subgroups of the main desert categories (*hot and dry, semiarid, coastal,* and *cold*).

HOT AND DRY DESERT

The major North American hot and dry deserts are the Chihuahuan, Mojave, Great Basin, and Sonoran. The Southern Asian, South and Central American, Ethiopian, and Australian deserts also fall into the hot and dry class.

These deserts are almost always warm during the day (year around) and are searing hot in the summer. Winters bring little rainfall. Temperatures show daily extremes due to low humidity. Desert surfaces receive over twice as much solar radiation as humid areas and lose nearly twice as much heat at night. Annual mean temperatures range from 20 to 25°C. The extreme maximum temperature ranges are commonly from 44 to 50°C. Minimum temperatures can sink to –18°C.

In hot and dry deserts, rainfall is scarce and dropped in short bursts between long dry spells. Evaporation rates are greater than rainfall rates. It is possible for rain to fall and evaporate before hitting the ground! Rainfall is lowest in Chile's Atacama Desert where 1.5 cm of rain annually is about average. Some years it gets no rain at all. Deep in the Sahara, the situation is the same, with rainfall around 1.5 cm per year.

Soils in these deserts are gravelly and thin. They may or may not have good drainage, and there is no subsurface water. The extreme environment allows wind to carry dust and sand particles away, leaving heavier pieces behind.

There is no forest canopy in hot and dry deserts. Low shrubs and short woody trees make up most of the larger vegetation. Leaves are *replete* (supplied with nutrients) and conserve water. They are small, thick, and covered with a thick outer layer to conserve moisture. A cactus is a good example of this. The leaves have become spines, and photosynthesis happens in the fleshy stems. Plants open their stomata (microscopic holes in a leaf's surface) for gas exchange only at night, when evaporation is lowest. Some of these desert plants include yuccas, prickly pears, agaves, and brittlebush.

Trade Wind Deserts

The *trade wind deserts* are found in two bands closest to the equator. Hot and dry trade winds scatter cloud cover as they get closer to the equator, allowing more sunlight to heat the land. Most of the major deserts of the world are found in regions crossed by the trade winds. The world's largest desert, the Sahara of North Africa, is a trade wind desert. It contains kilometers of linear dunes that are sometimes separated by as much as 6 km. Fig. 10-1 shows the pattern of trade winds around the globe.

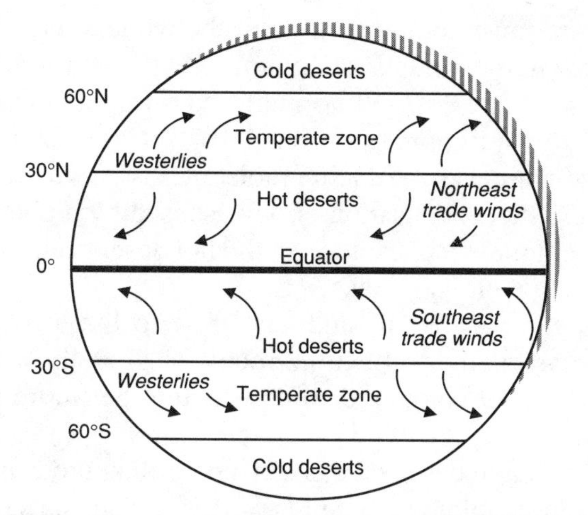

Fig. 10-1 Wind flows are horizontal in relation to the Earth's rotation.

There are other wind belts at different latitudes. The Westerlies play a major role in the more northern latitudes of the temperate zones.

The *Coriolis effect* was first noted by French mathematician Gaspard Gustave de Coriolis in 1844. In his studies of coordinate rotation, Coriolis hypothesized about circulating air and water patterns affected by the Earth's rotation. He found that wind and water currents were deflected eastward in the Northern Hemisphere and westward in the Southern Hemisphere. This rotational effect takes place between the equator and the poles of the earth.

Midlatitude Deserts

Midlatitude deserts are found between 30°N and 50°S. They occur between the subtropical high-pressure zones and the poles. These deserts, common in interior drainage basins far from oceans, have a wide temperature array. The Sonoran Desert of southwestern North America is a midlatitude desert.

SEMIARID DESERT

The U.S. deserts of Utah, Montana, Nevada, and southeastern Oregon are *semiarid deserts.* Worldwide they include the top-of-the-world landscapes of North America, Newfoundland, Greenland, Russia, Europe, and northern Asia.

The summers are fairly long and dry and the winters have little rain, mostly in the form of snow in the mountains. Summer temperatures are more moderate, averaging between 21 and 27°C. It normally doesn't get hotter than 38°C, with nighttime temperatures dropping to around 10°C. Cool nights make life easier for both plants and animals by reducing moisture loss from transpiration, sweating, and breathing. The condensation of dew, caused by night cooling, can meet or top the rainfall amounts received. As in the hot desert, rainfall is not common. It averages from 2 to 4 cm per year.

Semiarid soils can vary from sandy and fine to loose rock bits, gravel, or sand. In mountain areas, the soil is thin, rocky, or gravelly with good drainage. There is not a lot of salt compared to deserts that get more rain (with higher salt concentrations).

Most plants in semiarid deserts are prickly, providing protection in a dangerous environment. The large numbers of spines shade a plant surface's enough to cut down water loss. Other desert plants have a lot of fine hairs for the same reason. Some plants have silvery or glossy leaves to help them reflect greater amounts of radiant energy. These plants often have an unfavorable odor or taste. A few semiarid plants include creosote bush, white thorn, cat claw, mesquite, brittle bush, and lycium.

Semiarid desert plants differ by area. In some very dry areas, sparse plants may only cover 10% of the landscape. In other areas, sagebrush covers over 85% of the land. Desert plants grow between 15 and 122 cm in height.

Animals in these areas include jackrabbits, kangaroo rats, kangaroo mice, pocket mice, grasshopper mice, and antelope ground squirrels. They create underground burrows where they are insulated from the hot, dry climate. This burrowing method is also used by carnivores like the badger, kit fox, coyote, burrowing owl, and several lizards. Deer are found at lower elevations only in the winter. Grasshoppers and ants, as well as lizards and snakes, are also common.

COASTAL DESERTS

Coastal deserts are found on the western edges of continents between 15° and 30° latitudes near the Tropics of Cancer and Capricorn. These are also the largest of the climatic zones, covering nearly half of the Earth's area. They are located on both sides of the equator.

> The latitudes directly north and south of the equator are known as the **Tropic of Cancer** and **Tropic of Capricorn** respectively.

Coastal deserts are affected by cold ocean currents that follow the coastline. Since regional winds impact trade winds, these deserts are more changeable than others. Winter fogs created by rising cold currents often cover coastal deserts and block solar radiation.

Coastal deserts are complex because they exist where land, ocean, and atmospheric systems all mix. South America's coastal Atacama Desert is the driest desert on Earth. The Atacama Desert gets rainfall only once every 5 to 20 years.

Coastal deserts are found along the coasts of Chile and Australia. The chilly winters of coastal deserts are followed by long, warm summers. The average summer temperature ranges from 10 to 25°C; winter temperatures are 5°C or less. The greatest annual temperature is about 35°C and the least is about –4°C. In Chile, the temperature ranges from –2 to 5°C in July and 21 to 25°C in January. Remember, when it's summer in the Northern Hemisphere, it's winter in the Southern Hemisphere.

The average rainfall of these deserts is around 10 to 14 cm in most areas. The highest yearly rainfall over several years has been around 35 cm, with the lowest being around 5 cm.

The soil, fine with a medium salt content, is permeable and has good drainage. Plants have broad, shallow root systems so they can take advantage of any rain that comes their way. All coastal plants have thick, fleshy leaves that grab available water and store it for future use. In some plants, the surfaces are tightly ridged and grooved. When it rains, the stem bulges and grooves flatten to store water. As the water is used, the stem shrinks slowly back to its tight form. Coastal desert plants include the salt bush, buckwheat bush, rice grass, and black sage.

Dry climates force some types of toads to seal themselves in burrows and remain inactive for up to nine months. When a heavy rain comes along, soaking the earth, they get moving again. Some insects lay eggs that stay dormant for long periods, until the environment is suitable for hatching. Other coastal desert animals include, insects, coyotes, rabbits, toads, owls, eagles, lizards, and snakes.

Rain Shadow Deserts

Rain shadow deserts are formed when tall mountain ranges stop moisture-rich clouds from reaching areas on the far protected side of the mountains. These deserts are often found near coastal regions, but can be near other bodies of water. As moist air rises over a mountain range, water is dropped as rain or snow and the air loses its moisture. A desert is created on the far (dry) side of the range.

In the western United States, northern Nevada sits in a rain shadow of the Sierra Nevada Mountains. Snowfall on the western mountain slopes (California

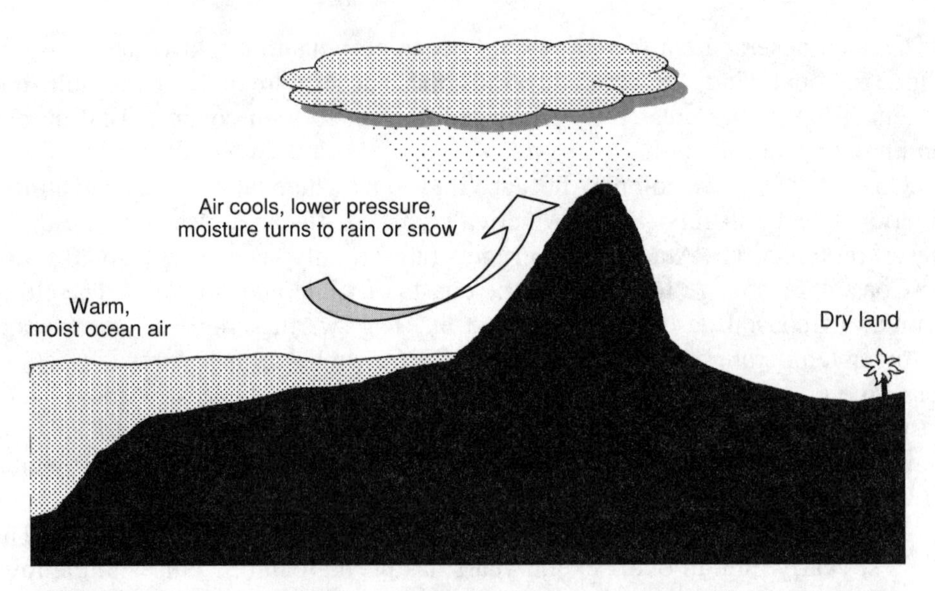

Fig. 10-2 Rainfall is much higher on one side of the mountain in a rain shadow desert.

side) is greater than that on the eastern slopes (Nevada side). Additionally, Nevada is nearly all located in the Great Basin area, which is fairly arid overall.

In the Turpan Depression of China, a rain shadow desert stretches for many kilometers of sand. There are a few oases and some mountain vegetation, but snow separates the Tian Shan vegetation from the rain shadow desert. Fig. 10-2 shows how rainfall drops on the ocean side of a mountain due to temperature, pressure, and topography changes.

Monsoon Deserts

When there is a large temperature difference between a continent and a nearby ocean, a *monsoon* results. This is a seasonal happening. In fact, monsoon comes from the Arabic word *mausim,* meaning "season."

Monsoons originally described the winds of the Arabian Sea, which blow for six months from the northeast and six months from the southwest. Now, monsoon generally describes winds with specific seasonal turnarounds. The southeast trade winds of the Indian Ocean dump heavy rains on the coast of India, losing their moisture on the eastern slopes of the Aravalli Range. Deserts of interior India and Pakistan are part of a monsoon rain shadow desert on the west side of the Aravalli Range.

It's odd that the wettest places in the world get no water in the winter at all. This is typical monsoon weather. When seasonal monsoon winds blow from one direction for six months, they bring torrential rains. Then, when they switch and blow from the opposite direction for the next six months, little to no rain falls. During the wet season moist air is cooled as it blows over rising land, letting lots of rain fall on the windward side of mountain ranges. Unfortunately, because of widespread conifer forest destruction (land clearing) that once protected the soil, the ground can't absorb the heavy rainfall during the monsoon season, resulting in erosion and mudslides. Monsoon conditions are also found in northern Australia, western, southern, and eastern Africa and Chile.

POLAR DESERTS

Polar or *cold deserts* are places with an annual rainfall amount of less than ¼ cm and a mean summer temperature of less than 10°C. Polar deserts cover nearly 5 million km^2 and are mostly bedrock or gravel plains. Sand dunes are not often seen in these deserts, but snow dunes are common where local rainfall is heavy.

Temperatures in polar deserts criss-cross the freezing point of water. This freezing/thawing fluctuation forms complex surface textures. The Dry Valleys of Antarctica have been ice-free for thousands of years.

Polar deserts experience cold winters with snowfall and high total rainfall all winter and sometimes in the summer. They are found in the Antarctic, Greenland, and close to the Artic region. Polar deserts have short, wet, and fairly warm summers with long, cold winters. The mean winter temperature is between –2 to 4°C, and the mean summer temperature is between 21 and 26°C.

In winter, polar deserts get a lot of snow. The mean annual rainfall is around 15 to 26 cm, although annual precipitation has reached a high of 46 cm and a low of 9 cm in some places. Heavy spring rains come in April or May, although, some places get heavy rains in autumn. The soil is dense, silty, and salty. It is fairly porous and drainage is good. A lot of soil salinity is leached out by rainfall.

Desert Features

Sand covers only around 20% of the Earth's deserts. Most of the sand is in sand sheets and sand seas that look like huge areas of rising and falling dunes.

The movement of sand and particles by the wind is called *eolian movement* after the Greek god of wind, Aeolus. Geologists have found that nearly 50% of desert surfaces are plains, where eolian movement of fine-grained material has uncovered loose gravels, pebbles, and stones.

The remaining arid land surfaces are composed of exposed outcrops, desert soils, and under-river deposits including sediment fans, basins, desert lakes, and oases. Bedrock outcrops are commonly seen as small mountains surrounded by heavily eroded plains.

OASES

Many oases are made by humans. Oases are sometimes the only places in the desert where crops can be grown and people are able to live.

An **oasis** is a fertile, green place in a sandy desert containing a spring, well, or irrigation.

In Egypt, a land that many people associate with the desert, there are six inhabited oases. The Al-Kharga Oasis is west of the Nile Valley town of Asyut. Past plate tectonic activity can be seen in the escarpment walls. Lengthwise, the oasis measures 185 km, while its width is roughly 20 km. Like Saudi Arabia, only 1% of Egypt's total area is used for farming.

DUNES

Most people think of sand dunes when they hear the word *dune*. However, a dune is any landmass created by deposited material that forms a low mound or ridge of sediment. Dunes can be formed by sediments of different sizes or even snow. They are formed when wind blows sediments, mostly from the same direction, until it encounters an obstruction of some sort. The wind slows and drops the heavier particles, which then build up and add to the size of the original obstruction. This becomes a cycle, with the blockage getting larger and more sand gathering.

The formation of a sand dune is progressive. The wind causes a layering effect. This is called a *slip face movement*. As the wind builds a mound, it tends to keep blowing some sand over the top in a leading edge. This leading edge eventually gets too heavy, creates an unstable slope, and slides down the other side in a layer. The wind continues to blow and builds another leading edge that eventually slips into a new layer. See Fig. 10-3 for the slope layering process and growth of sand dunes.

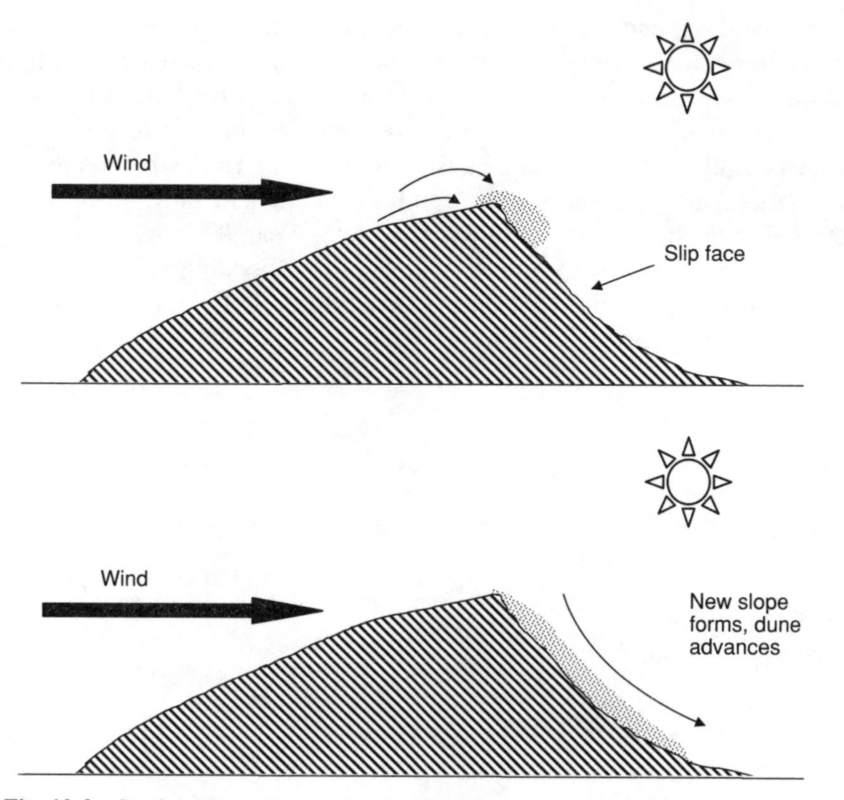

Fig. 10-3 Sand dunes are formed by the building and collapse of sand slope layers.

There are five main types of sand dunes. These are the *barchan, transverse, blowout, linear,* and *composite dunes.* Although it is sometimes easier to see different dune types from the air, some deserts have only one predominant type.

The *barchan* dune is a horseshoe-shaped dune with the front curve facing into the wind. Barchan dunes are often found in groups, but occasionally alone. They often move over a flat surface of pebbles or bedrock. The slip face of a barchan is away from the wind.

Transverse dunes form long lines of ridged dunes that are perpendicular to the wind direction. They have steep slip faces at the back sides of the ridges. These wavy dunes form in areas with plenty of sand and not much plant life. The sand dunes found behind beaches are often transverse dunes formed by strong ocean winds. The farther inland you travel, the more plant life you encounter.

Blowout dunes are basically opposite in shape from barchan dunes. In a blowout dune, the horseshoe-shaped curve faces away from the wind. The slip face is away from the wind. Blowout dunes have vegetation that stabilized the sand at one point, but has since become covered by sand to form a mound, creating the dune.

Linear dunes are mostly parallel to the wind and form long, straight ridges. These dunes can be over 100 meters tall and go on for many kilometers. Geologists think linear dunes are caused by winds that blow from one direction in one part of the year (northwest), and then shift and blow from a different direction (southwest) during another part of the year. The overall dune movement is easterly with a long, thin shape. The sand bedding is criss-crossed and the slopes are unequal. Fig. 10-4 diagrams the different types of sand dunes most commonly seen by geologists.

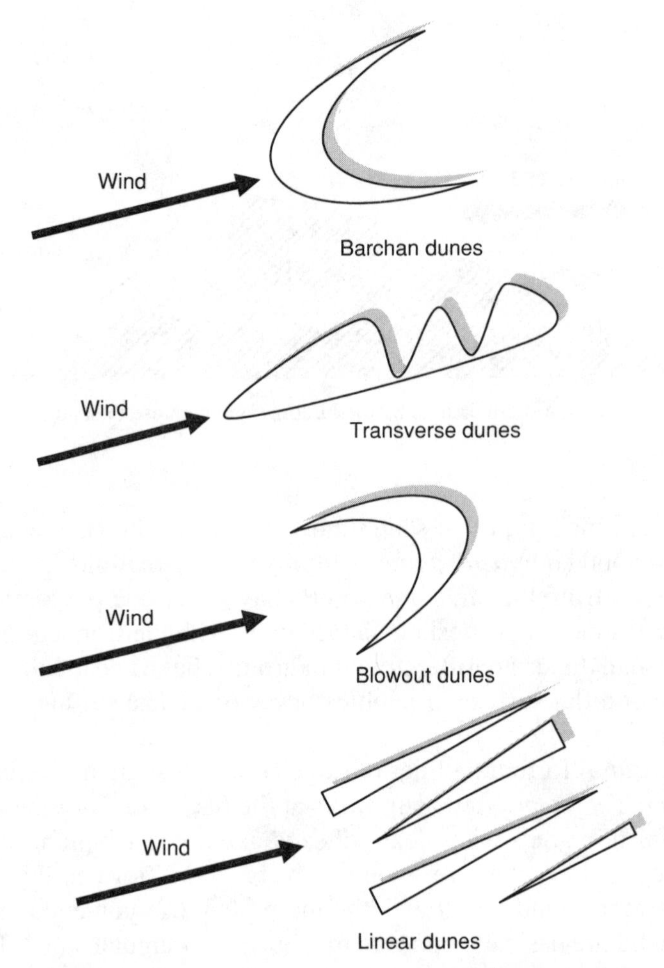

Fig. 10-4 There are four main types of sand dune formations.

A last dune type is really a combination of two or more types. *Composite dunes* form extremely large, tall, hilly forms known as *draas.* They are a mix of mostly transverse and linear dunes that get to be over 400 meters high. These giant dunes move much more slowly—sometimes only ½ meter per year—than their smaller cousins. They form wide dune fields called *ergs.* Some ergs can cover as much 500,000 km^2. That is one big sandbox!

SOILS

Desert soils are mineral-rich with low organic content. The cycle of rainfall and evaporation causes soils to form distinct salt layers. Calcium carbonate settling out of solution tends to cement sand and gravel into a hard material called *calcrete.* Calcrete can form layers up to 50 meters thick.

Caliche is a reddish-brown to white layer found in many desert soils. Caliche is found as nodules or as a coating on mineral grains. It is created by the interaction between water and carbon dioxide released by plant roots or by decaying organic matter.

Fine dust swept from the soil and mountains can be carried many kilometers across the desert before being released to the ground. When this windblown dust gathers into a silt layer, it is known as *loess.*

Loess, made up of rock particles between 0.01 and 0.06 cm in diameter, is blown around by the wind, sometimes kilometers away from its original location.

During the *dust bowl* years of the 1930s, drought, erosion, and poor farming methods caused huge dust storms. Over 100 million acres of southeastern Colorado, New Mexico, western Texas, and the panhandle areas of Oklahoma and Texas were affected. Fields and meager crops were covered by loess.

Many areas of the world have had similar problems. Large areas of China have been covered by loess blown from the Gobi desert.

PLANTS

Nearly all desert plants are drought-resistant or salt-tolerant. Some store water in their leaves, roots, and stems, then use it during the long dry times. Other desert plants have long tap roots that enter the water table, hold soil, and resist erosion.

When water evaporates and leaves minerals behind or minerals are washed from higher elevations, salt builds up in the soil. Chlorine is particularly retained and causes problems for plants. *Halophytes* are desert plants that adjust to high salt levels in several different ways. They are able to survive in the harsh desert climate where most other plants would quickly die.

Halophytes are plants that are able to tolerate extremely high levels of salts and still flourish.

Deserts are made up of halophytes that are *salt accumulators* (like *Salicornia*) or *salt excretors* (like the Tamarisk tree) and *nonhalophytes* (like *Zygophyllum*). They are truly the tough guys of the plant kingdom, living in hot and dry conditions that would shrivel most other plants in a few hours.

Deserts usually have a plant cover that is light but very diverse. It is thought that the Sonoran Desert in the United States has the most complex desert vegetation on the planet. Giant *saguaro cacti* offer nesting sites for desert birds and serve as trees of the desert. Saguaros are slow-growing but can live for 200 years. It takes about 75 years for a saguaro to grow tall and develop their first branches, but when fully grown, saguaro can reach 15 meters in height and weigh up to 10 tons. They populate the Sonoran Desert and give it a classicly "western" look to the dry landscape.

Although cacti are probably the first plants that come to mind when you think of deserts, plants in the pea and sunflower families are also well adapted to dry climates. Cold deserts also have specialized grasses and shrubs as the main vegetation.

WATER

There is rainfall in the deserts. In fact, desert storms can be fierce. Large storms in the Sahara have been known to dump up to 1 millimeter of rainwater per minute in a rapid burst. When this happens, dry stream beds, known as *arroyos* or *wadis,* become coursing torrents. Flash floods are a real danger.

As desert channels receive rainfall and runoff from high plateaus or mountains, they get a mixture of mud and sediment, too.

Most deserts are in low basins with little or no drainage, but a few deserts are intersected by rivers. The water from these large rivers sinks into soils and huge amounts of water are lost on their way through the deserts. The Colorado, Nile, and Yellow Rivers all flow across deserts, transporting sediments as they go.

Desert lakes are short-lived, shallow, and salty. When small lakes evaporate, they leave a *salt crust,* or *hardpan.* The flat area of clay, silt, or sand encrusted with salt that is left behind is called a *playa.* There are estimated to be over 100 playas in North American deserts.

Most big lakes are left over from the last Ice Age (12,000 years ago). Lake Bonneville was 52,000 km^2 and almost 300 meters deep. Huge during the last Ice Age, it stretched across Utah, Nevada, and Idaho. The only enduring remnants of Lake Bonneville are Utah's Great Salt Lake, Utah Lake, Sevier Lake, and the Bonneville Salt Flats.

The most unpredictable trait of all desert regions is the wide swing in annual rainfall levels. In the Namib desert in West Africa, for example, the average rainfall is 0.15 centimeters, but can be as little as 0 centimeters. Plants are amazingly tough and adaptable with regard to these seasonal, once-in-a-blue-moon desert rains.

Desertification

Deserts are formed from a variety of geological factors. Global temperatures, rainfall rates, and tectonic processes all contribute to the forming of the landmasses.

Desertification is the downgrading of rich soil and land into dry barren lands. Today, a lot of desertification is influenced by human activities and climatic changes. Desertification takes place because dry land environments are vulnerable to overdevelopment and poor land use. The following factors, which ruin a land's richness, can all increase desertification:

- Poverty
- Political instability
- Deforestation
- Overgrazing
- Bad irrigation practices

Estimates calculate that over 250 million people worldwide are directly harmed by desertification. Moreover, it is thought that the land supporting as much as one billion people in nearly a hundred countries is at risk. These populations are poor and have few resources to stem the desertification problem.

DEGRADATION OF DRY LANDS

When formally fertile land is degraded, there is a loss of *biodiversity.* As a result, the economic yield and complexity in croplands, pastures, and woodlands is damaged. *Degradation* is due mainly to climate mutability and unsound human activities, like overgrazing and deforestation. While drought is often linked with land degradation, it is a natural event. It happens when there is much less than normal rainfall in a region over a long period of time.

RAINFALL

Deserts have little if any freshwater supplies. In desertification, rainfall varies seasonally during the year, with wide swings happening over years and decades, leading to drought. Over time, dry land ecology has become used to this moisture fluctuation.

The biological and economic resources of dry lands—like soil quality, water supplies, plants, and crops—are easily damaged. For centuries, native people have learned to protect these resources with time-tested methods like crop rotation and nomadic herding.

Unfortunately, in recent decades these practices have changed due to economic and political conditions, as well as a leap in population levels. When people don't take climate and soil conditions into account and respond accordingly, desertification results.

LAND OVERUSE

Land overuse can come from economic circumstances, poor land laws, and cultural customs. Some people exploit land resources for their own gain with little thought for the land or neighboring areas. Some people in poverty have little choice but to overuse their meager resources, even to the extent of wearing out the land.

Trade and exploitation of a country's natural resources often leaves the land restoration in the hands of local people without the funding to have much of an impact. In the same way, an economy based on the sale of crops can cause farmers to ignore the overexploitation of the land.

Wars and national emergencies also destroy rich land by overburdening it with refugees and other displaced people. Natural disasters like floods and droughts can do the same thing.

POPULATION DENSITY

Most people blame desertification on overpopulation. However, it's possible for large populations to practice good conservation and land management methods and avoid desertification. Think of the midwestern "grain belt" of the United States. It is said that we could feed most of the world with the grain produced and stored in that one geographical region. However, the aquifer that provides most of the water for irrigation, the Ogallala aquifer, has been overmined for years. It may not be possible to continue irrigating crops in another 25 to 40 years.

Desertification is a complex problem and the relationship between population and desertification is not all that clear. We have seen, for example, that a decrease in population can cause desertification because there are less people to take care of the land. In some countries, when young people in villages go into the city to find work, their aging parents can't keep up with the land's needs and cultivation.

SOIL AND VEGETATION

Topsoil can be blown away by the wind or washed away by rainfall. However it happens, the soil's physical structure and biochemical makeup are changed. Cracks develop and nutrients are lost. If the water table rises from poor drainage and irrigation methods, the soil gets saturated and salts increase. When soil is stomped down by cattle, it's much harder for plants to grow in the compressed soil. This increases erosion.

Less plant growth is both a result and a cause of desertification. Loose soil can damage plants, bury them, or leave their roots exposed.

Weakened land may also allow downstream flooding, reduced water quality, increased river and lake sedimentation, and the buildup of silt in reservoirs and navigation channels. It can also cause dust storms and air pollution, resulting in damaged machinery, reduced visibility, and a lot of dirt. Windblown dust can increase health problems, including allergies, eye infections, and upper respiratory problems.

ECONOMIC

More and more people and governments are seeing the link between desertification, displaced people, and military conflicts. In the past 30 years, many African people have had to relocate or been forced to move to other countries because of war, drought, and land degradation. The environmental resources in and around cities and camps, where these people settle, come under severe pressure. Difficult living conditions and the loss of cultural identity further undermine social balance.

There is little detailed data on the economic losses resulting from desertification, although a World Bank study suggested that the depletion of natural resources in Senegal and Nigeria were equivalent to as much as 20% of their annual gross domestic product (GDP). At the global level, it is estimated that the annual income lost in areas immediately affected by desertification is approximately $42 billion each year. The indirect economic and social costs suffered outside the affected areas, including the influx of "environmental refugees" and losses to national food production, may be much greater.

TRENDS

Loss of soil, low moisture, and high temperatures are the parameters of several desertification studies. Interest areas include sites in Brazil, Namibia, South Africa, and the southwestern United States/northern Mexico. These locations have been the object of many years of active research into the continuing processes of land degradation and sand movement.

Larger global deserts are also monitored for geologic mineral mapping and the determination of active sand transport corridors.

In the past 70 years, the use of remote sensing has become an important tool in the study of dynamic features of deserts, like dunes. The ability to look at changes over time allows analysis of current climatic systems and marginal land areas at risk for future desertification.

In November 2004, NASA and the World Conservation Union, the world's largest environmental umbrella group, signed an agreement to use the space agency's satellite systems to monitor global environmental change in hopes of preserving the planet. NASA's mission is to understand developing (in many cases worsening) environmental problems and protect our home planet by using space-based observation techniques.

Remote sensing can give scientists an overall look at an entire area over different time periods. This provides a way to project past activity and monitor current climate changes. Drought-prone areas on the edges of wind pathways and sand seas are prone to dune advance and possible desertification. Human activity in these regions can also be examined for climatic impacts.

Findings presented at the 2005 American Meteorological Society (AMS) annual meeting described how the number of global land areas experiencing very dry conditions rose from 10–15% (1970s) to 30% (2002). Almost half of that change was found to be from increasing global temperatures rather than rainfall/snow decreases according to a study by the National Center for Atmospheric Research (NCAR) in Boulder, Colorado.

During three decades, extensive drying took place over large parts of Europe and Asia, Canada, western and southern Africa, and eastern Australia. Escalating global temperatures appear to be a major factor, says Aiguo Dai, the NCAR scientist who presented the results.

The study's details were published in the December 2004 issue of the *Journal of Hydrometeorology* by NCAR's Kevin Trenberth and Taotao Qian.

Clearly, evidence is growing to support global warming trends and impacts.

Paleodeserts

Information on ancient sand seas, changing lake basins, archaeology, and vegetation show that climatic conditions have changed drastically over vast areas of the Earth in the geologic past.

During the last 12,500 years, parts of some deserts were drier than they are today. About 10% of the land between 30°N and 30°S is now covered by sand seas. Around 18,000 years ago, sand seas in these two vast belts occupied almost 50% of this land area. Just like today, tropical rain forests and savannahs were located between the two belts.

Fossil desert sediments, some of them 500 million years old, have been discovered in different parts of the world. Surprisingly, tropical rain forests are today growing in places where ancient dunes were once found. Sand dune–like patterns have been recognized in current wet environments. These ancient dunes now get from 0.08 to 0.15 cm of rain annually.

The Nebraska Sand Hills is a dormant 57,000 km^2 dune region of central Nebraska. In the largest sand sea in the Western Hemisphere, the sand is now held in place by vegetation. On top of that, it receives over ½ meter of rain each year. Dunes in the Sand Hills measure as much as 120 meters in height.

When we think of dry, hot deserts and western dust storms, most people don't think they are an immediate problem. They are right on one level; it's not like lava rushing down a mountain toward you or a hurricane-force wind ready to rip away the barn roof. However, deserts do impact people and the environment. Desertification will only worsen without good land management and well thought out farming and water distribution methods.

Quiz

1. Deserts' daytime temperatures can go up to
   (a) 32°C
   (b) 45°C
   (c) 57°C
   (d) 62°C

2. Ancient sand deserts are called
   (a) paleodeserts
   (b) old dunes
   (c) bad picnic spots
   (d) beaches

3. Halophytes are
   (a) angels
   (b) animals that can see in the dark
   (c) plants that are able to tolerate extremely high levels of salts
   (d) fish that are bioluminescent

4. Peveril Meigs divided desert regions into three categories, according to their rainfall:
   (a) dry, drier, and driest
   (b) dry, moist, and wet
   (c) arid, temperate, and tropical
   (d) extremely arid, arid, and semiarid

5. Deforestation, overgrazing, and bad irrigation practices are all factors that contribute to
   (a) juvenile delinquency
   (b) desertification
   (c) glacier formation
   (d) metamorphism

6. Deserts cover roughly what percentage of the Earth's surface?
   (a) 20%
   (b) 30%
   (c) 40%
   (d) 45%

7. United States' deserts of Utah, Montana, Nevada, and southeastern Oregon are
   (a) polar deserts
   (b) semiarid deserts
   (c) coastal deserts
   (d) paleodeserts

8. A fertile, green place in a sandy desert containing a spring, well, or irrigation is called
   (a) an ice cap
   (b) a neutral zone
   (c) an oasis
   (d) a deciduous forest

9. Desert rainfall is
   (a) less than 50 cm/year
   (b) greater than 50 cm/year
   (c) not a problem, there is always lots of rain
   (d) only seen by people in oases

10. The forest canopy is
    (a) 2 meters high
    (b) 5 meters high
    (c) 10 meters high
    (d) nonexistent in hot and dry deserts

CHAPTER 11

Geochemical Cycling

We have studied the Earth from many different angles. We have talked about everything from the fiery core, semisolid mantle, and igneous and metamorphic rock layers to plate movement, ocean depths, and the atmosphere. We've looked at the effects of super hot volcano eruptions to gliding, super cold glaciers. Our planet is never boring. There is always some kind of geological happening going on somewhere!

A *geochemical cycle* is like an element's life cycle. As the element moves from one place to another, it takes on different forms. For example, a major geochemical cycle involves an element moving from magma to igneous rocks to sediments to sedimentary rocks to metamorphic rocks and then back to magma. A minor cycle might only include transport from sediments to sedimentary rocks to weathering and then back to sediments again.

When elements move from one Earth storage form to another, it is known as a **geochemical cycle.**

In this chapter, we will take a look at calcium and carbon's geochemical cycles. Since these two major Earth elements lead interesting and connected lives, they give us an idea of the intricate and complex interrelationships that exist at all levels of diverse ecosystems.

First, to understand the carbon and calcium cycles, we need to understand *residence time.*

Residence time is the length of time that an element hangs around in any one area. For example, in Chapter 5, when we studied the hydrologic cycle we saw how water is stored in reservoirs, sometimes for thousands of years.

Residence time equals the average amount of time an element, like carbon or calcium, spends in a geological reservoir.

Carbon dioxide, which makes up about 5% of the atmosphere, has a residence time of 10 years. Sulfur dioxide, a very minor atmospheric player has a residence time of hours to weeks. In contrast, oxygen which makes up around 21% of the atmosphere has a residence time of 6000 years. Amazingly, nitrogen, in amounts that equal 79% of the total atmospheric gases has a residence time of 400 million years. Dinosaurs probably breathed some of the same nitrogen molecules we are breathing today!

Some of the places that calcium spends its residence time are the atmosphere, oceans, crust, and mantle. Carbon spends its residence time in the atmosphere, oceans, different rock forms, and the biosphere. Nitrogen is mostly in the atmosphere, soil, and biosphere.

Calcium

Calcium makes up roughly 3.4% of the Earth's crust and has been around since the formation of our planet. Calcium sediments are carried from the mountains by streams and rivers to the land, lakes, and oceans. When calcium sediments sink to the bottom of these water volumes, they are hardened into calcium-containing sedimentary rock. Calcium-containing sediments may lie at the bottom of a body of water for a very long time.

Plate tectonics brings about crustal uplifting at plate margins. Geological uplift of the crust pushes calcium-containing rocks to the surface where they are weathered. We saw in Chapter 4 that over the Earth's history the rise can be so extreme that mountains are built up. Geologists have found calcium-rich rocks with fossilized sea life, high on mountain slopes in the middle of mountain ranges far from modern seas.

Water helps calcium move from the land to the oceans. High concentrations of dissolved calcium, and/or magnesium in fresh water causes what is known as *hard water.* When these minerals are in high quantities in water, around 89 to 100 parts per million, they don't react well with soap. In fact, bath tubs form mineral rings and laundered clothes take on a gray color from undissolved soap scum. Undissolved minerals in hard water are also deposited in plumbing, coffee pots, and steam irons. Frequently, people living in hard water areas use water softeners, which remove calcium and magnesium ions in an exchange with sodium ions.

Hard water can have benefits. In the aquatic environment, calcium and magnesium help keep fish from absorbing metals like lead, arsenic, and cadmium into the bloodstream through their gills. Therefore, the harder the water, the less easy it is for toxic metals to absorb into their gills.

In seawater, calcium concentrations are 100 to 1000 times higher than land levels and even greater concentrations are found in deeper, colder waters where there is little circulation. Calcium can average roughly one million years in the ocean before it appears on land again. Calcium ions stay in ocean water until they are precipitated as calcium carbonate.

Although upper ocean levels are highly saturated with calcium and carbonate ions, the saturation level depends on location and saturation conditions. Warm, shallow water contains lower levels of carbon dioxide because of photosynthesis and temperature. These conditions allow calcium carbonate to precipitate either inorganically or by aquatic organisms.

When calcium carbonate is used by marine inhabitants to build shells, it is called *biomineralization.* As these organisms die, their hard parts and shells sift down to the ocean's floor and gather or dissolve depending on depth, temperature, and pressure. Shells that sink to the deepest parts of the ocean are most often redissolved because of the higher levels of carbon dioxide in the colder, deeper waters.

> The dividing line that separates an area where calcium carbonate dissolves and accumulates is called the **lysocline.**

The calcium carbonate deposited by microorganisms is often mixed with other ocean sediments or washed from the land, depending on its location. Birds, animals, and humans eat seafood and shellfish, discarding the shells. This returns calcium back to the earth in a fairly rapid cycle.

However, most calcium takes a different path. It follows the main geological process of plate tectonics. Crustal plates and continental land masses, with their mountain-building movement, help stored calcium carbonate deposits get up to or near the surface in the form of limestone or marble (if changed by pressure and temperature). Fig. 11-1 illustrates the different compounds of the calcium cycle.

Limestone is a bedded sedimentary rock made up mostly of calcium carbonate. It's the most important of the carbonate rocks, consisting of sedimentary carbonate mud, calcium-based sand, and shells.

A good part of Europe and some areas east of the Mississippi River in the United States have a deep groundwork of limestone. The white "chalk" cliffs of Dover, England are made of limestone. Portions of coastal New England states have bands of marble between layers of metamorphosed sandstone and shale, pointing to ancient ocean sediments and higher sea levels. Continental calcium also collects during evaporation of brackish inland seas and reef production.

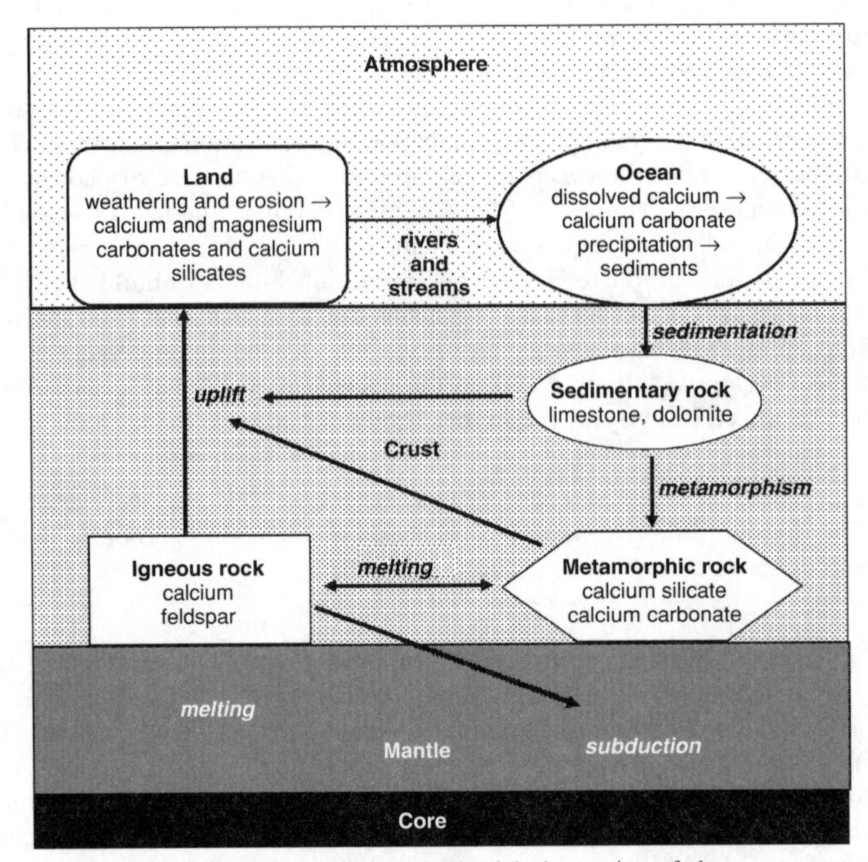

Fig. 11-1 Calcium enters and exits sinks in a variety of places.

Limestone is fairly insoluble. However, when plant roots and soil organisms of all sizes give off carbon dioxide that combines with groundwater, carbonic acid is formed. Carbonic acid dissolves limestone, releasing calcium. One thing that farmers consider when adding limestone to soil is particle size. When particles are small, the total reactive limestone surface area is larger, allowing for quicker dissolution.

Sometimes limestone contains magnesium as well as calcium in the form of carbonates and oxides. This type of limestone is known as dolomite. High levels of magnesium come from plants since it is the central element in the chlorophyll molecule.

SOIL

Calcium is taken up by the roots of plants either directly from the soil or after it has moved into the groundwater. When calcium is picked up by roots from the soil due to membrane permeability, both active and passive transport of ions takes place. Calcium is then transferred to the leaves, where it is stored.

Studies have found that calcium is taken up by trees and bushes, then returned to the ground with the leaf fall to be recycled by microorganisms that break down leaf litter.

Calcium is an important part of a plant's physical structure. Plant calcium also serves as an enzyme cofactor and has a strong effect on cell membranes. When there is a shortage of calcium, membranes get mushy.

Plant calcium goes back to the soil when leaves fall. It is stored in woody parts of a plant until it falls, decays, is burned, or is consumed by an animal.

When wood is burned, its volume is reduced to about 0.1 to 3.0% of its original volume as ash. Of this, approximately 30 to 60% is in the form of calcium oxide. This is the reason why gardeners use wood ashes as a calcium source. It's a good way to raise soil pH. An acid soil has a lot of the plant ion-binding sites occupied by hydrogen ions. A rich farming soil has about 60 to 70% of these sites filled with calcium cations, 10 to 20% filled with magnesium cations, 10 to 15% with hydrogen cations, 3 to 5% with potassium cations, and the rest with lesser level nutrients.

Besides equalizing soil pH available for plants, good calcium levels enhance soil structure, make phosphorus more available and improve the microbial environment. Calcium is said to aid the growth of symbiotic and nonsymbiotic nitrogen-fixing bacteria. This is why liming is important for the growth of legumes, whose roots host nitrogen-fixing bacteria in nodules.

CALCIUM IN FOOD

The highest levels of calcium are found in seafood. In fact, ocean kelp has the highest percentage of calcium found in any complete plant food. It is 1.09% calcium, with the calcium concentration of other seaweeds ranging from 0.30% to 0.90%.

When studying bacteria, plants, and animals, it's important to notice that the calcium concentration increases as you move up the food web. For example, calcium makes up around 1.5% of the average human body.

Animals get calcium from the foods they eat. Herbivores such as cows must get all their calcium from pasture, hay, and grains. Carnivores or omnivores can get their calcium from animals or from plants. Fermentation in the digestive tract changes calcium into a more absorbable form. It also suggests a connection to the digestive process of cows and other ruminants where microbial fermentation in the rumen (part of the cow's digestive system) is an essential early step in digestion. Ruminants like goats, sheep, camels, and cows provide most of the world's milk.

In humans, a person that weighs 68 kilograms has around 1.02 kg of body calcium, of which 99% is stored in the bones and teeth. Of the calcium found in foods eaten by humans, as little as 10 to 30% is absorbed. This could explain why, when we age, digestive processes slow, hormone levels change, and bone calcium levels drop.

Nitrogen

As we saw in chapter 3, nitrogen (N_2) makes up 79% of the atmosphere. All life, like proteins, requires nitrogen compounds to survive. However, they can't generally use nitrogen in the gaseous form.

To be used by an organism, nitrogen must be combined with hydrogen and oxygen. Nitrogen is taken out of the atmosphere by lightning or nitrogen-fixing bacteria. During storms, large amounts of nitrogen are oxidized by lightning and mixed with water (rain). This falls and is converted into nitrates. Plants take up nitrates to form proteins.

Plants are consumed by herbivores or carnivores. When these die (organic matter), the nitrogen compounds are broken down into ammonia. Ammonia can be taken up by plants again, dissolved by water, or remain in the soil to be converted to nitrates (*nitrification*). Nitrates stored in soil can end up in rivers and

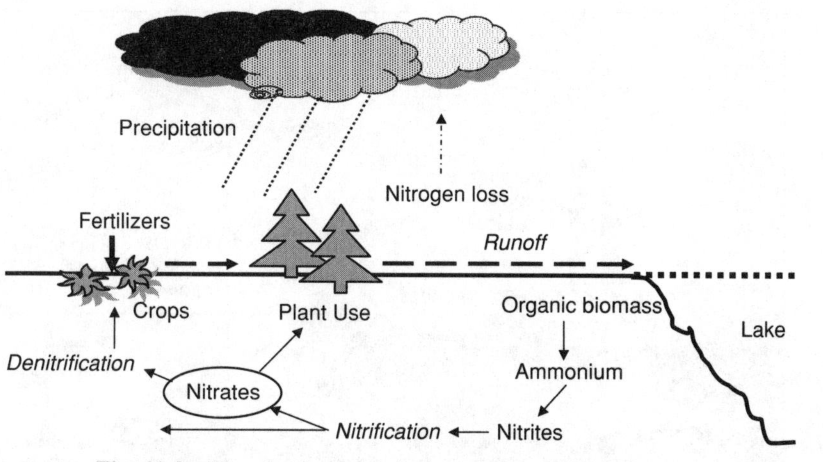

Fig. 11-2 The nitrogen cycle is essential to all living systems.

lakes through runoff. They can also be changed into free nitrogen and returned to the atmosphere. Fig. 11-2 gives you an idea of the nitrogen cycle.

As we saw in Chapter 8, extra nitrogen in water (from fertilizers and other runoff chemicals) can lead to eutrophication. High nitrate levels, along with phosphate, can cause the overgrowth of aquatic plants and algae, causing high dissolved oxygen consumption, killing fish and other aquatic organisms.

Carbon

Carbon is the fourth most abundant element in the universe after hydrogen, helium, and oxygen. At last count, there were more than 2 million known organic compounds, nearly 20 times more than all the other known chemicals combined.

Carbon is known as the *building block of life* and is the foundational element of all *organic* substances, from molds to mosquitoes to fossil fuels. Carbon cycles through the land, ocean, atmosphere, and the Earth's interior in a major biogeochemical cycle.

Organic matter must contain carbon from living or nonliving material in order to be considered organic.

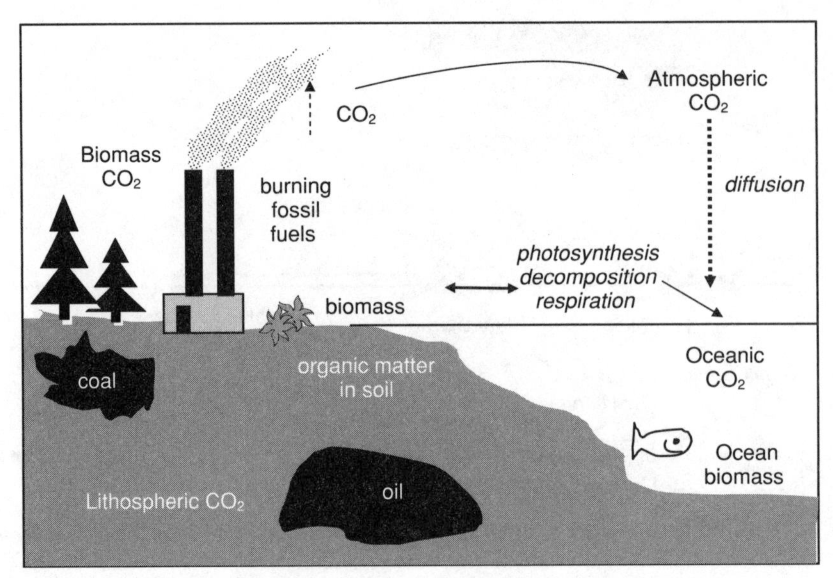

Fig. 11-3 The carbon cycle has a big industrial component.

The transport of carbon takes place in the atmosphere, biosphere, oceans, and landmasses. All of carbon's different lives are described by the carbon cycle. The carbon cycle has many different storage spots, also known as reservoirs or *sinks,* where carbon exchanges take place. The carbon cycle is shown in Fig. 11-3.

The global carbon cycle is divided into two types, the *geological carbon cycle,* which has been going on for millions of years, and the *biological carbon cycle,* which stretches from days to thousands of years.

When the Earth was first formed, small chunks of rock orbiting the Sun clumped together and formed the planets. At the same time, carbon-containing meteorites showered rocks down onto our planet's surface. This process allowed the carbon content of the Earth to rise. Geologists believe the total amount of carbon that cycles through today's Earth systems was around at the formation of the solar system.

GEOLOGICAL CARBON CYCLE

In the *geological carbon cycle,* carbon moves between rocks and minerals, seawater and the atmosphere through weathering. Table 11-1 shows all the various rock forms that contain carbonate.

Table 11-1 Carbonate is found in many different forms.

Carbonate rock forms	Composition
Micrite (microcrystalline limestone)	Very fine-grained; light gray or tan to nearly black
Oolitic limestone	Sand-sized oolites
Fossil-laden limestone	Fossils in a limestone matrix
Coquina	Fossil mash cemented together; may resemble granola
Chalk	Microscopic planktonic organisms such as coccolithophores
Chert	Silica skeletons of sponges, diatoms, and radiolarians
Crystalline limestone	Larger-grained than micrite
Travertine	Stalagtites and stalagmites ($CaCO_3$)
Coal	Converted land plant remains
Other	Intraclastic limestone, pelleted limestone

Carbon dioxide in the atmosphere reacts with water and minerals to form calcium carbonate. Calcium carbonate rock (limestone) is dissolved by rainwater through erosion and carried to the oceans. There, it settles out of the ocean water, forming sedimentary layers on the sea floor. Then, through plate tectonics, these sediments are subducted underneath the continents. With the extreme heat and pressure deep beneath the Earth's surface, the limestone melts and reacts with other minerals, freeing carbon dioxide. This carbon jumps back into the carbon cycle, returning to the atmosphere as carbon dioxide during volcanic eruptions.

The balance between weathering, subduction, and volcanism controls atmospheric carbon dioxide concentrations over geologic time. Some geologists have found that the oldest geologic sediments point to atmospheric carbon dioxide concentrations over 100 times current levels.

Conversely, ice core samples from Antarctica and Greenland make glaciologists think that carbon dioxide concentrations during the last Ice Age were only about ½ of today levels.

The amount of carbon recorded and exchanged at each step in the cycle controls whether a certain sink is increasing or decreasing. For example, if the ocean absorbs 2 gigatons more of carbon from the atmosphere than it releases in any one year, then atmospheric storage will decrease by the difference. Additionally, the atmosphere interacts with plants, soils, and fossil fuels. Everything is intimately interconnected.

The geological carbon cycle is intricate. Each player in the cycle impacts the other players. If the amount of carbon in one reservoir drops, then the amount in one or more of the others will rise. *Nature is all about balance.* However, it may take many years of large buildups in one area before things shift to another storage place. This might be the case with increases in atmospheric carbon dioxide. It may take a long time for the oceans to increase their uptake of carbon dioxide. The problem gets messy when everything—land, sea, and air—is overloaded. Then nature has a tough time keeping up.

The carbon cycle is a closed system. Original carbon amounts are squirreled away on the planet somewhere. Geologists are trying to figure out if everything really does balance out at the end of the carbon cycle equation. When all the sinks are estimated and added up, both sides of the equation should be equal. As population increases and global resources are challenged, experiments in this area are going to be more and more important.

BIOLOGICAL CARBON CYCLE

The biosphere and all living organisms, including you and me, play a big role in the movement of carbon in and out of the land and ocean through the processes of photosynthesis and respiration. On this planet, nearly every living thing depends on the creation of sugars from photosynthesis and the metabolism (respiration) of those sugars to support growth and reproduction.

> The **biological carbon cycle** occurs when plants absorb carbon dioxide and sunlight to make glucose and other sugars (carbohydrates) to build cellular structures.

Plants and animals use carbohydrates during respiration, the opposite of photosynthesis. Respiration converts this biological (metabolic) energy back to carbon dioxide. As a process pair, respiration and decomposition (respiration by bacteria and fungi) restore the biologically fixed carbon to the atmosphere. Yearly carbon levels taken up by photosynthesis and sent back to the atmosphere by respiration are 1,000 times higher than carbon levels transported through the geological cycle each year.

We've seen how photosynthesis and respiration play a big part in the long-term geological cycling of carbon. Land plants pull carbon dioxide from the atmosphere. In the oceans, the calcium carbonate shells of dead phytoplankton sink to the sea bed and form sediments. When photosynthesis is higher than respiration, organic matter gradually builds over millions of years and forms coal and oil deposits. These biologically regulated activities characterize atmospheric carbon dioxide removal and the storage of carbon in geologic sediments.

Total organic carbon (TOC) is used by hydrologists to check the health of fresh water. As we have seen, organic matter plays a big role in aquatic systems. It affects biogeochemical processes, nutrient cycling, biological availability, chemical transport, and interactions. It also directly affects municipal choices for wastewater and drinking water treatments. Organic content is commonly measured as total organic carbon and dissolved organic carbon, which are essential components of the carbon cycle.

BALANCE

All life is based on the element *carbon,* the main chemical ingredient of organic matter. There should be T-shirts that read, "Carbon is Life." Surprisingly, though, carbon is not hugely plentiful within the Earth's crust. The lithosphere is only 0.032% carbon by weight compared to oxygen and silicon that make up 45.2% and 29.4% of the Earth's surface rocks, respectively.

Carbon is stored on our planet in one or more of the following major storage reservoirs:

- Organic molecules in living and dead organisms found in the biosphere;
- Atmospheric carbon dioxide;
- Organic matter in soils;
- Lithosphere fossil fuels, sedimentary rock (limestone, dolomite, and chalk);
- Dissolved atmospheric carbon dioxide in the oceans; and
- Calcium carbonate in marine creatures' shells.

Table 11-2 illustrates the Earth's major carbon stores.

Table 11-2 Carbon is stored in various areas above and below the Earth.

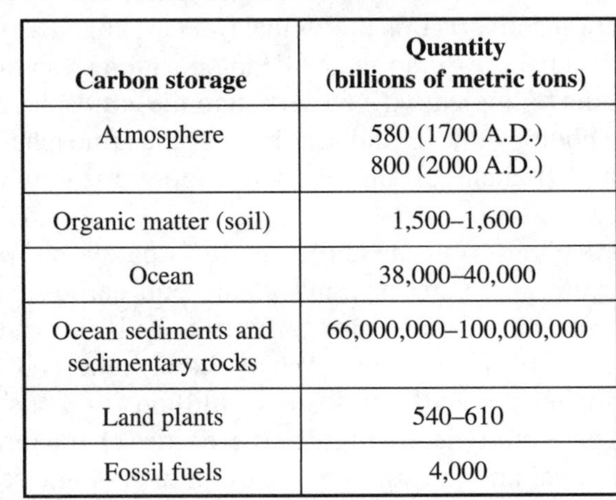

Carbon storage	Quantity (billions of metric tons)
Atmosphere	580 (1700 A.D.) 800 (2000 A.D.)
Organic matter (soil)	1,500–1,600
Ocean	38,000–40,000
Ocean sediments and sedimentary rocks	66,000,000–100,000,000
Land plants	540–610
Fossil fuels	4,000

Carbon in the lithosphere is stored in both inorganic and organic forms. Inorganic carbon deposits include fossil fuels like coal, oil, natural gas, oil shale, and limestone. Organic forms of lithospheric carbon include leaf and plant litter, organic matter, and other soil fractions.

Over geological history, the quantity of carbon dioxide found in the atmosphere has decreased. It is hypothesized that when the Earth's temperatures were a bit higher millions of years ago, plant life was plentiful because of the greater concentrations of atmospheric carbon dioxide. As time went on, biological mechanisms slowly locked some of the atmospheric carbon dioxide into fossil fuels and sedimentary rock. This carbon balancing process has kept the Earth's average global temperatures pretty much the same over time. However, human input in the carbon cycle (increased greenhouse gases) is recent.

CARBON IS NO. 1

Geologists are interested in carbon because it is such a versatile element. Not only does it exist in the air, land, and sea, but humans are made of approximately 50% carbon by dry weight. We are truly "carbon-based units."

Some environmental chemists study different ecosystems through carbon balancing accounts that include crop productivity, food chains, and nutrient cycling measurements.

> The **carbon cycle** involves the Earth's atmosphere, fossil fuels, oceans, soil, and animal and plant life of terrestrial ecosystems.

In addition, carbon dioxide is the major atmospheric greenhouse gas thought to be a result of human activities. We saw in Chapter 4 that these gases can lead to global warming.

Until alternative power sources like solar power are developed and used to a greater extent, atmospheric carbon dioxide increases will result mostly from the use of fossil fuels.

Geologists trying to understand seasonal carbon drops and gains in atmospheric carbon dioxide concentrations look for patterns. We've seen how global photosynthesis and respiration have to balance or carbon will either accumulate on land or be released to the atmosphere. Measuring year-to-year changes in carbon storage is tough. Some years might have more volcanic eruptions and have more carbon in the air, while other years or decades might have less carbon.

However, some measurements are clear. The clearing of forests for croplands is well documented, both historically and with satellites. It is a good measure of carbon storage changes. If forests get a chance to grow back on cleared land, they pull carbon from the atmosphere and start saving it up again in trees and soils. The change between total carbon released to the atmosphere and the total pulled back down governs whether the land is a supplier or reservoir of atmospheric carbon.

Atmospheric carbon = fossil fuels + land use changes – ocean uptake – unknown carbon deposit

When considering the global carbon equation between the atmosphere, fossil fuels, and the oceans, the global carbon tally is not completely known. Research goes on to discover the location of unknown carbon reservoirs.

Karst Formation

Karst is a specialized topography where the land is shaped by water's dissolving action on carbonate bedrock, like limestone, dolomite, or marble. It is an example of a calcium and carbon cycle on a small scale. Karst development takes place over thousands of years and results in unique land and subsurface structures such as *sinkholes,* vertical tunnels, vanishing streams and springs, interconnected underground drainage systems, and caves.

Karst formation is called a *carbon dioxide cascade.* It takes place as rain falls through the air, grabbing and dissolving carbon dioxide as it goes. Weak carbonic acid (H_2CO_3) is formed.

The mildly acidic water seeps into rock cracks or fissures where it starts to dissolve carbonate bedrock. This causes openings in the bedrock that get bigger until an underground drainage system begins to form, letting even more water come through. This speeds up the karst formation further. Over time, underground caves are carved out.

The top level of karst topography is called *epikarst,* which includes a crossing system of intersecting crevices and holes that gather and transfer surface water and minerals to the underground drainage system.

SINKHOLES

Karsts can be a big problem for homebuilders and business owners, especially in the Gulf Coast states of North America since they may form *sinkholes.* Sinkholes come in all sizes. Areas that are ripe for sinkholes can often be recognized by aerial or satellite photography. From these elevated views, characteristic karst circular patterns of ground cracks and depressions or lakes point to subsurface mineral dissolution. In karst regions, groundwater flow is speedy due to the high porosity and permeability of the underlying rock.

When groundwater is pumped out of these open underground water stores the water table drops, leaving a cavern. This open air space may then collapse when weighted down from above by ground construction.

The entire Florida peninsula contains solution-weathered limestone, with some cavities over 31 meters deep. Limestone in the area is covered by consolidated clays of the Miocene age and unconsolidated sands of the Pleistocene age. Groundwater is within 12 to 25 centimeters of the surface.

In 2002, the largest sinkhole in recent U.S. history opened up in Orlando, Florida, and swallowed large oak trees, park benches, and a sidewalk. Over 45 meters across with a depth of 18 meters, the sinkhole threatened nearby apartment buildings. City officials lined the sinkhole sides with plastic to keep rainfall from causing even more damage.

In Europe, geologists have discovered temperature inversions in limestone sinkholes of the eastern Alps. The Gstettner-Alm is an upland plateau area in a limestone subrange of the eastern Alps about 100 km southwest of Vienna, Austria. The Gstettner-Alm area, like many limestone areas, is characterized by sinkholes, or *dolinen.* The largest sinkhole in the Gstettner-Alm area, the Gruenloch, is about 1 km across. Recorded temperatures at the bottom of the Gruenloch are as low as –52.6°C, making it one of the coldest spots in central Europe. Researchers from the United States, the University of Vienna, and the Central Institute for Meteorology and Geodynamics are studying mechanisms that affect strong temperature inversions in sinkholes of different sizes near the Gstettner-Alm region.

KARST LIFE

Karst ecosystems often support odd or rare plant and animal species, both on the surface and underground. Some ferns and mosses need a limestone base on which to grow. Other ferns can grow in cool, wet, and dim cave entrances.

When visiting any cave, it's important to watch out for any wildlife living there, especially bats and other creatures, which are affected by even slight disturbances of their environment.

A lot of animals use different karst structures as a home. Caves are used from time to time by large carnivores like mountain lions and bears, for shelter or hibernation. Birds and other small mammals, like wood rats, nest in caves and other holes. Elk and deer often sleep near cave entrances during the summer when cave air is cooler and during the winter when cave air is warmer than outside temperatures. Since caves have fairly stable climates, they are also good homes for bat species that use them for roosting and hibernation.

Some karst-dependent species known as *troglobites* have evolved to living exclusively in the total darkness of caves. In British Columbia, Canada, a freshwater crustacean troglobite has been identified from the underground cave pools on Vancouver Island.

More commonly, *troglophiles* live inside and outside caves. Some troglophiles complete their entire life cycle within a cave, while other members of the same species live outside caves. Other Canadian troglophiles include certain species of salamanders, spiders, and crickets.

Some studies show that karst streams have a chemical impact on downstream aquatic environments. Minerals from a karst system benefit and raise fish populations in the following ways:

- Leaching of calcium carbonate from bedrock has important buffering effects on acidic streams;
- Groundwater associated with a karst results in cool, even stream temperatures throughout the year;
- Storage capacity in karst stream systems regulates seasonal flow rates;
- Karst streams supply more nutrients and encourage algae and moss growth;
- Aquatic insect populations within karst streams are larger and more diverse; and
- Karst stream systems provide more protective sites for fish to rest, reproduce, and avoid predators.

Karst aquifers also add to groundwater supplies and watersheds worldwide. However, there is one problem. The natural cleaning and filtering processes of surface streams are absent in karst underground systems. As a result, pollutants in karst waters may dangerously affect karst environments, including human water supplies and natural surface streams.

Calcium and carbon are essential to nearly all naturally occurring Earth processes. However, only through ongoing research can we hope to better understand our world and all of its magnificent workings.

Quiz

1. The dividing line between where calcium carbonate dissolves and accumulates in the ocean is called the
   (a) thermocline
   (b) biocline
   (c) geocline
   (d) lysocline

2. When calcium carbonate is used to build the shells of sea creatures, it is called
   (a) opalization
   (b) the bivalve process
   (c) biomineralization
   (d) shellization

3. Organic matter always contains
   (a) neon
   (b) dark chocolate
   (c) iodine
   (d) carbon

4. Oxygen's residence time in the atmosphere is approximately
   (a) 12 weeks
   (b) 300 years
   (c) 1000 years
   (d) 6000 years

5. High concentrations of calcium and/or magnesium in fresh water is commonly called
   (a) calciferous
   (b) hard water
   (c) soft water
   (d) white wash

6. Scientists have studied the carbon cycle in all of the following geochemical reservoirs, except the
   (a) soil
   (b) Earth's core
   (c) oceans
   (d) fossil fuels

7. When land is shaped by water's dissolving action on carbonate bedrock, it's called
   (a) karst
   (b) tephra
   (c) teflon
   (d) feng shui

8. By dry weight, approximately what percentage of carbon are humans composed of?
   (a) 20%
   (b) 50%
   (c) 70%
   (d) 90%

9. The balance of weathering, subduction, and volcanism controls
   (a) atmospheric carbon dioxide concentrations over geologic time
   (b) the formation of cumulonimbus clouds in the winter
   (c) the duration of the summer solstice
   (d) stock prices

10. Karst formation is also called a
   (a) rabbit hole
   (b) nitrous oxide cascade
   (c) carbon dioxide cascade
   (d) wetlands

CHAPTER 12

Solid and Hazardous Wastes

The simple act of living creates waste. Whether an organism is plant or animal, the use of raw materials, like food or water, results in waste. Human households create garbage. Most industrial processes produce solid and hazardous waste.

Hazardous waste in fresh water, oceans, or soil can kill fish and marine organisms or be taken up by plants that are later eaten by animals. This is how waste enters and contaminates the food web. In fact, predators (including humans) are most at risk from contaminants since toxins often concentrate in predators' tissues (*bioaccumulation*). Hazardous wastes can cause liver and kidney damage, as well as disruption to the nervous, reproductive, and immune systems. They can produce cancer and death in young and adult animals either from direct skin contact or gradual systemic poisoning.

These hazards are often stealthy and take years for their effects to be known. Similarly, they can take decades to be swept from the environment through natural circulation. In the United States, the Office of Solid Waste (OSW) regulates all hazardous waste under the Resource Conservation and Recovery Act (RCRA).

Since the Iron Age, when greater amounts of metal processing began, leftover waste had to be disposed of. The Industrial Revolution added to this waste buildup with widespread use of fossil fuels, magnifying the problem. Pollutants have been poured into the atmosphere and once pristine waterways, damaging them with harmful chemicals.

Early environmentalists sounded the alarm as early as the 1950s, but it was not until the 1962 publication of the book *The Silent Spring* by marine biologist Rachel Carson that environmental contamination came into broader public view. She pointed out the hazards of careless application and spraying of herbicides and pesticides.

Carson focused on *dichlorodiphenyltrichloroethane* (DDT). Her efforts brought to light DDT's harmful effect on fish-eating birds (e.g., eagles), a major factor in its ban from use in the United States.

In *Silent Spring,* her impassioned narrative explained that

> … over 200 basic chemicals have been created for use in killing insects, weeds, rodents and other organisms…. sold under several thousand different brand names. Can anyone believe it is possible to lay down such a barrage of poisons on the surface of the earth without making it unfit for all life?

Carson, an ecologist and then editor-in-chief of all U.S. Fish and Wildlife Service publications, was voted by *Time* magazine as one of the "100 Most Influential People of the Century" because of her important work in bringing the reckless use of chemicals and pesticides to policy makers' notice. She has also been called the *Mother of the Modern Environmental Movement.*

Fifty years ago people didn't recognize how the dumping of chemical wastes could affect public health and the environment. Thousands of properties, like contaminated warehouses and landfills, have become abandoned waste sites. Public awareness and alarm over the huge scope of hazardous dumping in the United States motivated the government to take action. The United States Environmental Protection Agency (EPA) was established and tasked with investigating, identifying, and cleaning up the worst hazardous waste sites nationwide.

In 1980, U.S. President Jimmy Carter signed a law called the Comprehensive Environmental Response, Compensation, and Liability Act (CERCLA) and established the first national hazardous waste identification and remediation program. Commonly called the Superfund Act, this law outlined regulations for disposing of hazardous waste and for cleaning up previously contaminated sites that had been abandoned. The Superfund Act is important because it is the only law that holds responsible all persons associated with contamination of a site, throughout that site's history.

The original fund set aside $1.6 billion for 5 years. Due to the serious hazards at many sites, the EPA was ordered to clean up first and collect fines later. A

National Priorities List (NPL) was established containing 120 sites and by 1989, had expanded to 1200 sites with 418 sites designated as "most hazardous."

On March 17, 2004, the EPA announced a milestone achievement. Cleanup work at the nation's first (and one of the worst) Superfund sites was completed. The Niagara Falls, New York site known as Love Canal took nearly 20 years and $400 million to clean up. The damage this multichemical waste dump brought upon the health of local inhabitants, animals, plants, and ecosystems is incalculable.

The dumping of hazardous wastes into Love Canal had a long history. Chemical waste dumping began in 1920 when the city of Niagara Falls, New York used the canal area as a landfill. Later, the U.S. Army used it to dispose of chemical warfare material. An accumulation of dumping took place during the 1940s and 1950s when the Hooker Chemical Company filled the canal with around 21,000 tons of organic solvents, acids, pesticides, and their byproducts.

It wasn't until the 1970s that Love Canal residents found out how much this thoughtless dumping cost them. Cancer rates and miscarriages were higher than average, people were getting sick from eating fruits and vegetables grown in gardens, and in some areas plants didn't grow at all.

But Love Canal was not the only site where decades of chemical and biological dumping made people sick. During the past 50 years, people in New Jersey, Pennsylvania, California, and Tennessee discovered their own problems.

Since 1982 the U.S. Army Corps of Engineers has initiated cleanup on many Superfund properties for the EPA, reducing future threats to public health and the environment. In 1982, the EPA allocated $12 million of Superfund work to the Corps. By 1988 that figure had risen to over $300 million and has remained high ever since. Today, the Corps is working on 143 projects nationwide under the Superfund program. Table 12-1 lists the steps taken to identify and clean up an EPA Superfund site.

Over 13,000 Superfund sites in the United States have been identified to date, with over 35,000 potential hazardous waste sites screened for federal action. Of these sites, 1295 have been proposed for addition to the NPL. The EPA estimates that between the years of 2004 and 2033, 294,000 hazardous waste sites will be identified in the United States alone, with over $250 billion paid out for cleanup by property owners and companies responsible for the contamination.

Many hazardous sites have been cleaned up or are in the process of cleanup. Getting rid of the hazardous waste and the contaminated soil takes a lot of time, effort, and funding. However, the reclaimed environment and human, animal, and plant habitats are well worth the challenge.

Table 12-1 The lifecycle of a Superfund site has specific stages.

Cleanup progress	Number of sites
Site discovery	39,542
Preliminary assessment	37,357
Site inspection	18,129
National priorities list	1233 (49 proposed)
Remedial inspection/ feasibility study	1033
Record of decision	1033
Remedial design	702
Remedial action	422
Construction complete	293
Delisted	83

Hazardous Waste

Solid, gaseous, or liquid waste, singly or combined, can create serious problems for humans and the environment if they are not treated, stored, transported, and managed safely. Whether singly or mixed, hazardous and toxic chemicals are generated by industry, agriculture, homes, and the environment.

Hazardous waste is highly flammable, corrosive, reactive (explosive or unstable), or toxic.

What kinds of modern waste are considered hazardous? There are three major classes of hazardous waste: *biological, chemical,* and *radioactive.* These types of waste are generated by many different industries and services. We will take a closer look at a few of these and their potential dangers.

BIOLOGICAL AND BIOHAZARDOUS WASTES

Biological waste is often called *organic waste* because it is composed of organic molecules containing carbon, hydrogen, and water. This can include everything from kitchen scraps (fruit/vegetable peels and rinds) to animal (bedding and carcasses) and human waste (leftover food, patient care waste, etc.).

Biological waste is defined as *infectious waste, pathological waste, chemotherapy waste,* or the containers and supplies created during its handling and/or storage. It is further described as waste that because of its quantity, type, or makeup requires special handling. Noninfectious waste can be handled as nonhazardous and can be disposed of through standard methods of trash disposal.

> **Biohazards** are infectious agents or hazardous biological materials that present a risk or potential risk to the health of humans, animals, or the environment.

Biohazardous materials include certain types of recombinant DNA; organisms infectious to humans, animals, or plants (parasites, viruses, bacteria, fungi, prions, rickettsia); and biologically active agents (toxins, allergens, venoms) that cause disease in other living organisms or significantly impact the environment. The risk can be direct (through infection) or indirect (through environmental damage.)

In a hospital or laboratory, biological waste is handled separately from chemical waste. Infectious agents are treated first, followed by chemical or radioactive contamination handling and disposal storage.

Biological + Radiation = Radiation Waste
Biological + Hazardous Chemical = Chemical Waste

A hazardous waste program that manages biological waste in a research, teaching, clinical laboratory, or clinical area is designed to 1) protect the people who handle, transport and dispose of waste; 2) protect the environment; and 3) minimize regulatory liability. Any attempt to ignore the special handling of biological waste puts people, and institutions at risk.

Infectious waste, classified by OSHA's Bloodborne Pathogen Standards List, is divided into seven major waste categories:

1. *Cultures and stocks:* agents infectious to humans, waste from biological production, live and dead vaccines, and anything used to contain, mix, or transfer agents. This includes, but is not limited to, petri dishes, pipettes, pipette tips, microtiter plates, disposable loops, vials, and toothpicks;

2. *Human blood, blood products, and infectious body fluids:* blood (not in a disposable container), serum, plasma, and other blood products or nonglass containers filled with discarded fluids. It also includes any substance or fluid that contains visible blood, semen, vaginal secretions, cerebrospinal fluid, synovial fluid, peritoneal fluid, pericardial fluid, pleural fluid, amniotic fluid, and/or saliva. Glass containers filled with such discarded fluids shall be considered sharps;
3. *Sharps:* needles, scalpel blades, hypodermic needles, syringes (with or without attached needles) and needles with attached tubing regardless of contact with infectious agents are considered by the EPA to be regulated medical waste. Other sharps: pasteur pipettes, disposable pipettes, razor blades, blood vials, test tubes, pipette tips, broken plastic culture dishes, glass culture dishes, and other types of broken and unbroken glass waste (including microscope slides and cover slips) in contact with infectious material. Items that can puncture or tear autoclave bags;
4. *Research animal waste:* contaminated carcasses, body parts, and bedding of animals exposed to infectious agents during research or testing. Animal carcasses and body parts not exposed to infectious agents during research or testing are disposed of as nonhazardous waste;
5. *Isolation waste:* biological waste and discarded material contaminated with human or animal body fluids and isolated because they are known to be infected with a highly communicable disease;
6. *Spills:* any material collected during or resulting from the cleanup of a spill of infectious or chemotherapy waste; and
7. *Mixed waste:* any waste mixed with infectious waste not considered to be chemical hazardous waste or radioactive waste.

These various categories are used by every major hospital, company, and university to protect anyone handling potentially hazardous biological wastes.

CHEMICAL WASTES

Scientists began to examine human impact on the planet as the world's population grew towards 6 billion. Chemists have become environmental sleuths. They have the huge task of understanding complex elemental interactions found in wastewater and smokestack gases. It's no surprise that cities have higher metal and acid levels in their wastes and air than rural areas. Scientists are working to piece together the total environmental interrelationship picture. The interconnectedness of all life forms also affects the complexity of environmental waste and hazardous pollution.

Many elements today were discovered using cutting edge technology and equipment. Since the 1960s, many elements added to the Periodic Table (chart of all the known elements) have been manufactured and not found in nature. These atoms have unheard-of uses that many research and applications scientists are just beginning to understand.

Chemists working in the plastics industry came under heavy criticism when landfills got overloaded with disposable plastic containers and a softer compound called *styrofoam.* Environmentalists sounded the alarm for consumers to think before they bought products, especially fast food, that came in these containers.

> Molecules that can be broken down into simpler elements by microorganisms are called **biodegradable.**

In order to meet the new concern, chemists doubled their interest in the biodegradability of plastic products. They found that by adding large complex *carbohydrates* $(C_6H_{10}O_5)_n$ to plastics microorganisms were able to break plastics down.

> **Carbohydrates** make up a large group of organic compounds containing carbon, oxygen, and hydrogen.

Chemical wastes are usually inorganic (without carbon). They include metals like mercury and lead, found to be extremely toxic in high levels to living systems. These refined materials are added to metals, paints, and other products. However, there are also naturally occurring inorganic hazardous compounds like mercury or uranium that are mined and released in large amounts.

There are several sources of hazardous chemical waste. They include batteries, kiln dust, construction debris, crude oil, natural gas, fossil fuel combustion, industry waste, pesticides, fertilizers, medical facilities, and used oil from vehicles and machinery.

Unrecognized early hazards came from a chemical family known as *dioxins.* Called the "most potent animal carcinogens" the EPA had ever evaluated in 1984, dioxins are one of the supreme "bad guy" pollutants. Composed of nearly 100 related organic compounds, the most well known and toxic being *2,3,7,8-*

tetrachlorodibenzo-paradioxin (TCDD), dioxins are known to cause weight loss, birth defects, kidney and liver problems, cancer, and death. *Agent orange,* an herbicide used in the Vietnam War to clear wide areas of jungle foliage, contained dioxin. Veterans of that war suffered many short- and long-term ill effects from contact with the herbicide.

Soil

One of the big chemical waste problems is the possibility of soil integration and migration of hazardous wastes that humans have stored underground. The surrounding soil is contaminated when hazardous wastes in the soil fail to biodegrade in landfills; waste barrels and containers rust, break, and leak; or deep wells or pits develop cracks and fissures. These stored waste contaminants are vulnerable to release since no one can predict an earthquake in years to come. It's possible that no one would still be living or would have records that describe what might be stored beneath or within the soil years into the future.

Water

As we learned in Chapter 8, dumped or buried wastes can leak into the soil and make their way to underground reservoirs. Once there, they contaminate the water as well as seep into surrounding rock. Water taken near or directly from these underground reservoirs is unsafe for use.

RADIOACTIVE WASTES

Land, water, and air can be affected by radioactive contamination. Depending on the wind or water flow, radioactive levels remain in place or are spread over a wide region. Radioactive wastes from uranium mining, production of energy (land-based power plants and nuclear submarines), or weapons development (missiles) are hot environmental issues. Public concern wants responsible *long-term* storage of radioactive wastes until they are safe.

Radioactive elements eventually decay or break down to form harmless materials, but these elements have very different *decay rates.* A few radioactive elements decay in a matter of hours or days, but there are some elements that take thousands of years to decay.

Radioactive decay is referred to in *half-life* periods; the time it takes for one-half of an element's original mass to decay and become harmless. Table 12-2 lists the half lives of several radioactive elements.

Table 12-2 Radioactive decay rates.

Element	Decay rate (half-life)	Element	Decay rate (half-life)
Rhodium-106	30 seconds	Krypton-85	10 years
Tellurium-134	42 minutes	Hydrogen-3	12 years
Rhodium-103	57 minutes	Curium-224	17.4 years
Lanthanum-140	40 hours	Strontium-90	28 years
Radon-222	4 days	Cesium-137	30 years
Xenon-133	5 days	Plutonium-238	87 years
Iodine-131	8 days	Americium-241	433 years
Barium-140	13 days	Radium-226	1622 years
Cerium-141	32 days	Plutonium-240	6,500 years
Niobium-95	35 days	Americium-243	7,300 years
Ruthenium-103	40 days	Plutonium-239	24,400 years
Strontium-89	54 days	Technecium-99	2×10^6 years
Zirconium-95	65 days	Iodine-129	1.7×10^7 years
Ruthenium-106	1 year	Uranium-235	7.1×10^8 years
Cerium-144	1.3 years	Uranium-238	4.5×10^9 years
Promethium-147	2.3 years	Rubidium-87	48.8×10^9 years

While elements are decaying, they give off radioactive energy. This is where the problem of radioactive waste comes in. Strontium (Sr^{90}) and Cesium (Cs^{137}) have half-lives of about 30 years (½ the radioactivity of a given amount of Sr^{90} will decay in 30 years). Plutonium (Pl^{239}) has a half-life of 24,000 years—not easy to handle or store!

Radioactive decay takes place when certain elemental isotopes (element forms) react/collide and there is an emission of energy in the form of radiation (alpha, beta, and gamma particles).

The 3 main types of radiation given off during the breakdown of radioactive elements are the *alpha* (α) and *beta* (β) particles, and *gamma* (γ) rays. Gamma rays are high energy electromagnetic waves like light, but with a shorter, more penetrating wavelength. Though alpha and beta particles are dangerous to living things since they penetrate cells and damage proteins, gamma rays are much more penetrating and harmful, stopped only by thick, dense metals like lead.

Nuclear (or **radioactive**) **waste** is a byproduct from nuclear reactors, fuel processing plants, and institutions such as hospitals and research facilities.

The storage of nuclear wastes during the time they take to decompose to safe materials is an area of high concern and study for governments; they are trying to figure out how to dispose of radioactive wastes from nuclear power plants and atomic weapons.

Radioactive waste also comes from reactors and other nuclear facilities being decommissioned or permanently shut down. The Nuclear Regulatory Commission divides wastes into two categories: high-level or low-level waste.

HIGH-LEVEL RADIOACTIVE WASTE

High-level radioactive waste (e.g. uranium) used in a nuclear power reactor eventually becomes *spent fuel* and no longer efficient in generating power for electricity. Spent fuel, thermally hot as well as highly radioactive, requires remote handling and shielding.

A nuclear power reactor contains Uranium (U^{235}) fuel, in the form of ceramic pellets inside metal rods. Before these fuel rods are used, they are only slightly radioactive and may be handled without special shielding. During the nuclear reaction, the fuel undergoes *fission,* where the nucleus of an atom of uranium splits, releasing two or three neutrons and a small amount of heat. The released neutrons then smack into and split other atoms and a domino effect takes place. This releases huge amounts of heat that is used to generate electricity at nuclear power plants.

The splitting of heavy uranium atoms during reactor operation creates radioactive isotopes of several lighter elements, such as Cesium (Ce^{137}) and Strontium (Sr^{90}), called *fission products.* These daughter products cause the heat and penetrating radiation in high-level waste.

Some uranium atoms also capture neutrons from surrounding uranium atoms to form heavier elements like *plutonium.* These heavier-than-uranium, or *transuranic,* elements produce less heat and penetrating radiation than fission products, but they take a lot longer to decay. *Transuranic wastes* (TRU) are responsible for the majority of radioactive hazard still present in high-level wastes after a thousand years.

High-level wastes are extremely dangerous to humans and other life because their high radiation levels produce fatal doses during short periods of direct exposure. For example, 10 years after being taken out of a reactor, the surface dose rate given off by a typical spent fuel assembly is greater than 10,000 rem/hour (radiation unit of measure). A fatal whole-body dose for humans is around 500 rem in a singe exposure. Reprocessing of high-level waste divides leftover uranium and unreacted plutonium from the fission products. Uranium and plutonium can be reused as reactor fuel. Most high-level waste (other than spent fuel) in the past 35 years has come from fuel reprocessing from government-owned plutonium reactors, as well as naval, research, and test reactors. However, no commercial waste fuel reprocessing is taking place in the United States currently. Most existing commercial high-level waste comes from spent fuel.

LOW-LEVEL RADIOACTIVE WASTE

Low-level radioactive waste includes everything except high-level and uranium recovery/mining wastes. These wastes are stored in surface facilities rather than in the deep geologic sites required for high-level wastes. There have been seven U.S. commercial facilities licensed to bury low-level radioactive wastes. These operations are found in:

- West Valley, New York;
- Maxey Flats near Morehead, Kentucky;
- Sheffield, Illinois;
- Beatty, Nevada;
- Hanford, Washington;
- Clive, Utah; and
- Barnwell, South Carolina.

Today, only Hanford, Clive, and Barnwell are still receiving waste for burial. The other sites have permanently stopped accepting wastes. Long decaying transuranic waste storage is limited at all of the sites. Transuranic wastes (material contaminated with neptunium, americium, and plutonium) contain artificially made elements from spent fuel reprocessing and nuclear weapons.

Protecting populations from accidents in handling or terrorist threats will continue to push much research focus toward the reactivity and degradation of radioactive compounds and elements.

Nonhazardous Waste

Municipal solid waste is the garbage generated by homes, businesses, and institutions. Other kinds of solid wastes include sludge from wastewater treatment plants, water treatment plants, and air pollution control facilities. Additional discarded materials includes solid, liquid, semisolid, or containerized gaseous materials from industrial, commercial, mining, and agricultural operations, and community activities.

> When solid waste does not pose a threat to the environment or human health, it is classified as **nonhazardous waste.**

Although an unknown quantity of solid waste is managed by individuals and organizations, a recent Environmental Research and Education Foundation (EREF) study listed the amount of solid waste managed off-site in the United States at 544.7 million tons (reported by waste facilities). Of these 544.7 million tons of solid wastes, 63% was in municipal solid waste landfills, 21% in material recycling facilities, 6% in incinerators, 5% in construction and demolition landfills, and 5% in compost facilities. Private companies own 5% of the off-site facilities that manage solid waste and the public sector owns 47%. Fig. 12-1 shows a few of the nonhazardous wastes treated at waste disposal facilities.

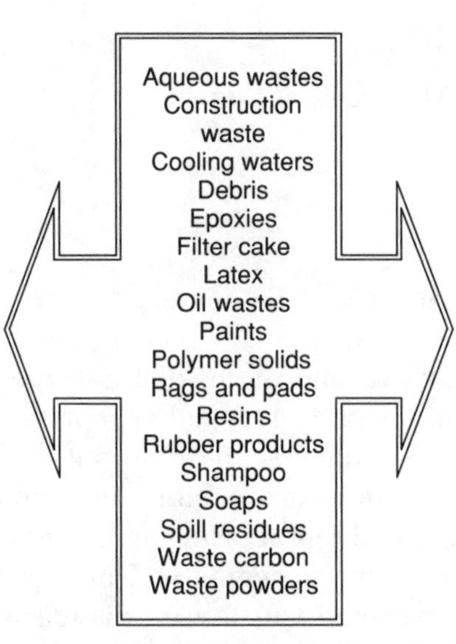

Fig. 12-1 There are a number of wastes that are considered to be nonhazardous waste.

Responsible waste processing facilities handle nonhazardous waste as carefully as hazardous waste. Nonhazardous waste arrives at facilities in a variety of containers including fiber and plastic drums, original packaging, bags, and shrink-wrapped or steel-strapped pallets, along with steel drums, cubic-yard boxes, railcars, tank trucks, roll-off boxes, and dump trailers.

Disposal methods depend on the waste type, but commonly include incineration, sludge-dewatering, wastewater treatment, waste-to-energy, secure landfill, and others.

Making a Difference

Sometimes it seems like the buildup of hazardous waste is beyond our individual efforts. However, by demanding responsible environmental processing, treatment, and storage, individuals can have a positive impact.

There are plenty of things we can do about nonhazardous environmental problems. These fall into the larger category of *reuse and recycling*.

> **Reuse** and **recycling** keep materials and products out of the waste disposal cycle for a longer time, while preserving natural resources.

As individuals, we can buy recycled paper products so that original wood fibers are used again and again. The same thing can be done with recycled glass and plastics. And it's easy! Many cities have recycling programs that allow people to recycle right at their door or driveway.

Ink cartridges from computer printers and electronics parts can be recycled. Used oil from vehicles, paints, and other products can be taken to approved recycling and disposal sites.

In order to make garbage disposal costs fair, some municipal areas have adopted "pay-as-you-throw" policies. Those people who don't want to recycle and produce high trash and waste have to pay more than people who are lowering their waste output through recycling.

↑ Household Waste = ↑ $$$
↑ Reuse + ↑ Recycling = ↓ $$

Some people argue that reuse and recycling is a bother, but if you think about the destruction of essential global forests that are cut down for paper to print bills or advertisements (junk mail), the need comes into sharper focus. In fact, in today's electronic information age, print media is becoming less and less important.

Waste can be reduced in terms of space. Trash *compaction* is another easy way of reducing landfill volume and disposal costs. Although trash is highly compacted at landfill sites, the initial compacting at home helps make landfill sites more productive, as well as lowering processing costs and pollution from bull dozers and machinery used at landfill sites.

Composting by collecting organic waste (potato, banana and orange peels, watermelon rinds, etc.) and reusing the degraded material as fertilizer and soil nutrients is another good way to recycle.

There is a popular slogan that helps focus environmental efforts: "reduce, reuse, and recycle." It reminds us that our global resources are not infinite as people once thought.

Reusing paper bags and plastic containers is easy and helps lower the resource hit of creating new ones. During past severe economic times, like the U.S. Great Depression (1930s), money was scarce and everything was reused and recycled. Many people today still remember those difficult times and continue to recycle many products.

Quiz

1. When solid waste does not pose a threat to the environment or to human health it is classified as
   (a) hazardous waste
   (b) fluid waste
   (c) nonhazardous waste
   (d) transuranic waste

2. The Comprehensive Environmental Response, Compensation, and Liability Act is commonly called the
   (a) Supermarket
   (b) Super duper
   (c) Clean Air Act
   (d) Superfund

3. The 1962 environmental book *The Silent Spring* was written by marine biologist
   (a) Rachel Carson
   (b) Jack Showers

(c) Marvel Calkin
(d) Ralph Nader

4. Cesium (Ce^{137}) and Strontium (Sr^{90}) are

(a) often added to toothpaste
(b) fission products
(c) both elements that burn yellow in a flame
(d) nonhazardous waste products

5. Highly flammable, corrosive, reactive (explosive or unstable), or toxic wastes are considered

(a) nonhazardous
(b) low priority
(c) hazardous
(d) lively

6. These types of waste are responsible for the highest radioactive danger found in high-level wastes after a thousand years:

(a) silicate
(b) transuranic
(c) biological
(d) landfill

7. Sharps are classified as what kind of hazardous waste?

(a) Biological
(b) Meteorological
(c) Claustrophobic
(d) Transuranic

8. Molecules broken down into simpler elements by microorganisms are said to be

(a) quantum strings
(b) biodegradable
(c) inorganic esters
(d) buckyballs

9. In order to make garbage disposal costs fair, some municipal areas have adopted
   - (a) higher taxes
   - (b) mobile landfill facilities
   - (c) "pay-as-you-throw" policies
   - (d) larger trash cans
10. Nuclear waste comes from all of the following sources except
    - (a) nuclear reactor byproducts
    - (b) research facilities
    - (c) hospitals
    - (d) hair salons

Part Three Test

1. The difference between mass wasting and erosion is that erosion happens
   (a) over much greater distances
   (b) over short distances
   (c) only with water
   (d) only with wind

2. The slow or sudden movement of rock down slope as a result of gravity is called
   (a) intrusion
   (b) topography
   (c) extrusion
   (d) mass wasting

3. When rock disintegrates and is removed from the surface of continents, it's called
   (a) superimposition
   (b) denudation
   (c) lithification
   (d) sonification

4. Slower, long-term soil or rock movement with a series of sliding surfaces and plasticity is known as
   (a) creep
   (b) slip
   (c) slither
   (d) a rolling stone gathers no moss

5. The repeated freeze-thaw cycle of water in extreme climates causes
   (a) lava flows
   (b) increased plant growth
   (c) frost wedging
   (d) colds and flu

6. The popular slogan that focuses environmental efforts is
   (a) chop, slash, and burn
   (b) location, location, location
   (c) all pesticides all the time
   (d) reduce, reuse, and recycle

7. Overplanting, overgrazing, deforestation, and poor irrigation methods are all factors that can lead to
   (a) earthquakes
   (b) tsunamis
   (c) desertification
   (d) fossilization

8. Dry streambeds are also known as
   (a) channels
   (b) plateaus
   (c) wadis
   (d) saguaros

9. What has the greatest effect on mass wasting?
   (a) Water
   (b) Gravity
   (c) Wind
   (d) The sun

10. A raindrop falls through still air and hits the soil at about
   (a) 25 km/hour
   (b) 32 km/hour

(c) 38 km/hour
(d) 42 km/hour

11. High concentrations of dissolved calcium and/or magnesium in fresh water cause what is commonly known as
    (a) soft water
    (b) hard water
    (c) recycled water
    (d) mineral water

12. Remote sensing has been important as a tool to study dynamic features of deserts, like dunes, for the past
    (a) 25 years
    (b) 40 years
    (c) 50 years
    (d) 70 years

13. The flat area of clay, silt, or sand covered with salt in a dried desert lake is called a
    (a) continental shelf
    (b) playa
    (c) abyss
    (d) scarp

14. What measurement do hydrologists use to check the health of fresh water?
    (a) Catch limit
    (b) Temperature
    (c) Total organic carbon
    (d) Salinity

15. All of the following are classified by OSHA as infectious waste except
    (a) cultures and stocks
    (b) sharps
    (c) recycled geothermal water
    (d) human blood

16. Who is known as the Mother of the Modern Environmental Movement?
    (a) Shirley Temple Black
    (b) Florence Nightingale
    (c) Marie Curie
    (d) Rachel Carson

17. Depending on composition and texture, weathering that occurs at different rates in the same geographical region is known as
    (a) drought
    (b) global warming
    (c) differential weathering
    (d) sequential weathering

18. Pedocal, pedalfer, and laterite are 3 types of
    (a) pesticide
    (b) hydroelectric power plants
    (c) soil
    (d) erosion

19. Biological, chemical, and radioactive are three major classes of
    (a) hazardous waste
    (b) solar energy
    (c) sedimentary rock
    (d) facial peels

20. Molecules that can be broken down into simpler elements by microorganisms are said to be
    (a) halophytes
    (b) nonreactive
    (c) infectious
    (d) biodegradable

21. When fine soil grains are carried as dust during windy conditions, it is known as
    (a) dust devils
    (b) smoking paddocks
    (c) tornadoes
    (d) tsunamis

22. Frost wedging is primarily a form of
    (a) chemical weathering
    (b) cosmic weathering
    (c) biological weathering
    (d) physical weathering

23. Most hot and dry deserts have humidity levels around
    (a) 10%
    (b) 30%

(c) 50%
(d) 70%

24. When land is shaped by water's dissolving action on carbonate bedrock, like limestone, dolomite, or marble, it forms a
(a) rabbit hole
(b) geothermal geyser
(c) volcano
(d) karst

25. Rock slides take place
(a) when large amounts of rock free-fall from very steep areas of a slope
(b) only on the ocean floor
(c) across sandy beaches
(d) at the same time as snowstorms

26. The carbon cycle, where carbon exchanges take place, has many different storage spots, known as reservoirs or
(a) ditches
(b) caves
(c) buckets
(d) sinks

27. The law that established the first national hazardous waste identification and remediation program is commonly known as the
(a) Clean Air Act
(b) Superfund Act
(c) National Parks Act
(d) Love Canal Fund

28. A reddish-brown–to–white layer found in many desert soils is called
(a) rojo
(b) desert dung
(c) caliche
(d) caliente

29. The breakdown of large rocks into smaller bits that have the same chemical and mineralogical makeup is called
(a) rotational weathering
(b) mechanical weathering
(c) chemical weathering
(d) gravel

30. The Nuclear Regulatory Commission divides waste into two categories:
    (a) hot and cold waste
    (b) high-level and low-level radioactive waste
    (c) dry and wet waste
    (d) biological and technical waste

31. When calcium carbonate is used by marine inhabitants to build shells, it is called
    (a) biomineralization
    (b) chitinization
    (c) carbonization
    (d) conchization

32. Desert plants that adjust to high salt levels are known as
    (a) shrubberies
    (b) neophytes
    (c) geophytes
    (d) halophytes

33. The most important natural acid, formed when carbon dioxide dissolves in water, is called
    (a) acetic acid
    (b) carbonic acid
    (c) hydrochloric acid
    (d) ascorbic acid

34. All of the following help to control erosion, except
    (a) limiting tillage and direct-drilling practices
    (b) controlling wind access to the soil
    (c) increasing the frequency of field use
    (d) rotating livestock grazing

35. When elements move from one Earth storage form to another, it is known as a
    (a) geochemical cycle
    (b) rock slide
    (c) hurricane
    (d) point source

36. Which dune is horseshoe-shaped?
    (a) Transverse
    (b) Barchan

(c) Linear
(d) Billow

37. The soggiest type of soil, found in tropical and subtropical climates, is high in organic matter and known as
    (a) mud
    (b) gumbo
    (c) pedalfer
    (d) laterite

38. What physical weathering type is an important rock-breaking force in wet climates?
    (a) Laterite
    (b) Biological decay
    (c) Frost wedging
    (d) Typhoons

39. Sand covers around what percentage of the Earth's deserts?
    (a) 20%
    (b) 30%
    (c) 40%
    (d) 50%

40. Hazardous and toxic chemicals are generated by all of the following except
    (a) households
    (b) sand dunes
    (c) agriculture
    (d) industry

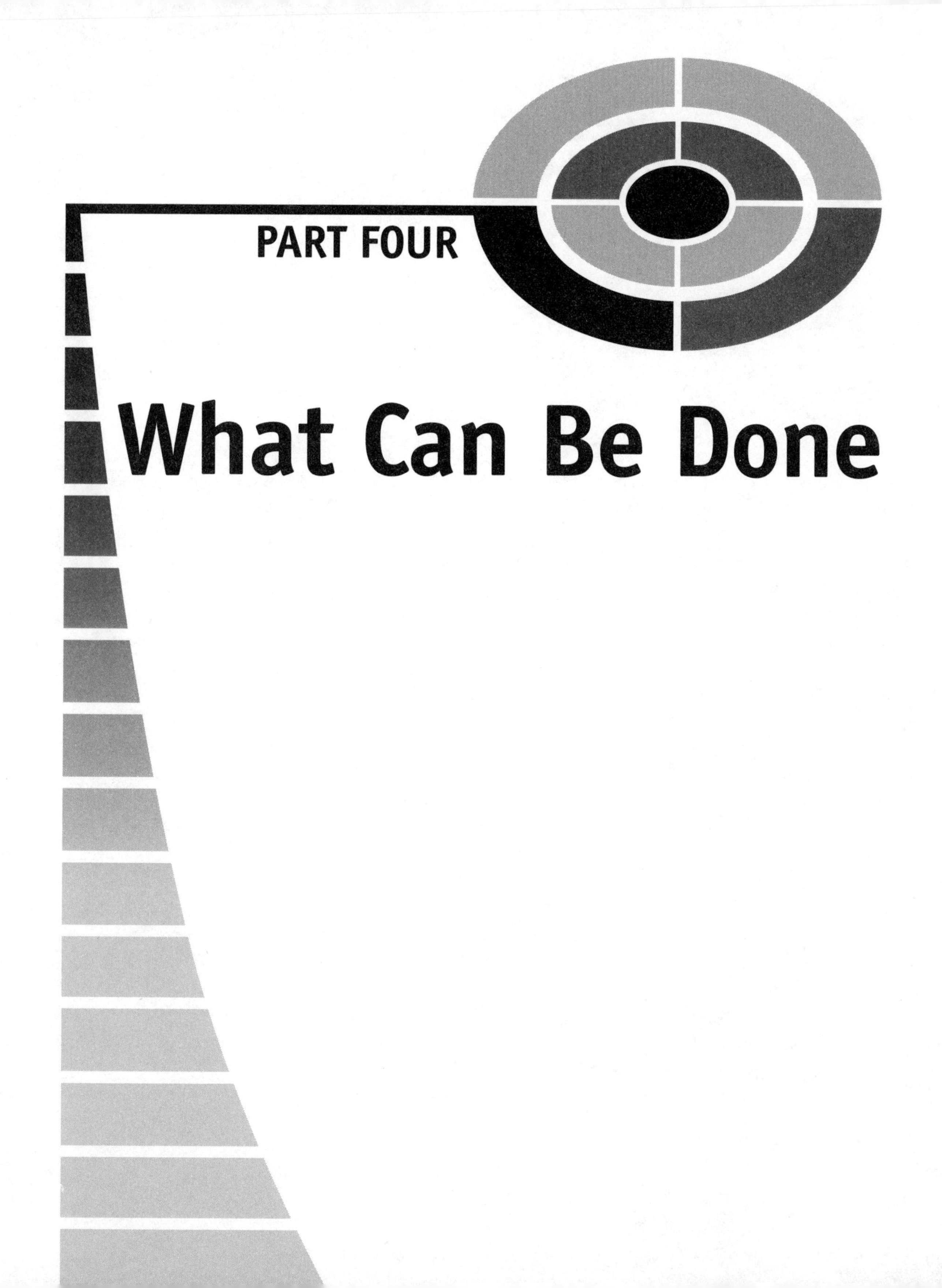

PART FOUR

What Can Be Done

CHAPTER 13

Fossil Fuels

Everyone uses the term "fossil fuels," but what are fossil fuels anyway? Very simply, they are fuels made from mining or pumping out of the ground prehistoric dead stuff. However, all ancient organic material (fossils) didn't turn into oil. Specific geological factors must be present within oil-rich rocks. Solid rock (for containment) and a seal (salt or clay) that keeps the oil from rising to the surface must be present. It is estimated that, under these conditions, only 2% of all organic material gradually becomes oil. Oil can be as thin as gasoline or as thick as tar.

> **Petroleum** (oil) is considered a nonrenewable energy source because it takes millions of years to form.

First Oil Use

Humans have used oil since ancient times. The ancient Chinese and Egyptians burned oil for lighting.

In 1839, Abraham Gesner, a governmental geologist in New Brunswick, Nova Scotia, discovered *albertite* (a solid coal-like material). After immigrating to the United States, Gesner developed and patented a process for manufacturing *kerosene*. He is often called the *Father of the Petroleum Industry*.

Before the 1800s, however, light was produced from torches, tallow candles, and lamps that burned animal fat. Because it burned cleaner than other fuels, whale oil was popular for lamp oils and candles. However, it was expensive. A gallon cost about $2.00, which today would cost around $760 per liter ($200/gallon). And we think our oil is expensive!

As whale oil became more and more expensive, people started looking for other fuel sources. In 1857, Michael Dietz invented a clean-burning kerosene lamp. Almost overnight, whale oil demand dropped. Most historians and ecologists believe that if kerosene had not come onto the market, many whale species would have quickly become extinct from overhunting.

Kerosene, known as "coal oil," was cheap and smelled better than animal fat when burned. It also didn't spoil like whale oil. By 1860, around 30 U.S. kerosene plants were in production. Kerosene was used to light homes and businesses before the invention of the electric light bulb by Thomas Edison in 1879.

Oil Wells

The first oil well in North America was the Oil Springs, Ontario, well in 1855. Between 1858 and 1860, nearly 1.5 million liters of crude oil was shipped to buyers in the United States and elsewhere.

In the United States, oil discovery peaked in 1930 with the discovery of the East Texas field at Spindletop that produced 12,000 m^3 of oil/day (80,000 barrels/day). U.S. peak production came in 1970. Similarly, worldwide oil discovery peaked in 1964, and peak production is estimated to occur by 2010.

One of the biggest problems with fossil fuels, besides their effect on global warming, is our dependence on them: Oil provides around 40% of the world's energy needs and about 90% of its transport fuel. When it's used up and eventually runs out, our way of life and travel will grind to a stop.

> **Fossil fuels** are solids, liquids, and gases created through the compression of ancient organic plant and animal material in the Earth's crust.

Fossils fuels are hydrocarbons formed into coal (solid), oil (fluid), and natural gas (mostly methane). These can be burned and used as fuels by themselves

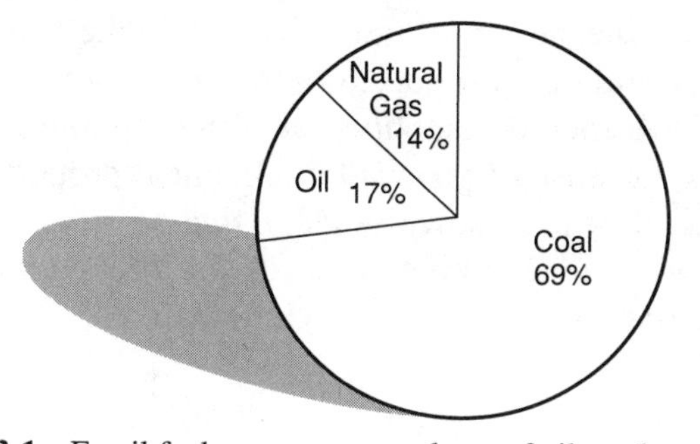

Fig. 13-1 Fossil fuel reserves are made up of oil, coal, and gas.

or processed to produce purer products like propane and gasoline. Fossil fuels (oil and natural gas) are also used in the petrochemical industry to make chemicals, plastics, and fertilizers. The colossal problem facing the countries of the world in the next 10 to 30 years is that fossil fuels are running out and we need one or more efficient, nonpolluting replacements. Fig. 13-1 provides a snapshot of the U.S. Department of Energy's (DOE) estimated global fossil fuel energy resources available as of January 1, 1996. Since we have continued, and in many cases, increased fossil fuel use, quantities are even smaller.

The burning of fossil fuels is probably the biggest single source of air pollution in the industrialized world. Reducing the amount of smoke, ash, and combustion products from fossil fuel burning is critical to the future of life on Earth. Besides the fact that the use of fossil fuels is a huge factor in air pollution, it causes big problems (slicks, tar) in the water as well.

In the first part of this chapter, we'll look at current global fossil fuel levels and energy alternatives. In the second part of the chapter, we'll study ongoing fossil fuels problems (spills) and ways to help.

Energy

Fossil fuels were first seen as a convenient, virtually limitless source of energy. Some oil bubbled up out of the ground on its own without any human effort at all! This was a lot easier than mining coal from deep mines with the related dangers of cave-ins and toxic dust. Fossil fuels were found around the world in the form of liquid petroleum that could be used and transported by planes, train, automobiles, ships, and pipelines.

The most common fuels used for transportation in the United States are gasoline and diesel fuel, but several additional energy sources are able to power motor vehicles. These include alcohols, electricity, natural gas, and propane. When vehicle fuels, because of physical or chemical properties, create less pollution than gasoline, they are known as *clean fuels.*

In this chapter we will examine some of the fuels available today, while exploring in greater detail potential energy alternatives.

Oil Demand

Since World War II, petroleum has taken the place of coal as the United States' leading energy source. Today, it provides more than 39% of the energy used in the United States (coal 22% and natural gas 23%).

Americans use nearly 2,646 million liters (700 million gallons) of petroleum per day. And that number is growing. In fact, we use 12% more transportation oil than we did in 1973, even though today's vehicles get more miles per gallon than 1970s models. There are 50% more vehicles in operation today than in the 1970s.

So how is the United States oil industry keeping up with demand? Foreign oil from oil-producing countries like Saudi Arabia has filled the gap. In 1994, the United States bought 45% of its petroleum from other countries and that percentage increases between 1 and 2% each year. In 2004, the United States imported nearly 12 million barrels of oil/day from foreign oil sources (up 2% from 2003). The 2005 worldwide demand for oil is estimated at 21 million barrels of oil/day (an increase of over 3% from 2004). But these foreign oil supplies aren't limitless.

HUBBERT'S PEAK

Fossil fuels have been exploited worldwide for decades following many technological advances. However, it wasn't until large cities became sooty, dirty places from fossil fuel burning that people began to think there might be a better way. They debated whether it was easier to stick with fossil fuels and their problems or switch to a better energy source. Scientists began to study the world's energy needs in greater depth.

In 1956, Shell Oil Company geophysicist M. King Hubbert calculated that the oil well extraction rate in the United States (lower 48 states) would peak around

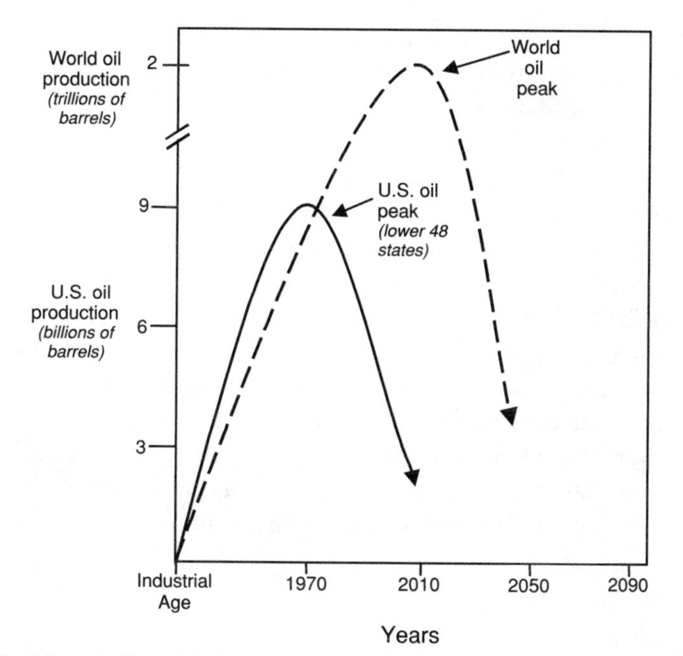

Fig. 13-2 United States' oil production peaked in 1970, as Hubbert predicted from his calculations.

1970 and drop from then on. At the time, people didn't believe him. Hubbert was highly criticized by oil experts and economists, but it turned out he was right. Oil production peaked at around 9 billion barrels/day in 1970, with today's production at less than 6 billion barrels/day and falling. Fig. 13-2 gives an idea of the United States oil production peak that Hubbert calculated. The projected world production peak is also shown.

In 2001, coal, oil, and natural gas provided 86% of the energy used by all the people of the world. The *International Energy Outlook 2004* (IEO 2004), published by DOE, contains a projection that, in 2025, fossil fuels will provide 87% of the energy consumed globally. The IEO 2004 also projects that global energy consumption will steadily increase by an average of 1.8% per year from 2001 to 2025.

Because the industrialized world depends on fossil fuels in a big way, everyone is trying to figure out what quantitites of our fossil fuels are left. Plus, they want to know the number of years left before all known reserves of fossil fuel (coal, oil, and natural gas) are used up, based on world consumption rates.

If fossil fuels can only provide an additional 1% of global energy needs by 2025, where will the additional energy needed come from? Scientists are not just figuring out the lifetime of fossil fuels (how long it is economically and physically recoverable), but studying Hubbert's calculations to discover the worldwide peak production date.

Where most people get confused is that even if we use all the fossil fuels available, there will always be some left in the Earth. The big problem is that we can't get to it. There are technological as well as cost limitations.

Geologists, performing calculations with updated global oil data, have discovered that global oil production will peak by 2010. The numbers are clear. Of the approximately 2.5 trillion barrels of the Earth's total supply of stored oil (United States Geological Survey estimate), nearly 50% of it has already been used. Since the peak is at the halfway point and fossil fuel consumption is still rising around the world, the downhill slide could be quick. The time for energy alternatives is now! Just like when you drive your car, you can't wait until you run out of gas to stop at a station and get more. Fossil fuel drain could be slowed if some of the demand is lowered by new energy sources and more efficient methods.

ELECTRICITY

Battery-powered electric vehicles are a great option for many commuters. Practically pollution-free, they offer one of the best options for lowering vehicle emissions in polluted cities. Unfortunately, power plants that create electricity do pollute. But efficient emission control measures are more easily installed and maintained on individual power plants than on millions of vehicles.

Even though the driving range of electric cars is limited by low-power vehicle batteries, cutting-edge research is being done in this area. By creating batteries that take minutes instead of hours to recharge and run for longer distances on one charge, electric power could become a widespread, "clean fuel" transportation option for future decades.

ETHANOL

Ethanol (grain alcohol) is the main vehicular fuel in Brazil, and ethanol/gasoline blends, known as *gasohol,* have been used in the United States for many years. Pure ethanol fuel gives great performance, and has the added plus of low hydrocarbon and toxic emissions. Ethanol can be produced from corn or other crops as well as from wood or paper wastes. Since these crops pull carbon dioxide out of the atmosphere as they grow, they decrease greenhouse gas buildup.

Ethanol is now used in the United States as a fuel additive (oxygenate) to reduce incomplete combustion emissions. However, it may have very serious repercussions as a groundwater contaminant.

The problem with ethanol, a fairly nonpolluting alternative, is its ability to increase the polluting effects of other compounds, like benzene. Ethanol can act as a solvent and has been found to slow the breakdown of benzene, toluene, and other chemicals in the soil and groundwater. This is a problem because the longer highly toxic compounds, like benzene, stay in the soil, the greater the chance of public contact and health risk.

With current technology and pricing, ethanol is more expensive than gasoline, but as fossil fuels decrease and new processing is developed, the price differences will be eliminated. Eventually, oil and gasoline will be so expensive that any alternative will be a bargain.

METHANOL

Methanol (wood alcohol), like ethanol, is a high-performance liquid fuel that releases low levels of toxic and ozone-forming compounds. It can be made for about the same cost as gasoline from natural gas, wood, and coal. All major auto manufacturers have produced cars that run on M85, a blend of 85% methanol and 15% gasoline. Cars that burn pure methanol (M100) offer much greater air quality and efficiency advantages. Many auto manufacturers have developed advanced M100 prototypes. Methanol is the fuel of choice for race cars because of its superior performance and fire safety characteristics.

COAL

In 2003, coal provided 24% of the energy used by the world's people. *The International Energy Outlook 2003* (IEO 2004) projects that in 2025 coal will provide only 23% of the energy consumed globally. Global coal use, however, is projected to increase by 1.5% per year between 2001 and 2025.

Global reserves of coal equal 1083 billion short tons (Bst). The United States has 271 Bst of this total. Since coal is plentiful in the United States, some people think of it as our backup energy source. Currently, many electrical generating plants burn coal. Unfortunately, burning coal creates a lot of atmospheric pollution and particulates, and increasing its use will only worsen the global greenhouse problem. It's kind of a win-lose situation.

BIOMASS

When the sun's energy is stored within materials like plant and animal matter (organic compounds), it's known as *biomass.* Biomass is considered renewable because it is made more quickly than the fossil fuels that take millions of years to form. There are a wide variety of biomass fuel sources, including agricultural residue, pulp and paper mill remains, construction waste, forest residue, energy (corn) crops, landfill methane, and animal waste.

Energy in the form of electricity, heat, steam, and fuel can be obtained from these sources through conversion methods, like direct combustion boiler and steam turbines, anaerobic digestion, co-firing, gasification, and pyrolysis. The *co-firing method,* which mixes coal and biomass, may be biomass' best economic alternative, particularly in mixed heat and power applications.

NATURAL GAS

According to DOE estimates (IEO 2004), natural gas provided 23% of all energy used worldwide in 2003. It was projected that in 2025, natural gas will provide 25% of the energy consumed globally, with global natural gas consumption expected to grow by 2.1% per year from 2001 to 2025.

Global natural gas reserves are roughly 6076 trillion cubic feet (Tcf). Geologists estimate that the United States has around 2400 Tcf of natural gas resource reserves left. Of this amount, natural gas is mostly located in low-permeability formations, shales, coal beds, and existing fields.

Natural gas (methane) is plentiful and widely used for home heating and industrial processes. It is easily transported through pipelines and costs about the same or less than gasoline. Compressed natural gas (CNG) vehicles give off low levels of toxins and ozone-forming hydrocarbons. However, since CNG fuel must be stored under pressure in heavy tanks, the cost of building these specialized tanks can be limiting.

There are significant tradeoffs for CNG vehicles among emissions, vehicle power, efficiency, and range; however, natural gas is already used in some fleet vehicles and appears to have a great future as a motor vehicle fuel.

PROPANE

Propane, or liquefied petroleum gas (LPG), is a byproduct of petroleum refining and natural gas production. It burns more cleanly than gasoline, but is limited in supply. Propane-fueled vehicles are already common in many parts of the world.

NEW GASOLINE TYPES

The petroleum industry is beginning to market gasoline mixes that give off fewer hydrocarbons, nitrogen oxides, carbon monoxide, and toxics than standard gasoline. These new gasolines can be used without huge changes to existing vehicles or fuel distribution systems.

According to the US DOE and EPA, less than 15% of the energy in vehicle fuel is used to move a car down the road or run accessories like air conditioning or power steering. The rest is lost to waste heat, the friction of moving engine parts, or pumping air into and out of the engine. Of the energy in 1 gallon of gasoline, 62% is lost to engine friction, engine pumping losses, and to waste heat. In urban stop-and-go driving, another 17% is lost in traffic. Vehicle accessories like the water pump or air conditioning use another 2%.

With fossil fuels being used up, environmental scientists are spending more and more time on energy use forecasts, but pollution reduction is still critically important.

Spills

Everyone who has ever made oil and vinegar salad dressing knows that oil and water don't mix, but rather separate and layer because of their different densities. Oil spills in lakes, rivers, open seas, and the world's oceans cause the same thing to happen. Waste and/or dumped oil spreads across water and coats shorelines, creating a sticky mess. Spills foul the water for everything from single-celled organisms and birds to sea mammals and humans.

Oil spills into rivers, bays, and oceans are caused mostly by crashes and crunches involving tankers, barges, pipelines, refineries, and storage facilities. This happens when oil is transported to end users. Spills can be caused by:

- Human error or carelessness;
- Vandals;
- Equipment malfunction;
- Natural disasters like hurricanes;
- Terrorists acts;
- Acts of war between countries; or
- Illegal dumping.

Although oil floats on salt water (oceans) and fresh water (rivers and lakes), super heavy oil sometimes sinks in fresh water. This creates its own set of problems at multiple water depths, not only for fish and marine mammals, but bottom-dwelling creatures as well.

Unfortunately, spilled oil spreads quickly across the water's surface to form a thin layer, called an *oil slick.* As the oil continues to spread, the slick layer gets thinner and thinner, finally becoming a very thin layer called a *sheen,* which reflects rainbow colors in certain lights. After a rain, sheens are common in parking lots and on roads, where oil has previously dripped from vehicles.

Following an oil spill, a variety of local, state, and Federal agencies as well as volunteer organizations respond to the incident, depending on who's needed. Size, location, and extent of a spill determine the tools needed to clean up spilled oil. Some of these include:

- Booms (floating oil barriers placed around a tanker leaking oil to collect the oil);
- Skimmers (boats that skim spilled oil from the water's surface);
- Sorbents (huge sponges used to absorb oil);
- Chemical dispersants and biological agents (enzymes or microorganisms that clump or break oil down into its simpler chemical components);
- In-situ burning (burning of freshly spilled oil from the water's surface);
- High- or low-pressure washing of oil from beaches with water hoses;
- Vacuum trucks (vacuuming spilled oil from beaches and/or the water's surface; or
- Shovels and heavy road equipment (to collect oily beach sand and gravel for disposal or cleanup).

Oil spill methods and tools depend on several factors. The weather, type and amount of oil spilled, currents, distance from land, types of bird and marine habitats in the area, and population of an area are all taken into consideration. Assorted cleanup methods work on different beach types and with different kinds of oil. For example, heavy road machinery works well on sandy beaches, but not in marshes or on rocky beaches with big boulders.

In-situ burning (ISB) involves the controlled burning of oil that has spilled from a vessel or facility, or burning oil on the vessel itself, before it spills into the environment.

In some cases, work stations are set up to clean and rehabilitate wildlife. Although it doesn't happen often, there are times when agencies don't respond to a spill at all, because responding is pointless or would cause even more spill damage.

RUN OFF AND LEAKAGE

Oil end users, not ships and pipelines, account for 85% of the petroleum pollution in North American oceans, according to a 2004 research study released by the National Research Council (NRC). The NRC is the operating agency of the congressionally charted National Academy of Sciences (NAS) in Washington, D.C.

The NRC report, "Oil in the Sea: Inputs, Fates, and Effects," reports that 109,800 kiloliters of petroleum pour into North American waters each year. Much of it comes from land-based run-off, polluted rivers, boats, and other recreational watercraft. Huge oil slicks and black, sticky beaches caused by shipping spills and accidents get more press, but point sources leaking smaller amounts of petroleum daily are the greater villains of sea and ocean oil pollution. Oil spills are responsible for 8% of the annual dump of petroleum into the seas, while oil drilling and extraction adds another 3%.

In addition to manmade oil spills, the study determined that nearly 178,000 kiloliters of oil from natural geologic formations on the sea floor seep into the ocean yearly. The upside is that the NRC found that there is less total petroleum in the oceans today than reported in a similar 1985 NRC study. Worldwide, the report says, 795,000 kiloliters of oil from manmade sources flowed into the oceans, with another 681,000 kiloliters from natural seepage.

When the Iraqi army (occupying Kuwait in January 1991) began destroying tankers, oil terminals, and oil wells, nearly 9 million barrels of oil were spilled into the Arabian Gulf, forming a 600-square-mile oil slick. Four hundred miles of the western Gulf shoreline was slicked with an unknown amount of sinking oil. Tar layers as much as 18 centimeters thick formed on beaches. Cleanup operations in April 1991 estimated that roughly 1 million barrels of oil were removed from Arabian Gulf waters and shores.

Wildlife Impact

Have you ever played on a beach that was littered with tar balls? Even sticky dime- and quarter-sized black blobs of tar can be tough to get off your feet when you get home. Can you imagine swimming in waters that smell of gasoline and that have the consistency of thick molasses or honey coating your entire body? Welcome to the world of birds, mammals, fish, and shellfish affected by an oil spill.

Nearly all nearby biological populations are impacted by oil spills. Those susceptible are plant communities on land, marsh grasses in estuaries, and kelp beds in the ocean; microscopic plants and animals; and larger animals, such as fish, amphibians, reptiles, birds, and mammals. They can't escape contact, smothering, toxicity, and long-term effects from the physical and chemical nature of spilled oil.

Oil destroys the insulating ability of fur-bearing mammals like sea otters and the water-repelling abilities of bird feathers. Damage appears to be directly related to how important the fur and blubber are to staying warm, a process known as *thermoregulation.* Without this protection, these animals are exposed to the cold. Mammals that have been affected in the past include river otters, beavers, sea otters, polar bears, manatees, seals, sea lions, walrus, whales, porpoises, and dolphins. These mammals need clean fur or hair to stay warm.

If oil is swallowed, it can cause stomach and intestinal bleeding, as well as liver and kidney damage for species that groom themselves like birds. After a habitat has been slammed by a big spill, all of the following are present: dehydration from the elimination of clean water; disorders and destruction of red blood cells from oil ingestion; pneumonia from oil vapor inhalation; skin and eye irritation from direct contact with oil; and impaired reproduction.

Breathing hydrocarbon vapors results in nerve damage and behavioral abnormalities in all birds and mammals. A large spill can result in the injury and death of hundreds or thousands of birds and mammals. Additionally, eggs and young are especially susceptible to oil contamination. Tiny amounts of oil on bird eggs can also kill embryos.

Fish are exposed to spilled oil in three main ways: 1) direct contact that contaminates their gills; 2) exposure of eggs, larvae, and young to water that contains toxic and volatile chemicals; and 3) eating contaminated food. Fish exposed to oil suffer from changes in heart and respiratory rate, enlarged livers, reduced growth, fin erosion, a variety of biochemical and cellular changes, and changes in reproductive and behavioral responses. Chronic exposure to chemicals found in oil can cause genetic abnormalities and cancer in sensitive species.

If chemicals such as dispersants are used to treat a spill, there may be a higher impact on fish and shellfish by increasing the concentration of oil at different depths. This also affects humans in areas with commercial and recreational fisheries.

TAR BALLS

When crude oil drifts on the ocean's surface, its physical properties change. During the first few hours of a spill, the oil spreads into a thin slick. Winds and waves tear the slick into smaller patches that get scattered over a much wider area. Various physical, chemical, and biological processes change the appearance of the oil. These changes are known as *oil weathering.*

Winds and waves continue to pull and rip the spilled oil into smaller bits or tar balls. While some tar balls are as large as a plate, most are much smaller. These small blobs of petroleum tar last a long time in open waters and travel hundreds of miles.

> **Tar balls,** small brown-black blobs of oil that stick to your feet when you go to the beach, are leftovers of oil spills.

The number of tar balls found on a particular beach depends on many factors including tanker traffic, wind patterns, sea currents, the last occurrence of an oil spill, and how often the beach is cleaned.

Weathering forms tar balls that are hard and crusty on the outside and soft on the inside, like a chocolate candy with a caramel center. Wave or beach activity (birds, animals, or people) may break tar balls open, exposing their sticky centers.

Although research is ongoing with regard to tar ball formation, it is known that temperature is important to tar ball stickiness. Like honey, tar balls become more fluid as they are warmed. Particulates and sediments in the surrounding water can also stick to tar balls and make them more solid. The more sand and debris layers on the surface of a tar ball, the harder it is to break it open.

The National Oceanographic and Atmospheric Administration (NOAA) recommends that people especially sensitive to chemicals, like hydrocarbons found in crude oil and petroleum products, avoid beaches with tar balls as they may develop allergic reactions or rashes even from brief contact. Should contact occur, wash the area with soap and water, baby oil, or a safe cleaning compound like the cleaning paste sold in auto parts stores. Don't use solvents, gasoline, kerosene, diesel fuel, or similar products on the skin. These products, when applied to skin, present a greater health hazard than the smeared tarball itself. Plus, they are petroleum products too—allergens!

As we've seen, there are plenty of disadvantages to the continued use of fossil fuels as a energy source. The hazardous pollutants released from fossil fuel combustion and oil spills are slowly poisoning the Earth.

However, since the passage of the Clean Air Act in 1970, things have been getting better. Oil companies have started lowering air and water emissions produced by oil refineries. Gasoline mixtures burn more cleanly with less lead, nitrogen oxide, carbon monoxide, and hydrocarbons released into the air.

Without a doubt, our environmental situation is a challenge. Current and future generations must balance the growing oil demand with responsible protection of the environment. But luckily, we have many promising options.

Quiz

1. Fossil fuels were first seen as a
   (a) a quirk of nature
   (b) passing energy phase
   (c) sticky mess
   (d) convenient, virtually limitless source of energy

2. When vehicle fuels, because of physical or chemical properties, create less pollution than gasoline, they are known as
   (a) clean fuels
   (b) blue fuels
   (c) super fuels
   (d) strange fuels

3. Tanker traffic, wind patterns, sea currents, the last occurrence of an oil spill, and how often the beach is cleaned all affect the quantity and size of
   (a) sea gulls
   (b) tar balls
   (c) sun bathers
   (d) gum balls

4. Oil end users cause what percentage of petroleum pollution in North American oceans, according to a 2004 National Research Council research study?
   (a) 20%
   (b) 45%
   (c) 60%
   (d) 85%

5. A high-performance liquid fuel that releases low levels of toxic and ozone-forming compounds is
   (a) gasoline
   (b) decanol
   (c) methanol
   (d) molasses

6. Booms, skimmers, sorbents, chemical dispersants, and biological agents are all used to
   (a) contain and clean up oil spills
   (b) cook eggs
   (c) build hospitals
   (d) reduce wrinkles

7. Ethanol, the main vehicular fuel in Brazil, and ethanol/gasoline blends, are known as
   (a) tobacohol
   (b) CNG
   (c) jet fuel
   (d) gasohol

8. Less than what percentage of the energy in gasoline is used to move a car down the road?
   (a) 15%
   (b) 30%
   (c) 50%
   (d) 75%

9. When birds and fur-bearing mammals are coated with oil from spills, it most affects their body's capability for
   (a) flying
   (b) thermoregulation
   (c) swimming
   (d) coordination

10. What process causes tar balls to become hard and crusty on the outside and soft on the inside?
    (a) heating
    (b) sieving
    (c) weathering
    (d) washing

CHAPTER 14

Nuclear and Solar Energy

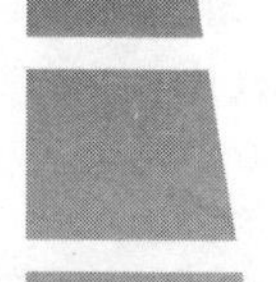

In order to significantly reduce greenhouse gas emissions and discontinue the industrialized world's dependence on fossil fuel consumption, some hard choices will have to be made. Even environmentalists who support the Gaia theory and want all-natural everything are beginning to wonder if nuclear energy is the key to our planet's future health. It is a clean, efficient energy source. When it's operating normally, it produces no greenhouse gases. Unfortunately, it produces high-level radioactive wastes and thermal pollution. Most people see nuclear energy as a Pandora's box—open the lid and a whole lot of uncontrollable stuff may be released.

Environmental groups, including Greenpeace and Friends of the Earth, lobby against clean nuclear energy and for the stop-gap Kyoto Treaty. Renewable energies, such as wind, geothermal, and hydroelectric, are part of the solution. However, nuclear energy is the only non-greenhouse gas–emitting power source that can effectively replace fossil fuels and satisfy global demand. It's available now, with even better control and safety technologies than were available just 20 years ago.

The human error factor in any nuclear equation is a problem. Science and technology can devise the greatest devices, processes, systems, or whatever, but if a significant mistake is made in manufacturing or operations, lives could be lost. With nuclear reactors, safety concerns are magnified.

Solar power is important and gets much better press coverage than nuclear power, but solar may be a case of "too little, too late." Even with efficiency and technology upgrades, solar power will not be able step right into the place of fossil fuels. The sun's energy is plentiful, but capturing, harnessing, and distributing its energy to millions of users is more complicated than it might appear.

In this chapter, we will examine the pros and cons of nuclear and solar power. As you learn about each, try to decide which you think is best.

Nuclear Energy

When people think of nuclear energy, they think of bombs and weapons-grade uranium (U^{235}). However, nuclear energy has its geological roots, as well.

Uranium is present in minute amounts in the Earth's crust, with only a 0.00016% chance that an average continental rock sample will contain the element. Natural uranium is 99.3% U^{238} and 0.7% U^{235}. The type of uranium that undergoes fission and releases huge amounts of energy is U^{235}. This accounts for only 1 in every 139 atoms of mined uranium. It is commonly found in small amounts of the uranium oxide mineral *uraninite* (commonly called *pitchblende*) in granite and other volcanic rocks. Uranium is also found in sedimentary rocks, seawater, coal, and granite.

Nuclear power is generated using the metal uranium, which is mined in various parts of the world.

Nuclear power comes from the fission of uranium, plutonium, or thorium, or the fusion of hydrogen into helium. The first large-scale nuclear power plant opened in 1956 at Calder Hall, Cumbria, England. Some military ships and submarines now have nuclear power plants for engines. Nuclear power plants use almost all uranium. The fission of an atom of uranium produces 10 million times the energy produced by the combustion of a carbon atom from coal. So, it's an incredibly efficient fuel resource.

A nuclear power reactor contains a core with a large number of fuel rods. Each rod is full of pellets of uranium oxide. Most nuclear power plants today use enriched uranium in which the concentration of U^{235} is increased from 0.7% U^{235} to around 4 or 5% U^{235}. An atom of U^{235} undergoes fission when it absorbs a neutron. Fission produces two fission pieces and other atomic particles that speed off at high velocity. When they finally stop, the energy is converted to heat (10 million times the heat of burning a coal atom).

By the mid 1990s, roughly 20% of electricity in the United States was provided by domestic licensed power reactors. The good thing about nuclear energy is that current reactors, which use natural U^{235}, can provide energy for several hundred years. Good!

Under appropriate operating conditions, the neutrons given off by fission reactions can "breed" more fuel from otherwise nonfissionable isotopes. *Breeder reactors,* which produce more energy than they use, would allow us to have energy for billions of years. The most common breeder reaction is that of plutonium (Pu^{239}), a byproduct of nonfissionable U^{238}. Uranium's most stable isotope, U^{238}, has a half-life of nearly $4^1/_2$ million years. However, since breeder reactors use uranium byproducts, which are extremely dangerous and can be used in nuclear weapons, there is a problem. Not good!

Currently, there are around 425 nuclear power plants worldwide, with 110 of these located in the United States. Fig. 14-1 gives a general overview of the parts of a nuclear power plant. Water, heated to steam, drives the turbine that generates electricity. These plants produce nearly 20% of the world's total electricity. In addition, they create huge amounts of reliable energy from small amounts of

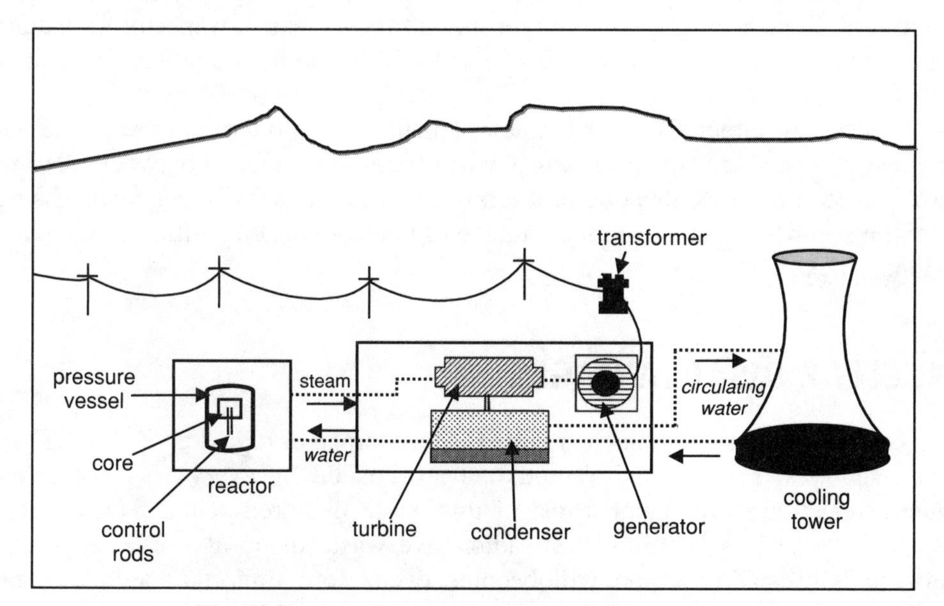

Fig. 14-1 A nuclear plant is composed of several parts.

fuel without the air pollution associated with burning fossil fuels. However, the reactors are only licensed for 40 years. Many early U.S. reactors have been or are in the process of being decommissioned.

In 1973, only 83 billion kilowatt hours (kWh) of nuclear power was produced. In 2001, nuclear power produced nine times as many kilowatt hours, with 104 licensed nuclear power plants generating 769 billion kWh of electricity. Even though there was rapid growth in nuclear power generation, there were big obstacles to nuclear power in the United States. Since 1977, no new nuclear power plants have been built and none is planned. We are beginning to look at nuclear power generation again. Its popularity is rising.

Worldwide, France has the greatest number of nuclear power plants on a per capita basis, and is second in installed nuclear capacity after the United States. Due to France's limited energy resources, energy security and imports are big concerns. The French government has strongly promoted increases in nuclear power generation over the past 30 years. To date, roughly 77% of France's electricity comes from the country's 58 nuclear reactors. In order to reduce their dependency on fossil fuels, the French invested heavily in nuclear. This was a big switch from 1973, when more than 80% of French power generation came from fossil fuels.

France now appears to be rethinking their nuclear push, however. Even though the French government planned to eventually have 100% nuclear power generation, strong environmental criticism seems to have slowed nuclear growth. When Germany chose to phase out nuclear power, French public concern grew. Currently, French opinion polls favor an end to nuclear power. Unfortunately, they have no clean, affordable substitutes that can handle the power demand at this time.

France must either replace old nuclear plants with modern ones or phase out nuclear power, since several reactors will need to be replaced between 2015 to 2020. Since France is also one of the few countries with a nuclear reprocessing plant, it's important to the public's image of nuclear waste handling to see what happens next.

NUCLEAR DISADVANTAGES

The serious problem with nuclear power is the storage of radioactive waste. Each year, about 30 metric tons of used fuel are created by every 1000-megawatt nuclear electric power plant. Most of this waste is stored at the power plants because of the lack of high-level radioactive waste disposal sites. Long-term storage is crucial now and will become even more important as additional radioactive waste accumulates.

Although not much waste is created at any one plant, it is extremely dangerous. It must be sealed up and buried for decades, even centuries, to allow time for the radioactivity to gradually disappear.

Geologists are studying the pros and cons of storing radioactive waste beneath the deep oceans and in stable (no tectonic activity) rock formations. The big concern is that sometimes groundwater can seep into land sites and become contaminated. Citizens of Nevada, a state that has been designated as a radioactive repository, are not happy about these significant safety hazards. Since the Earth is always surprising geologists, there is no way to guarantee that a chosen site will be stable and safe for radioactive storage of hundreds or thousands of years. The debate over nuclear energy will probably go on for a long time.

ACCIDENTS

Nuclear power is reliable, but lots of attention and funding must be focused on safety. If there is a problem, a nuclear accident can be a huge environmental disaster. On March 28, 1979, the nuclear accident at Three Mile Island, a two-unit nuclear plant on the Susquehanna River in Pennsylvania, caused great safety concerns and fears of radiation leakage.

At Three Mile Island, the reactor lost cooling water, overheated, and some of the fuel rods melted and ruptured. This resulted in a release of radioactive gases into the atmosphere and critical damage to the reactor. People within a 1-mile radius of the reactor were evacuated.

The world's worst nuclear plant accident occurred on April 26, 1986 at the Chernobyl, Ukraine (former Soviet Union), nuclear power plant. Nearly 200 emergency personnel who responded to the accident died within three weeks of the event from radiation poisoning in addition to the 31 who died the day of the accident. The Chernobyl plant, a graphite-regulated reactor not used in the United States, did not have the concrete containment dome that is mandatory on all U.S. nuclear plants.

The Chernobyl accident happened when two blasts, seconds apart, blew off the reactor roof, spewing radioactive gases into the atmosphere. Fires, poor containment design, several operator mistakes, and a power surge led to catastrophic and deadly circumstances. With the core continuing to heat up, Russian officials ordered aircraft to dump 5,000 tons of lead, sand, and clay onto the site to bring the temperature down.

The East/West politics of the time resulted in an early cover-up of the incident, until Swedish scientists detected the radioactive cloud moving across several countries and demanded an explanation.

In the aftermath of Chernobyl, 130,000 people had to be relocated from the contamination area. Ongoing health studies link increased birth defects and a greatly increased rate of thyroid cancer in children living in the area surrounding Chernobyl.

In addition to radioactivity, the spiraling rise in temperature during a meltdown is a huge safety issue. This is known as *the China Syndrome.*

The China Syndrome describes what happens when nuclear core temperatures escalate out of control, causing the core to melt through the Earth's crust and mantle to the core and then seemingly through to the other side of the Earth (China).

In 1979, a Hollywood movie by the same name described incidents that took place at the Rancho Seco power plant in California. However, unlike in *The China Syndrome* movie, the emergency system shut the plant down safely as it was designed to do.

REGULATIONS

Compared to the extreme Chernobyl accident, the Three Mile Island incident was not serious and no deaths occurred. In fact, the best thing to come out of Three Mile Island was much better plant and safety designs and the creation of the Institute of Nuclear Power Operations (INPO). The INPO established guidelines for excellence in nuclear plant operations and increased communication within the nuclear industry.

Strong regulatory impact in response to public outrage has been a big factor in the growth of nuclear power. In 2000 however, the Nuclear Regulatory Commission (NRC), in an encouraging nod to the U.S. nuclear power industry, granted the first-ever renewal of a nuclear power plant's operating license. The 20-year extension (2034 and 2036 for two reactors) was given to the 1700-megawatt plant in Calvert Cliffs, Maryland. As of March 2002, Exelon and Dominion Resources reportedly were looking at sites to build the first new nuclear power plants in the United States in nearly thirty years. Nuclear generation is thought to have increased by 0.6 to 0.7% in 2002 and 2003.

After the discovery of corrosion in a major nuclear plant section in Ohio, the NRC ordered safety information on 68 other units. After checking the problem thoroughly, the problem was found to affect only the Ohio unit.

In January 2002, Energy Secretary Spencer Abraham recommended Yucca Mountain, Nevada as the nation's permanent nuclear waste depository. In specially designed containers, over 77,000 tons of nuclear waste from both power plants and nuclear weapons would be buried in a series of tunnels dug 302 meters below the mountain's peak and 274 meters above the water table.

Even though the Yucca Mountain site has been studied as a radioactive waste site for over 20 years, the state of Nevada is firmly opposed to everything from the geological assumptions (the area is susceptible to earthquakes) to the nickel alloy spent fuel containers (promised to last 10,000 years, but projected at more like 500 years by state and environmental opponents). Permanent geologic disposal is critical for managing used nuclear fuel from commercial electric power generation.

Currently the construction schedule calls for completion of the Yucca Mountain site by 2010 with the first waste shipments to arrive at that time. The capacity of the Yucca Mountain project will be reached within 25 years of its projected opening at current nuclear waste production rates. Planning for the expansion of Yucca Mountain or construction of a second storage site must begin now to meet expected radioactive storage demand.

Another big industry and public controversy is centered on the hazards of transporting nuclear materials to the Yucca Mountain storage site by rail or truck. Since most nuclear power plants are in the eastern part of the United States, the waste would have to be transported west about 2000 miles to Nevada. Besides just the usual hazardous mishaps that could take place across that distance, the very real threat of terrorist sabotage hangs like a black cloud over the entire transportation plan. In any case, nuclear utilities are running out of radioactive waste storage capacity and if we continue to use nuclear power, we'll have to store the waste somewhere.

Solar Energy

The sun is an immense source of energy. It lights and heats our planet and supplies energy for plant photosynthesis. For thousands of years, humans wished they could capture and store energy from the sun. However, it has only been in the past half century that energy technology with which to harness solar power has been available.

Sunlight is a huge natural resource. The amount of solar energy reaching the earth annually is much greater than worldwide energy demands, although it changes with time of day, location, and the season.

Most solar power generation is based on the *photovoltaic* (PV) reaction that produces voltage when exposed to radiant energy (especially light). A solar PV cell converts sunlight into electrical energy.

Solar technologies use the sun's energy and light to provide heat, light, hot water, electricity, and cooling for homes, businesses, and industry. Accessibility to unblocked sunlight for use in both passive solar designs and active systems is protected by zoning laws and ordinances in many parts of the country.

CRYSTALLINE SILICON SOLAR CELLS

Traditionally, crystalline silicon was used as the light-absorbing semiconductor in most solar cells. It produces stable solar cells, using technology developed in the microelectronics industry with good efficiencies.

There are two types of crystalline silicon used in the solar industry. The first, *monocrystalline silicon,* is produced by slicing thin wafers (up to 0.150 cm in diameter and 350 microns thick) from a high-purity single crystal. The second, *multicrystalline silicon,* is made by cutting a cast block of silicon into bars, then wafers. Most crystalline silicon cell manufacturers use multicrystalline technology.

For both mono- and multicrystalline silicon, a semiconductor connection is made by diffusing phosphorus onto the top of a boron-coated silicon wafer. Contacts are applied to the cell's front and back, with front contact pattern designed to allow maximum light exposure of the silicon material with minimum cell resistance losses. The most efficient production cells use monocrystalline silicon with covered, laser-grooved grid contacts for maximum light absorption and current gathering.

Some companies skip parts of the crystal growth/casting and wafer-sawing method. Instead, they produce a silicon ribbon, either as a plain two-dimensional strip or as an octagonal post, by extracting it from a silicon melt.

Crystalline silicon cell technology supplies about 90% of solar cell market demand. The rest comes from thin film technologies.

Efficiency

An average crystalline silicon–cell solar module has an efficiency of 15%. An average thin film–cell solar module has an efficiency of 6%.

Solar cells function more efficiently under focused light. For this reason, mirrors and lenses are used to direct light onto specially designed cells with heat sinks or active cell cooling, and to disperse the high heat that is created. Unlike common flat plate PV arrays, concentrator systems need direct sunlight (clear skies) and don't operate under cloudy conditions. They generally follow the sun's path across the sky during the day using *single-axis tracking*. To follow the sun's changing height in the sky during the seasons, *two-axis tracking* is useful.

THIN SOLAR CELLS

The high cost of crystalline silicon wafers led the semiconductor industry to look for cheaper ways and materials to make solar cells. The most commonly used materials are *amorphous silicon* or *polycrystalline materials* [cadmium telluride, copper indium (gallium), diselenide]. These materials all strongly absorb light and are naturally only about 1 micron thick, so production costs drop. Unfortunately, they are environmentally unfriendly.

These wafers allow large area deposition (substrates of up to 1 meter) and high volume manufacturing. The thin film semiconductor layers are deposited onto coated glass or stainless steel sheets. Intricate thin film methods have taken about 20 years of development get from the initial research stage to the manufacture of a prototype product.

Staebler-Wronski Effect

Amorphous silicon is the best developed of the thin film technologies with a single sequence of layers, but amorphous silicon cells undergo a lot of power output degradation (15 to 35%) when exposed to the sun. Not a good thing for *solar* cells!

Solar cell degradation is called the *Staebler-Wronski Effect*. The effect describes how the best stability requires the thinnest layers. This is important to increase the electric field strength across the material. However, it also reduces light absorption and cell efficiency. The industry has developed in-line, stacked layers to reduce band gap and improve light absorption, but this technology leads to additional complexity, lower yields, and higher cost.

The best silicon solar cell methods produce thin film cells, laminated to produce a weather-resistant and environmentally sturdy material while achieving low manufacturing costs and high, stable efficiencies with the highest process yields.

ELECTROCHEMICAL SOLAR CELLS

While crystalline and thin film solar cells have solid-state light-absorbing layers, electrochemical solar cells are active in a liquid phase. Electrochemical solar cells use a *dye sensitizer* to absorb light and produce electron pairs in a nanocrystalline titanium dioxide semiconductor layer. This is sandwiched between a tin oxide–coated glass sheet (front cell contact) and a rear carbon contact layer with a glass or foil backing sheet.

Although these cells are made at lower manufacturing costs because of simplicity and cheap materials, the ability of companies to scale up large manufacturing processes and demonstrate reliable field operations is still to be proven. Small device prototypes powered by dye-sensitized nanocrystalline electrochemical PV cells are now coming on the market.

SOLAR TECHNOLOGY TYPES

We have seen how *photovoltaic cells* that convert sunlight directly into electricity are made of semiconducting materials. Photovoltaics can absorb solar energy almost anywhere on Earth (hot or cold climates)—as long as there is sunlight. Simple solar cells power watches and calculators while complex solar systems light homes and create power for the electric grid.

Around 50% of the world's solar cell production in 2003 occurred in Japan. The United States accounted for roughly 12% of solar PV. On the electricity supply side, the amount of solar-powered electricity produced worldwide reached 742 megawatts in 2003.

Japan has overtaken the United States as the largest net exporter of PV cells and modules. Four companies account for over 50% of solar cell production: Sharp, Kyocera, BP Solar, and Shell Solar. Of the top five manufacturers, Sharp is the largest, with the fastest growth over the last five years. Sanyo, fifth-largest, has shown the second-highest rate of growth over the same period.

Buildings designed for *passive solar* and day lighting use design features like large south-facing windows and construction materials that absorb and slowly release the sun's heat. No mechanical processes are used in passive solar heating. Incorporating passive solar designs can reduce heating bills as much as 50%. Passive solar designs can also provide cooling through natural ventilation.

Concentrated solar power technologies use reflective materials such as mirrors to concentrate the sun's energy. This concentrated thermal energy is then converted into electricity and can be used to power a turbine.

Solar hot-water heaters use either the sun to heat water or a heat-transfer fluid in the collectors. Commonly, this type of system will lower the need for traditional electric or gas water heating by as much as two-thirds. High-temperature solar water heaters have also been used to provide energy-efficient hot water and heat for commercial and industrial facilities.

Solar Now

Until the oil shortages of the early 1970s that made people see a need for alternate energy, the main use for photovoltaic cells was in providing electricity to orbiting satellites.

To produce electrical power for the International Space Station, the crew of the space shuttle *Endeavour* deployed the largest-ever solar panel array in space on December 5, 2000. The panels were the first set of four to be installed. According to NASA, the expansive solar panels are so reflective that the station is visible to the human eye—only the Moon and bright planets like Venus are brighter. The panels produce enough electricity to power around 30 households.

Groups, organizations, communities, states, and countries have been encouraging policy makers to provide tax benefits for using and developing alternative energy sources, including the following:

- Energy-efficient products
- Hybrid cars
- Ethanol
- Solar energy
- Wind energy
- Geothermal energy
- Hydroelectric power

Founded in 1954, the International Solar Energy Society (ISES) has been serving the needs of the renewable energy community with many ideas to promote solar energy. A United Nations organization, it is active in more than 50 countries. The ISES teaches its members the benefits of clean, renewable energy methods, performance, and education worldwide.

Data can be gathered on the amount of solar energy available to a specific collector, as well as how much it changes monthly, yearly, and in different locations. Collecting this information takes a national network of solar radiation monitoring sites.

Recent years have seen a lot of growth in the number of PV installations on buildings connected to the electricity grid. Demand has been encouraged by government programs (Japan and Germany) and by incentive pricing policies or electricity service providers (Switzerland and the United States). The main push comes from individuals and/or companies who want to get electricity from a clean, nonpolluting, renewable source and agree to pay a little extra for the option.

An **electricity grid** describes an electricity transmission and distribution system, usually supplying power across a wide geographical region.

An individual PV system, connected to a larger supply grid, can supply electricity to a home or building. Any extra electricity can be sent to the public grid. Batteries are not needed since the grid is able to meet any extra demand. However, for a home to be independent of the grid supply, battery storage is required to offer power during nighttime hours.

Grid-connected systems are independent power systems joined to a regional grid; they draw on the grid's reserve capacity when they need it and give electricity back during times of extra production.

Solar PV modules can be added to a pitched roof above the existing roof tiles, or the tiles can be replaced by specially designed PV roof tiles or roof-tiling systems. If you want to put a PV system onto a roof and connect it to the grid supply, local regulations and permissions are required.

One disadvantage of solar power is the amount of land needed. For enough solar panels to supply the electricity needs of an urban area, a lot of acreage is required.

COMMON USES

Photovoltaic systems are practical for beach, remote cabins, or vacation homes that don't have access to the electricity grid. These systems can be set up to meet power needs for less money. The fact that solar energy is highly reliable and needs little maintenance makes it a great choice for hard-to-reach locations with ample sunlight. Polar research stations are a good example.

Remote mountain or desert homes in sunny locations can get reliable electricity from solar generation for lighting, radio, or television. PV systems are made up of a PV panel, a rechargeable battery to store the energy captured during daylight hours, a regulator, and the necessary wiring and switches. These simple systems are known as solar home systems.

For home application, 50- to 100-watt modules may be required, although smaller panels of 10- to 15-watt will give enough power for a basic single-lamp system.

Central power applications use solar energy in the same way a traditional utility company operates a major power station. There are hub locations from which power is sent out to meet demand.

Power sent out in small amounts, usually near the point of electrical usage, is known as **distributed power.**

For many years, solar energy has been used for industrial applications where power was needed in remote locations. Many of these applications needed only a few kilowatts of power. Solar was used for powering microwave repeater stations, TV and radio, telemetry and radio telephones, as well as flashing school traffic lights.

Solar-gathered power was also found on transportation signaling (navigation buoys, lighthouses, and air strip warning lights). Solar has been used to power environmental and specific monitoring equipment, as well as corrosion safeguard systems for pipelines, well heads, bridges, or other structures. For greater electrical needs, a hybrid power system may link a PV system with a small diesel generator, for better reliability.

Apart from off-grid homes, other remote buildings such as schools, community halls, and clinics can all benefit from electrification with solar energy. Solar can power TV, video, telephones, and a range of refrigeration equipment (that meet World Health Organization standards for vaccine refrigeration), for example. In rural areas, rather than mount solar power panels on individual dwellings, it's also possible to configure central village power plants that can either power homes through a wired network or act as a battery-charging station where people can bring batteries to be recharged.

PV systems can be used to pump water in remote areas as part of a portable water supply system. Specialized solar water pumps are designed for submersible use (borehole) or to float on open water. Usually, the ability to store water in a tank means that battery power storage is unnecessary. Large-scale desalination plants can also be PV-powered. Larger off-grid systems can be constructed to power larger and more sophisticated electrical loads by using an array of PV modules and having more battery storage capacity.

To meet the largest power requirements in an off-grid location, the PV system is sometimes best configured with a small diesel generator. This means that the PV system no longer has to be able to meet the worst sunlight conditions presented during the year. The diesel generator can provide backup power, but is rarely used by the PV system during the year, so fuel and maintenance costs stay low and diesel (fossil fuel) use is minimal.

In large office buildings, a central atrium or skylight can be covered with semitransparent glass or glass PV modules that create shaded light. Industrial buildings with large roof areas can also use large solar modules. Solar energy can also be used to light associated parking lots, outlying facilities, and signs.

Another solar energy home/business option is to use sunshades or balconies with a PV system. Sunshades could (1) incorporate a PV system externally mounted to a building or (2) contain PV cells mounted between glass sheets of a window.

The bottom line is that the Earth gets more energy from the sun in one hour than the entire planet uses in a whole year. Since 2 billion of the world's people have no access to electricity, solar power is an excellent, renewable, and nonpolluting energy option. For many people, solar photovoltaic cells provide the cheapest electricity source, especially as fossil fuel prices soar. Solar power is a great choice for much of our energy needs, not just in industrialized countries, but around the world.

Quiz

1. A big concern facing geologists in the long-term storage of radioactive waste is that
   (a) people living in the area will begin to glow in the dark
   (b) the half-life will take even longer to pass
   (c) groundwater will seep into land sites and become contaminated
   (d) people will mistake it for fluorescent minerals

2. The isotope of uranium that undergoes fission and releases huge amounts of energy is
   (a) U^{190}
   (b) Pt^{225}
   (c) I^{60}
   (d) U^{235}

3. Nuclear power is generated using the metal
   (a) copper
   (b) uranium
   (c) mercury
   (d) actinium

4. Solar cell degradation is called the
   (a) Staebler-Wronski Effect
   (b) Bert-Ernie Effect
   (c) Watson-Crick Effect
   (d) Pasteur-Salk Effect

5. Solar energy can provide electricity for all of the following uses, except
   (a) signs
   (b) traffic lights
   (c) submarines
   (d) lighting for parking lots

6. A strong regulatory backlash to nuclear power occurred in response to what event?
   (a) Three Mile Island
   (b) the Trail of Tears
   (c) Mount Rushmore
   (d) the eruption of Mount St. Helens

7. No new nuclear power plants have been built or planned in the United States since
   (a) 1958
   (b) 1962
   (c) 1977
   (d) 1983

8. Solar panels convert sunlight directly into electricity using silicon-filled
   (a) mood rings
   (b) photovoltaic cells
   (c) rubber balls
   (d) plant cells

9. Cadmium telluride is a common material used to make
   (a) dark chocolate
   (b) nylon
   (c) clothes hangers
   (d) wafer-thin solar cells

10. Which country's government has strongly promoted increases in nuclear power use over the past 30 years?
    (a) Columbia
    (b) Madagascar
    (c) Holland
    (d) France

CHAPTER 15

Wind, Hydroelectric, and Geothermal Energy

There are many, many possibilities for wind, hydroelectric, and geothermal energy, not only in the United States, but around the world. These natural energy sources from the Earth have been used successfully for centuries. They are generally safe, clean, and use existing technology.

Wind Energy

The fact that wind energy is natural is obvious, but did you know that its created by the sun? Wind energy is created by the Earth's atmospheric circulation patterns, which are heated and influenced by the Sun's heat.

Ancient people used wind power directly to power ships for travel, trade, and to discover new lands. Long before electricity, windmills provided energy for stone mills to grind grain into flour and/or pump water. Think of the beautiful windmills that dot the countryside in the Netherlands and other countries.

Today, wind power can be converted into multiuse electricity through large, high-tech windmills and *turbines.* The electricity is transmitted through power lines to locations far removed from the generating wind tower. These large tower turbines look something like the giant version of a child's pinwheel toy. Fig. 15-1 shows how turbines use long blades to catch the wind.

Horizontal wind turbines, like windmills, are built high on a tower to capture the greatest amount of wind energy. At 30 meters or more above the ground, they can catch the faster and less turbulent wind. Turbines catch the wind's energy with propeller-like blades. Usually, two or three blades are mounted on a shaft to form a *rotor.*

A turbine blade acts a lot like an airplane wing. When the wind blows, a pocket of low-pressure air forms on the downwind side of the blade. This low-pressure air pocket then pulls the blade toward it, causing the rotor to turn. This is called *aerodynamic lift.* Lift force is much stronger than the wind's frictional force against the front side of the blade, which is known as *drag.* Frictional drag force is perpendicular to the lift force and slows the rotor's rotation slightly. To build an efficient wind turbine, the blade must have a fairly high lift-to-drag

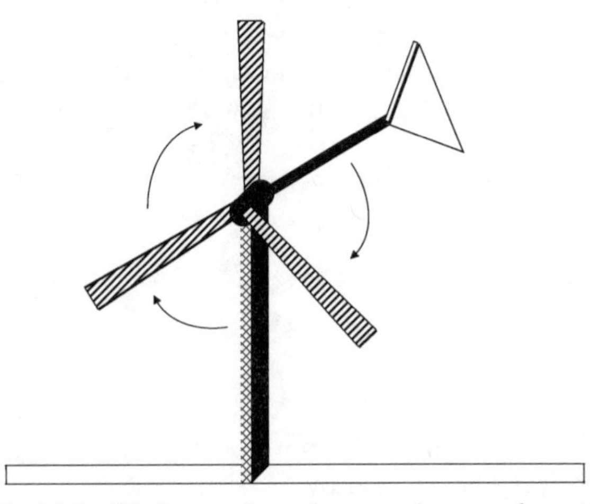

Fig. 15-1 Wind energy is another natural source of energy.

ratio. This ratio can be changed along the blade's length to boost the turbine's energy yield at different wind speeds.

The right combination of lift and drag makes the turbine spin like a propeller. The rotor is mounted to a shaft that spins a generator and makes electricity.

The United States has huge wind energy potential. Table 15-1 lists the top 20 states with excellent wind resources for generating electricity from wind energy, such as (1) coastal areas, since the wind always blows there, (2) flat plains of the Midwest, and (3) mountain passes that act as a funnel to increase wind velocity.

Wind farms currently only generate about 1/3 of the electricity that they could. Wind proponents project that with continued government funding and development, wind energy could provide 6% of the nation's electricity by 2020. This clean *green energy* source creates electricity without consuming natural resources or producing greenhouse gases.

Table 15-1 Wind power in the top 20 states.

Rank	State	Kilowatt hours (billions)	Rank	State	Kilowatt hours (billions)
1	North Dakota	1210	12	New Mexico	435
2	Texas	1190	13	Idaho	73
3	Kansas	1070	14	Michigan	65
4	South Dakota	1030	15	New York	62
5	Montana	1020	16	Illinois	61
6	Nebraska	868	17	California	59
7	Wyoming	747	18	Wisconsin	58
8	Oklahoma	725	19	Maine	56
9	Minnesota	657	20	Missouri	52
10	Iowa	551			
11	Colorado	481	**Total**		**10,470**

Source: An Assessment of the Available Windy Land Area and Wind Energy Potential in the contiguous United States, Pacific Northwest Lab, 1991.

In 1997, at the United Nations Convention on Climate Change in Kyoto, Japan, 160 nations reached a first-ever agreement to reduce emissions of greenhouse gases, including carbon dioxide. Most industrialized nations committed to lowering average emissions between 2008 to 2012 to approximately 5% below 1990 levels. To fulfill this multilateral agreement, many nations began to use wind power.

The European Wind Energy Association intends to provide over 10% of Europe's electricity by 2030. Denmark, the largest user of windmills, projects 10% of its electricity generation will come from wind power in the next 10 years.

Today, more than 13,932 megawatts (MW) of wind energy are being generated worldwide, and production is predicted to reach nearly 50,000 MW by the year 2005. With an energy cost of roughly 3.5 to 4 cents per kilowatt hour and dropping, wind is less expensive than coal, oil, nuclear, and most natural gas–fired generation.

The great thing about wind, besides being a clean energy source, is that it's free! Wind farms don't need fossil fuels. The land underneath turbines can be used for farming, and turbines provide a good way to get power to remote areas.

DISADVANTAGES OF WIND

One of the drawbacks of wind power is that it isn't always predictable; some days the air is still, without even a faint breeze. Another commonly cited problem is that some people believe wind turbines are ugly. Others, however, see them as a great nonpolluting way to produce cheap power. A big disadvantage of wind farms is that they need a lot of acreage to generate enough electricity for urban areas. Land for wind farms, especially in coastal areas, can be relatively expensive and difficult to buy from vacationers. Moreover, there is some perception that wind turbines may interfere with television reception—a truly horrible offense compared to polluting the planet!

The main drawback to wind energy is thought to be the noise created by the rotors. A wind generator makes a constant, low, humming noise day and night, which is either pleasant or annoying, depending on who you ask. Although wind turbines create only about 60 decibels (db) of noise when running efficiently, some opponents say it's too high especially when multiplied by the numbers needed to support a power grid. Table 15-2 lists several common sound decibel levels and shows where a wind turbine ranks on the scale.

Table 15-2 Wind turbines are relatively quiet compared to many modern noises.

Sound	Decibel level (dB)	Sound	Decibel level (dB)
Rustling leaves	20	Screaming child	90
Whispering	25	Passing motorcycle	90
Library	30	Subway train	100
Refrigerator	45	Diesel truck	100
Normal conversation	60	Jackhammer	100
Wind turbine	60	Helicopter	105
Washing machine	65	Lawn mower	105
Dishwasher	65	Sandblasting	110
Car	70	Live rock music	90–130
Vacuum cleaner	70	Auto horn	120
Busy traffic	75	Airplane propeller	120
Alarm clock	80	Air raid siren	130
Noisy restaurant	80	Gunshot	140
Outboard motor	80	Jet engine	140
Electric shaver	85	Rocket launch	180

Hydroelectric Power

Have you ever tried to cross a rushing creek? Even if the water is only a few inches deep, the force of quickly moving water can knock you over. Hydroelectric power doesn't pollute the atmosphere like the burning of fossil fuels (coal or natural gas). Moving water is powerful and since hydroelectric plants are fueled by water, it's a clean, generally available fuel source.

Flowing water creates energy that can be turned into electricity. This is called *hydroelectric power* or *hydropower*. The two major ways that water flow is used to make electricity include (1) huge amounts of water spinning giant turbines and (2) tidal diversion where water can be directed both up and down pipes linked to turbines. As water flow is restricted, it flows faster, spinning turbines and generating electricity.

Small hydroelectric power systems can provide enough electricity for a home, farm, or ranch. So for those people lucky enough to live near a river, this is a good way to get electricity.

TYPES OF HYDROELECTRIC PLANTS

There are three types of hydroelectric power plants: *impoundment, diversion,* and *pumped storage.* Some hydropower plants use dams and some do not.

Hydroelectric power is commonly generated from water stored in a reservoir at a power plant dam built on a river. Water released from the reservoir through the dam flows and spins turbines that produce electricity. This most common type of hydroelectric power plant is called a large *impoundment plant,* and uses a dam to store river water in a reservoir. When water is released from the reservoir, it powers a turbine that activates a generator to produce electricity. The water may be controlled to provide more or less electricity or to keep a constant reservoir level. Fig. 15-2 illustrates an impoundment (dam) hydroelectric power plant.

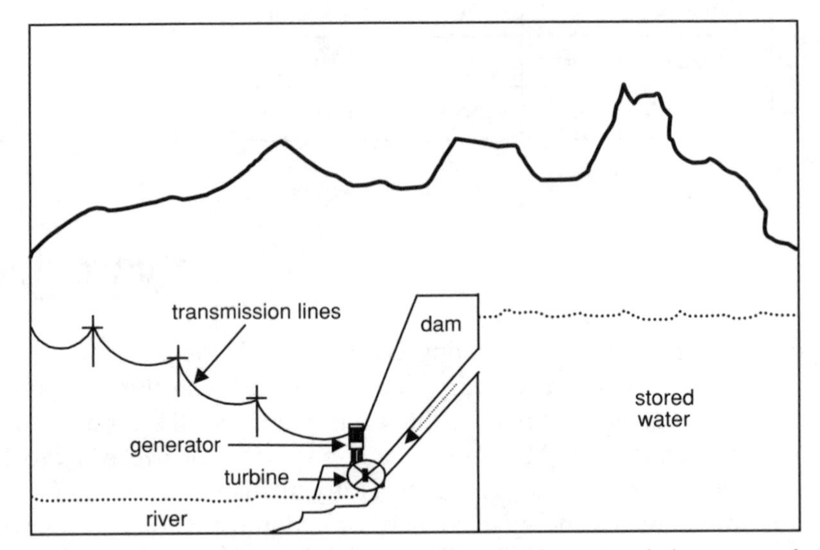

Fig. 15-2 An impoundment hydroelectric power plant is commonly known as a dam.

A **hydroelectric impoundment plant** holds water in a reservoir, then uses the stored potential energy to drive a turbine and produce electricity when the water is released.

Hydroelectric power doesn't always need a big dam. Some hydroelectric power plants use small canals to channel river water through turbines. This type of hydroelectric power plant is called a *diversion,* or *run-of-river,* plant. It channels a part of a river's flow through a canal or sluice and may not need a dam (depending on the local geography). Alaska's Tazimina plant is an example of a diversion hydropower plant. No dam is needed.

Another type of hydroelectric power plant, the *pumped storage plant,* stores power by sending it from a power grid into electric generators. These generators spin the turbines backward, which makes the turbines pump water from a river or lower reservoir to an upper reservoir, storing it. During periods of high electrical demand, the water is released from the upper reservoir back to its starting place. This spins the turbines forward, causing the generators to make electricity.

A lot of dams, built for other reasons (like irrigation) than power generation, had a hydroelectric power plant added later. In the United States, there are around 80,000 dams. However, only 2400 of these produce power. The majority of the dams are used for recreation, stock/farm ponds, flood control, water supply, and irrigation.

Hydropower plants range in size from small systems for a home or village to large plants that produce utility power for many users. Table 15-3 lists the different Department of Energy classifications of hydroelectric power plants.

Table 15-3 There are 3 main classifications of hydroelectric power plants.

Size of hydroelectric power plant	Power generated (megawatts)
Large	> 30
Small	0.1 to 30
Micro	100 kw–0.1

TURBINES

There are a number of turbine types used for hydroelectric power. They are chosen with regard to the amount of stored water, called *head*, available to drive them. The rotating part of the turbine is called the *runner*. The most commonly used hydroelectric turbines include:

- A *Pelton turbine,* with one or more water jets pushing on the buckets of a runner, like a water wheel. Pelton turbines are used for high-head sites (50–6000 feet);
- A *Francis turbine,* with a runner of 9 or more fixed vanes. The water enters the turbine at a perpendicular angle to the shaft, and is released in a direction parallel to the shaft. Francis turbines operate best on medium head sites (10–2000 feet);
- A *propeller turbine,* with a runner that has 3 to 6 fixed blades, like a boat propeller. The water passes through the runner and moves the blades. Propeller turbines can operate at low head sites (10–300 feet); and
- A *Kaplan propeller turbine* (propeller turbine), where the blades' pitch can be altered to increase peak operating output.

Table 15-4 lists the different energy levels (MW) generated by the various turbine types.

Table 15-4 In different applications, turbine configurations produce varying amounts of energy.

Hydroelectric turbines	Power generated (megawatts)
Pelton	200
Francis	800
Propeller	100
Kaplan	400

DISADVANTAGES OF HYDROELECTRIC

Hydroelectric power plants affect water quality and flow and can cause low dissolved oxygen levels in the water. Maintaining normal water flows downstream of a hydropower plant is important for the survival of local habitats and marine species. Many don't survive downstream of large dams independent of flow.

Fish can be impacted if they aren't able to migrate upstream past impoundment dams to spawning grounds or if they cannot migrate downstream to the ocean. Upstream fish movement can be helped using fish ladders or by collecting and hauling spawning fish upstream by truck. Downstream fish movement is assisted by rerouting fish from turbine intakes with screens, racks, or underwater lights and sounds, and by keeping a minimum flow near the turbine.

Drought can also influence hydroelectric plant operations. When water is low or not available, plants can't make electricity.

Newly constructed hydroelectric plants impact the local ecology and may compete with other site uses. Humans, flora, and fauna may lose their natural habitat or recreation spots. Local native cultures and historical sites may also be changed. As with every aspect of the energy puzzle, there are pros and cons to every alternative.

Geothermal Energy

Like solar energy, the Earth's thermal energy is a free gift. The word *geothermal* comes from the Greek words *geo* (earth) and *therme* (heat), and describes the earth's core heat. This interior heat came from our plant's original compression of dust and gas. The Earth's heat is continuously regenerated from the decay of radioactive elements that takes place in all rocks. Internal geothermal heat from core processes supplies energy for plate tectonics, mountain building, volcanic eruptions, and earthquakes.

CORE HEAT

As we learned in Chapter 1, the earth's core has temperatures between 4000 and 7000°C. This internal heat moves outward toward the crust through convection and tectonic activity as magma (lava).

Some magma reaches the surface and creates volcanoes. However, most of it stays underground, pooling into huge subterranean geothermal areas of hot rock, sometimes as big as an entire mountain range. Cooling can take between 5000 and more than million years. These geothermal regions have high temperature gradients.

The Pacific Ocean's *Ring of Fire* (oceanic plates are subducted beneath continental plates) has a high number of volcanoes and is high in geothermal potential. The Atlantic midocean or continental rift zones (where plates separate) also have high heat energy, like Iceland and Kenya. Specific hot spots (above magma columns), like the Hawaiian Islands and Yellowstone National Park in the United States, are also good sources.

WATER'S PART

In some areas with high temperature gradients, there are deep subterranean faults and cracks that allow rainwater and snow to seep underground. In these areas, the water is heated and circulates back to the surface. Once there, it becomes hot springs and holes, as well as geysers.

If rising hot water hits solid rock it becomes trapped and fills up the holes and cracks of the surrounding rock, forming a geothermal reservoir. Much hotter than surface hot springs, geothermal reservoirs can reach temperatures of over 370°C and are huge sources of energy.

Geothermal energy is also created when deep underground heat is transferred by thermal conduction through water to the surface. Think of a wet sauna where rocks are heated and steam is produced as water is poured over them. When ground water is heated it forms *hydrothermal reservoirs* like hot ponds, geysers, and steam.

To capture geothermal energy, holes are drilled into hot spots and the superheated groundwater is pumped to the surface. Scientists and engineers use geological, electrical, magnetic, geochemical, and seismic tests to help find the reservoirs. If they are near the surface, they can be reached by drilling exploratory wells, sometimes over two miles deep.

After a test well proves the existence of a geothermal reservoir, production wells are drilled. Hot water and steam shoot up at temperatures between 120 and 370°C.

The Earth's heat is used to produce electricity in geothermal power plants, while shallow areas of lower temperature (21 to 149°C) are used directly in health spas, greenhouse heating, fish farms, vegetable drying, industry, and heating equipment for homes, schools, and offices.

If a geothermal resource is used directly for heat generation, the hot water is usually fed to a heat exchanger before being injected back into the Earth. Used geothermal water is sent back to the underground reservoir to help keep up the total pressure. However, the returned water is at a lower temperature. Eventually, the hot water reservoir will be cooled if too much hot water is removed. Hot water and steam are found in lots of subsurface locations in the western United States.

Within the past five years, many more countries began using geothermal heat to produce electricity. Water reservoirs with temperatures of 80°C to 180°C have been tapped to provide reliable and inexpensive heat for homes, businesses, and industry. Reykjavik, Iceland's capital, is heated completely by geothermal energy from volcanic sources within the Midatlantic Ridge that intersects the country.

GEOTHERMAL POWER GENERATION

In 1999, 8217 megawatts of electricity were being produced from around 250 geothermal power plants operating in 22 countries around the world. These plants provide reliable power for over 60 million people, mostly in developing countries. Today, geothermal power provides about 10 percent of the United States' energy supply. Table 15-5 provides a list of countries that use geothermal energy and the megawatts of electricity that each produced in 1999.

Table 15-5 There are a number of different countries using geothermal energy.

Country	Megawatts	Country	Megawatts
United States	2850	Kenya	45
Philippines	1848	China	32
Italy	768.5	Turkey	21
Mexico	743	Russia	11
Indonesia	589.5	Portugal	11
Japan	530	Guatemala	5
New Zealand	345	France	4
Costa Rica	120	Taiwan	3
Iceland	140	Thailand	0.3
El Salvador	105	Zambia	0.2
Nicaragua	70	**Total**	**8217**

There is a large amount of geothermal capacity available from power plants in the western United States. It is estimated that geothermal energy provides around 2% of the electricity in Utah, 6% of the electricity in California, and nearly 10% of the electricity in northern Nevada.

The electrical energy generated in the United States from geothermal resources is more than twice that from solar and wind combined. However, it's important to know that geothermal energy is not easily transported and loses up to 90% of its heat energy if not used near its source.

TYPES OF GEOTHERMAL PLANTS

In geothermal power plants, hot water and steam are used to turn turbine generators that make electricity, but unlike fossil fuel power plants, no fuel is ignited. Geothermal plants release water vapor, not smoky pollution. There are four main types of geothermal power plants: (1) *flashed steam,* (2) *dry steam,* (3) *binary,* and (4) *hybrid.*

The most common type of geothermal power plant is the *flashed steam* power plant. Hot well water is sent through separators where, released from the pressure of the deep reservoir, some water instantly boils to steam. The steam's extreme force spins a turbine generator. To save water and keep up reservoir pressure, the residual water and condensed steam are sent back down an injection well into the reservoir to be recycled and reheated.

Dry steam plants are less common. They produce mostly steam and very little water. In dry steam plants, the steam shoots directly through a rock-catcher and into the turbine. It is thought that the first dry steam geothermal power plant was built in Larderello, Italy, in 1904. These Larderello power plants were destroyed during World War II, but have been rebuilt and expanded, and produce electricity today.

The Geysers (northern California) dry steam reservoir has been providing geothermal power since 1960. Known as the largest dry steam field in the world, it furnishes enough electricity to supply a city the size of San Francisco.

In a *binary* power plant the geothermal water is sent through one side of a *heat exchanger* (series of pipes), where heat is transmitted to a second (binary) liquid, known as a *working fluid,* in a separate, nearby set of pipes.

> A **working fluid** (isobutane or isopentane) boils and flashes to a gas at a lower temperature than water.

The working fluid boils to a gas, like steam, and powers a turbine generator. A working fluid can be concentrated repeatedly back to a liquid and reused. The geothermal water passes only through the heat exchanger and is immediately sent back to the reservoir.

Although binary power plants are more expensive to build than steam-driven plants, they have several advantages: (1) a *working fluid* (isobutane or isopentane) boils and becomes gaseous at a lower temperature than water (electricity can be obtained from lower temperature reservoirs), (2) a binary system uses reservoir water more efficiently (hot water in a closed system results in less heat/water loss), and (3) binary power plants have almost no emissions.

Hybrid power plants combine processes. In some power plants, flash and binary processes are combined to produce power. A hybrid system has been developed in Hawaii that provides nearly 25% of the electricity used in the state.

Other Geothermal Uses

Geothermal resources have been used around the planet to improve farming yields. Water warmed by geothermal reservoirs warms greenhouses to help grow flowers, vegetables, and other crops. For hundreds of years, farmers in central Italy have grown winter vegetables in fields heated by natural steam. In Hungary, it is estimated that mineral-rich geothermal waters provide 80% of vegetable farmers' needs.

Geothermal aquaculture (marine farming) uses naturally heated water to increase the growth of fish, shellfish, reptiles, and amphibians. This direct use of geothermal resources is growing as fossil fuel costs climb. In China, geothermal fish farms cover almost 2 million square meters. Eels and alligators are raised in Japan, while geothermal aquaculture in the United States (Idaho, Utah, Oregon, and California) grow catfish, trout, alligators, tilapia, and tropical fish for pet shops. In Iceland, the production of abalone may exceed 2½ million annually.

Worldwide, industrial geothermal water usage is also on the rise as an alternative to costly fossil fuels. These uses include the drying of fish, fruits, vegetables, and timber products, as well as washing wool, dying cloth, manufacturing paper, and pasteurizing milk. High-temperature water can be directed under sidewalks and roads to prevent icing in freezing weather. Thermal waters have also been used in gold and silver extraction from ore.

Beside hot springs baths, the oldest and most common use of geothermal water, geothermal waters have been involved in the heating of single buildings as well as commercial and residential districts.

A *geothermal district heating* system supplies heat by pumping geothermal water (60°C or hotter) from one or more wells drilled into a geothermal reservoir. This hot water is sent through a heat exchanger, which transfers the heat to water pumped into buildings through separate pipes. After going through the heat exchanger circuit, the used water is directed back into the reservoir to reheat.

The first district heating system in the United States was established in 1893 in Boise, Idaho. In the western United States there are over 275 communities that are close enough to geothermal reservoirs to take advantage of geothermal district heating. Currently, there are 18 district heating systems in use in the United States; the largest are in Boise, Idaho, and San Bernardino, California.

Clean, low-cost geothermal district heating is also used to warm homes in Iceland, Turkey, Poland, and Hungary. Reykjavik has the world's largest geothermal heating district system, as nearly all the buildings use geothermal heat. Before Icelanders switched to geothermal heat, Reykjavik was heavily polluted from fossil fuel burning. Now, it is one of the cleanest cities in the world.

DISADVANTAGES OF GEOTHERMAL

Geothermal energy has few polluting problems itself, but there are some drawbacks in its processing. For example, steam can sometimes bring up toxic heavy metals, sulfur, minerals, salts, radon, and toxic gases. If water or steam is vented above ground as in an open-loop system, it can pollute. Scrubbers can filter out toxic components, but they produce hazardous sludge that has to be disposed of later. As we learned in Chapter 12, hazardous waste contaminates soil and potentially groundwater. The bright side is that geothermal power plants create far fewer pollutants than fossil fuels. In a closed-loop (recycled) system, these problem pollutant don't come above the ground.

Another potential drawback is that a lot of geothermal resource areas, like Yellowstone National Park, are located in pristine, environmentally sensitive areas. The careless construction of geothermal plants in these areas could greatly impact local ecology.

In addition to industry and governments, individuals are becoming increasingly interested in using clean power, even when it costs more. The reality of the world's pollution and global warming situation is just beginning to hit home around the world.

In the next chapter, we'll look at green energy in even more detail and explore the latest cutting-edge research and energy options.

Quiz

1. What do marine farmers use to increase the growth of fish, shellfish, reptiles, and amphibians?
   (a) High-protein flies and worms
   (b) Highly saline water
   (c) Geothermally heated water
   (d) Marine steroids

2. Geothermal reservoirs can reach temperatures of
   (a) 370°C
   (b) 560°C
   (c) 740°C
   (d) 820°C

3. Impoundment, diversion, and pumped storage are all types of
   (a) solar cells
   (b) wind turbines
   (c) dams
   (d) hydroelectric power plants

4. The first geothermal district heating system in the United States was established in Boise, Idaho, in
   (a) 1864
   (b) 1893
   (c) 1902
   (d) 1910

5. Flowing water that creates energy and is turned into electricity is called
   (a) nuclear power
   (b) hydroelectric power
   (c) solar power
   (d) thermal energy

6. What force is perpendicular to the lift force of a wind turbine rotor?
   (a) Friction
   (b) UV radiation
   (c) Heat
   (d) Fusion

7. The following are all types of geothermal power plants, except
   (a) flashed steam
   (b) hybrid
   (c) trinary
   (d) dry steam

8. When deep underground heat is transferred by thermal conduction through water to the surface, it is called
   (a) solar energy
   (b) nuclear energy
   (c) wind energy
   (d) geothermal energy

9. Propeller-like turbine blades are used to generate electricity from
   (a) sunlight
   (b) rain
   (c) wind
   (d) snow

10. What percentage the United States' energy needs are met by geothermal power?
    (a) 50%
    (b) 35%
    (c) 20%
    (d) 10%

CHAPTER 16

Future Policy and Alternatives

No one argues that environmental interrelationships are very complex. Everything is connected to everything else. Natural climate regulation and chemical interactions among the atmosphere, water, and soil result in an intricate dance that affects the rise and fall of natural processes and biodiversity.

Environmental protection and problem solving, combined with technological progress and global energy requirements, are like strands in a mass of knotted string. Whenever one strand is pulled, a tightening takes place in another spot. Working the knots free takes time and patience, but with careful study and a little luck, it can be accomplished. Any long-term environmental or energy solution must involve balance and sustainability.

Sustainability

The Environmental Performance Measurement Project—a collaborative effort between the Center for International Earth Science Information Network

(CIESIN) at Columbia University, and the World Economic Forum—produces the Environmental Sustainability Index (ESI). The ESI is a compiled index that tracks a diverse set of socioeconomic, environmental, and institutional indicators that characterize and influence environmental sustainability at the national level. This index provides a valuable policy tool, allowing benchmarking of environmental performance by country and issue.

In the January 28, 2005, ESI, Finland scored highest in environmental sustainability out of 146 countries. The top five countries were Finland, Norway, Uruguay, Sweden, and Iceland. Their high ESI scores resulted from substantial natural resources, low population density, and successful management of environment and development issues.

ESI ranks countries on 21 factors of environmental sustainability, including natural resources, past and present pollution levels, environmental management efforts, protection of global resources, and the capacity to improve environmental performance over time.

The United States ranked 45th—after the Netherlands (44) and ahead of the United Kingdom (46). The United States got good scores on issues like water quality and environmental protection capacity. Low scores on other issues, however, such as waste generation and greenhouse gas emissions, pulled the overall U.S. rank down.

The lowest-ranked countries—North Korea, Taiwan, Iraq, Turkmenistan, and Uzbekistan—face many challenges, both natural and manufactured that are related to poor environmental policy and changing political structures.

The 2005 ESI showed income as a critical driver of environmental results. However, at all levels of economic development, some countries managed environmental issues well while others did not. Developed countries face environmental challenges, particularly pollution stresses and consumption-related issues, separate from those of developing countries where dwindling natural and economic resources make pollution control difficult.

GOING GREEN

When you hear the expression "going green," do gardeners, children who eat too many hot dogs, and avid golfers come to mind? Or maybe organic vegetable farmers or Costa Rican tree farms? In fact, *going green* and *green power* can be confusing to the general public or anyone not focused on today's environmental issues. Commonly, going green means choosing alternative processes and products that allow the environment to sustain itself. Instead of always taking, we take only what we need, and if possible, give back.

Energy (electricity) production is considered **green energy** if it is created without causing any harmful environmental impact.

With new technology, a ton of cooperation, and a little luck, going green doesn't have to be an impossible mission. Humans, who are used to facing challenges, have often overcome hurdles through science and engineering.

The discovery of penicillin in 1928, which wiped out global infectious killers, was a serendipitous event (i.e., discovered by accident). Those lucky events could still happen, but we can't count on one or two "tricks up the energy sleeve." We have to rethink, relearn, recycle—and redo our old fossil fuels ways with green alternatives.

Green Power

Clean energy is electricity produced from renewable energy sources, like wind, solar, hydroelectric, geothermal, and low-emission energy sources like landfill gas. Compared to common power plants (coal, gas, and oil), green power has an extremely low environmental impact. In other words, it's a very good thing for the environment. However, most of these technologies today can't meet global energy demands. Fig. 16-1 shows how the world's overall energy usage is distributed among various sources.

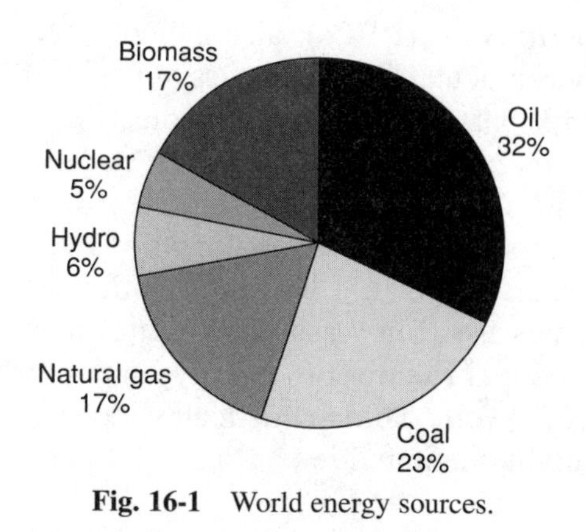

Fig. 16-1 World energy sources.

Most power plants add to greenhouse gases, acid rain, smog, and toxic mercury emissions. They also consume water resources, pollute water bodies, and impact local sites, as with solid waste disposal and fuel processing. Depending on a plant's age, size, emissions technology, and waste disposal procedures, environmental impact is high to low.

The good news is that things are changing. Along with consumer demand for clean renewable energy, deregulation of the utilities industry is beginning to drive green power growth. Alternatives like solar, wind, geothermal steam, biomass, and small-scale hydroelectric power sources are becoming more feasible and readily accepted. Small commercial solar power plants have begun serving some energy markets.

Green power is produced by a renewable "green" energy source, distinct from power produced by fossil fuel, nuclear, and other types of generators.

Marketing divisions of energy companies have begun offering renewable, green energy options as a new, environmentally savvy customer choice. In some states, like California, environmental standards are high and meet a strict definition of *green power.* Other states don't have to meet a state standard. An environmentally smart buyer will read the fine print and investigate the exact energy source mix offered by utilities.

RENEWABLE ENERGY CERTIFICATES

Renewable energy certificates (RECs), also called *green certificates* or *green tags,* describe the environmental characteristics of power from renewable energy projects and are sold separately from general electricity. Consumers can buy green certificates whether or not they have access to green power through their local utility. They can also buy green certificates without having to switch electricity suppliers. Today, over 30 organizations sell wholesale or retail green energy certificates. Table 16-1 lists information from the February 2004 green power certificates list (for Washington state) maintained by the Green Power Network, a National Renewable Energy Laboratory project of the U.S. Department of Energy. (*Note:* Some certificates are specific to certain states, while others are available nationwide.)

Table 16-1 Green certificates.

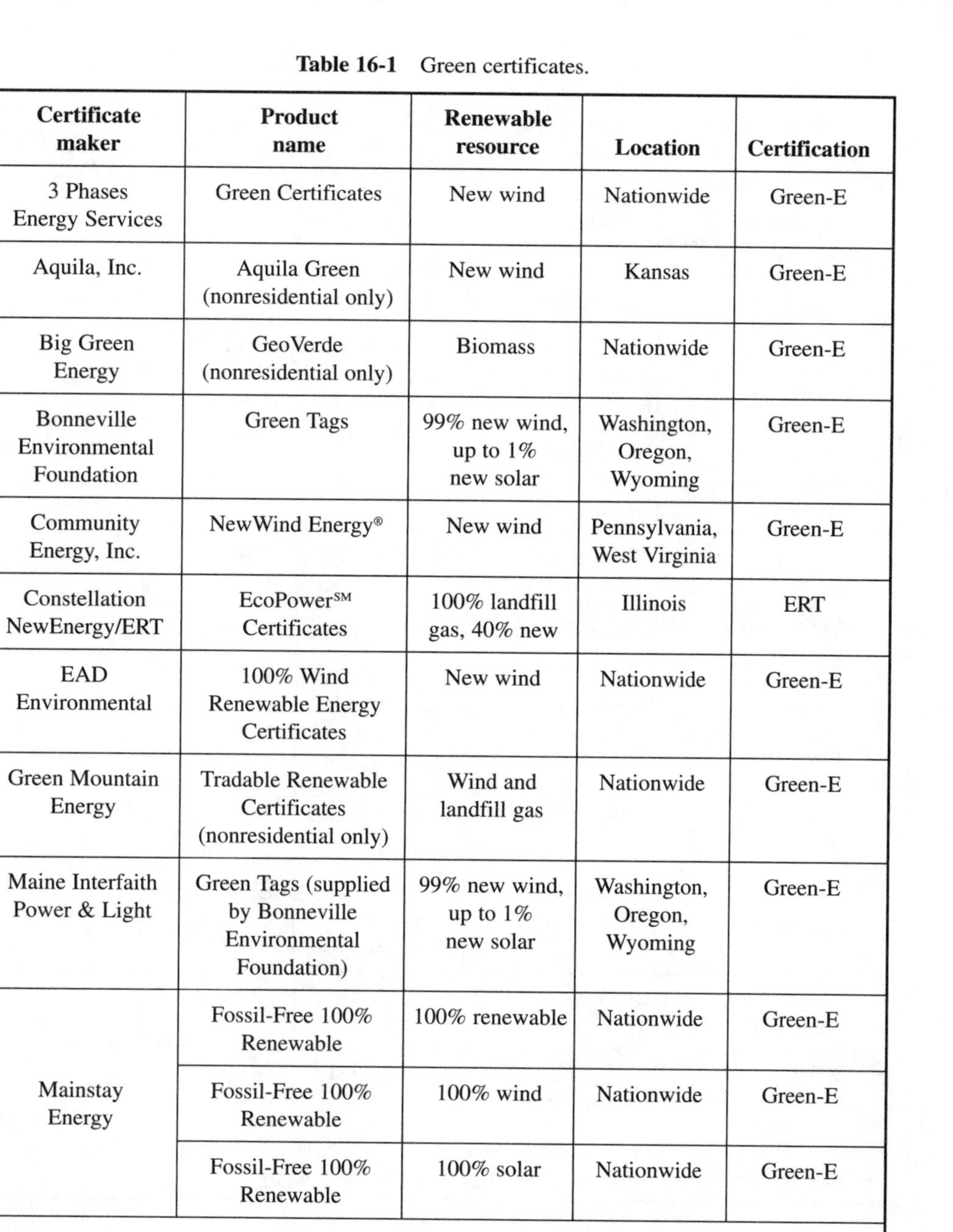

Certificate maker	Product name	Renewable resource	Location	Certification
3 Phases Energy Services	Green Certificates	New wind	Nationwide	Green-E
Aquila, Inc.	Aquila Green (nonresidential only)	New wind	Kansas	Green-E
Big Green Energy	GeoVerde (nonresidential only)	Biomass	Nationwide	Green-E
Bonneville Environmental Foundation	Green Tags	99% new wind, up to 1% new solar	Washington, Oregon, Wyoming	Green-E
Community Energy, Inc.	NewWind Energy®	New wind	Pennsylvania, West Virginia	Green-E
Constellation NewEnergy/ERT	EcoPower℠ Certificates	100% landfill gas, 40% new	Illinois	ERT
EAD Environmental	100% Wind Renewable Energy Certificates	New wind	Nationwide	Green-E
Green Mountain Energy	Tradable Renewable Certificates (nonresidential only)	Wind and landfill gas	Nationwide	Green-E
Maine Interfaith Power & Light	Green Tags (supplied by Bonneville Environmental Foundation)	99% new wind, up to 1% new solar	Washington, Oregon, Wyoming	Green-E
Mainstay Energy	Fossil-Free 100% Renewable	100% renewable	Nationwide	Green-E
	Fossil-Free 100% Renewable	100% wind	Nationwide	Green-E
	Fossil-Free 100% Renewable	100% solar	Nationwide	Green-E

**Data on green power options (February 2004) is from the Green Power Network (National Renewable Energy Laboratory for the U.S. Department of Energy).*

(*continued*)

Table 16-1 Green certificates. *(continued)*

Certificate maker	Product name	Renewable resource	Location	Certification
Mass Energy/People's Power and Light	New England Wind℠	New Wind	Massachusetts	Green-E
*Native*Energy	Wind Builders℠	New wind	South Dakota	—
	Vermont Cool/Home℠ (residential only)	New biomass (diary farm methane); and new wind	Vermont (biomass); South Dakota (wind)	—
PG&E National Energy Group	Pure Wind℠ Certificates	New wind	New York	—
Peoples Energy Services/ERT	EcoPower Certificates	100% landfill gas, 40% new	Illinois	ERT
Renewable Choice Energy	American Wind™	New wind	Nationwide	Green-E
Sterling Planet	Green America Program™	40% wind, 35% biomass, 15% geothermal, 5% low-impact hydro, 5% solar (all new)	Nationwide	Green-E
Sun Power Electric	ReGen	99% new landfill gas, 1% new solar	Massachusetts, Rhode Island	Green-E
Waverly Light & Power	Iowa Energy Tags™	Wind	Iowa	—
WindCurrent	Chesapeake WindCurrent	New wind	Midatlantic states	Green-E

**Data on green power options (February 2004) is from the Green Power Network (National Renewable Energy Laboratory for the U.S. Department of Energy).*

ENVIRONMENTAL RESOURCES TRUST

Environmental Resources Trust, Inc. (ERT) is a Washington, D.C.–based non-profit organization that uses market forces to protect and improve the global environment. Established in 1996, ERT uses energy markets to meet the challenges of climate change, secure clean and reliable power, and encourage sustainable land use.

ERT is made up of the National Audubon Society, the National Fish and Wildlife Foundation, Environmental Defense, and the German Marshall Fund. ERT energizes green power markets by supplying important auditing and verification services. These services support consumer confidence by verifying that new green power sources are actually being brought online.

ERT has three focused programs to carry out its mission. The GHG Registry^{SM} validates industrial greenhouse gas emission profiles by creating a market that can facilitate emission decreases. The EcoPower^{SM} program verifies and promotes blocks of clean power from new renewable energy sources. ERT's EcoLands^{SM} program works out plans to encourage and assist landowners in their land use decisions. The bulk of ERT's Clean Power program involves:

1. Verifying specific energy blocks as green and giving them the EcoPower^{SM} label;
2. Creating a Power Scorecard Rating System that categorizes the environmental attributes of various clean power blocks;
3. Advertising EcoPower^{SM} blocks and negotiating their sale from generators to consumers, like municipalities;
4. Developing an EcoPower^{SM} ticket program that guarantees consumers a definite claim to purchased power; and
5. Auditing EcoPower^{SM} energy blocks and preparing verification reports.

ERT's Power Scorecard Rating System, verification, and marketing have done a lot to raise public awareness of the clean power market. EcoPower^{SM} tickets make it possible for clean energy generated in one area to be available for sale elsewhere.

ERT is supported by philanthropic contributions and fee-for-service revenues. It is nonpartisan and addresses environmental issues through economic methods. Today, legitimate energy providers who advertise clean energy to consumers have each kilowatt hour verified as "new, certified, zero-emissions renewable power" (EcoPower or Green Power) by ERT.

GREEN-E PROGRAM

Within the Green-E Renewable Branding Program established in California, power suppliers must meet specific power criteria. Companies that meet these standards can use the Green-E logo if they also follow strict rules of professional conduct. These Green-E standards include the following:

- Power must contain 50% or more renewable energy content averaged over one year;
- The fossil fuel portion (if any) of Green-E power must have equal or lower air emissions than an equivalent amount in the power grid;
- Air emissions from renewable power produced from waste materials (like biomass) have to be equal to or less than air emissions normally produced during disposal of the waste; and
- Green-E power can't contain any nuclear power beyond that currently included in the power grid.

A Green-E standard has been developed that requires eligible energy products to get a certain part of their renewable electricity content from an alternative, renewable power source. An independent board of consumer, environmental, and other public interest groups governs the Green-E program.

However, like everything else worth having, energy alternatives won't be cheap, at least not initially. Consumers will have to pay more for a cleaner electricity blend, as much as 10 to 20% more per month. The good thing about it is that the money will be applied to even more development of clean energy sources.

Green Energy Highlights

The largest geothermal field and biggest resource of green power worldwide is The Geysers, California, generating 1000 MW of electricity for 1 million California households. This meets 70% of the power demand between San Francisco and Oregon. To add to this resource, Calpine Corporation and the city of Santa Rosa built a 41-mile pipeline that pumps 11 million gallons/day of reclaimed water from the city's water treatment plant to The Geysers. This recycled water, injected into the geothermal reservoir, increases steam production, but also cools the underground reservoirs.

The injected water, heated naturally by local geothermal resources, boils to provide added steam for power plants that convert it into electricity. Besides increasing power production, the water injection helps reduce the amount of naturally occurring gas that must be eliminated to meet air quality standards. The innovative Santa Rosa Geysers Recharge Project not only increased power generation by 85 MW, but redirected the city's treatment plant water that had been released into the Russian River. The project, costing nearly $250 million, began operations in December 2003.

Edison Source, a subsidiary of Edison International Co., has electricity options known as *Earthsource 50* and *Earthsource 100* that have earned the Green-E certification. These environmentally improved sources get 50% and 100%, respectively, of their generation from renewable resources, including wind, solar, geothermal, biomass, and hydroelectric power. In return for protecting the environment, most people's average monthly bill increased about $17.00—the cost of a large everything-on-it pizza.

Green Mountain Energy Resources (GMER) provides a Green-E certified wind generation program. It builds a new wind turbine at a Wyoming wind project for every 3000 customers who choose the program. The turbines create up to 10% of the electricity needed, with the rest coming from at least 75% renewable sources. The GMER program requires a three-year obligation that raises an average monthly bill about $11.55.

In December of 2004, several major corporations agreed to buy 62 megawatts of green power. As part of the World Resources Institute (WRI), the Green Power Market Development Group bought 62 megawatts of electricity from renewable energy sources in a 12-month period. It then established a corporate partnership interested in using green power. The partnership's members included Alcoa Inc., Cargill Dow LLC, General Motors Corporation, Johnson & Johnson, Delphi Corporation, Dow Chemical USA, DuPont, FedEx Kinko's, Interface Inc., Pitney Bowes, IBM, and Staples.

The 62 MW of green power was enough to power more than 80 corporate facilities in 18 states. It included 39 MW of certified renewable energy credits (21 MW of biomass power and 18 MW of wind power), 21 MW of power from landfill gas (supported by DuPont and Johnson & Johnson), and 2 MW of wind and solar power facilities at Johnson & Johnson and IBM. In addition, Staples is installing two 280 kW solar power systems at its California facilities. According to the WRI, five of the corporate green power partners now use renewable energy for 10% or more of their United States power requirements.

GREEN PRACTICES

It's important to remember that "going green" is not just about transportation (fuel) and energy needs, it's about balance. It's about using only what we need and leaving the rest. It's about rethinking the way we look at global resources. It's about respect for ourselves, our neighbors, and ultimately, our planet.

If industrialized nations don't soon realize that the free ride is over and begin to plan for the future, the consequences will be severe. Fouled air, water, and soil will be the inheritance of generations to come. Nations struggling to rise above baseline existence will never get there.

At this rate, the future looks dim, but it doesn't have to. Even now, brilliant minds around the world have turned their focus on environmental and energy issues. Policy makers and global energy companies are realizing that a change is critical. Advances are being made.

In order to assist individuals, communities, states, and industries in going green, the DOE division of Energy Efficiency and Renewable Energy (EERE) was established. This information outlet provides hundreds of Web sites and thousands of online energy efficiency and renewable energy documents that describe green concepts in the following areas:

- Energy Efficiency: buildings, industry, power, and transportation;
- Green Power Information: consumers, children, and states;
- Renewable Energy: biomass, geothermal, hydrogen, hydroelectric, tidal, solar, and wind;
- Financing/Investment; and
- Education.

As electricity markets change, everyone will have the chance to choose between electricity suppliers and various power generation sources. Additionally, energy deregulation, advancing technology, and other incentives make clean energy even more workable.

One of the major factors for using green power sources is the rising cost of oil and natural gas. Since long-term fossil fuel costs will only go up as supply dwindles, economic incentives for green energy will look better and better.

In deregulated states, residential market demand for clean power is between 1 and 3% of households. Roughly 1.0 billion kWHs of clean power/year is now sold residentially.

Nanotechnology

According to scientific advances in the past five years, nanotechnology and its growing potential could help revolutionize the energy industry, especially in the areas of storage and transmission. Advances in solar power cells made of plastics, as well as batteries that renew themselves, are definitely good for the environment.

Nanotechnology is the study of compounds at the single-atom level, or 10^{-9}-meter (nanometer) scale.

Nanotechnology is able to manipulate structures on the atomic and molecular levels, the nanometer scale (billionths of a meter). Because nanomaterials have a lot more surface area for chemical reactions or storage to take place, they act as *super-catalysts* for many different applications. This large surface area improves the strength, electrical, and thermal properties of nanomaterials. It can be compared to the differences between a regular man and Superman (the cartoon superhero from the planet Krypton who could leap tall buildings in a single bound). That's how different carbon nanomaterials are from regular forms of carbon like graphite. They are stronger, faster, and have characteristics that scientists only dreamed about a decade ago.

Nanotechnology advances offer the chance to leap beyond our current alternatives for energy supply by introducing technologies that are more efficient, inexpensive, and environmentally sound.

Research of the past 25 years has shown us an increasingly troublesome picture of continued fossil fuel use. Our worldwide energy problem requires groundbreaking new technology along with enhancement of current technologies. Energy transmission and storage efficiencies must be improved using inexpensive, environmentally safe materials if they are to be widely accepted as a fossil fuel replacement.

Nanotechnology is beginning to look like the solution of choice. Nanotubes and other nanomaterials provide unique ways to transport electricity efficiently and at a lower cost over long distances.

Of course, the reduction/elimination of the United States' reliance on foreign oil is a major goal, but another big challenge will be whether nanoscience can provide the advances to implement widespread collection, conversion, and transmission of affordable solar energy.

NANOTECHNOLOGY APPLICATIONS

MPhase Technologies in Norwalk, Connecticut, has joined with Lucent Technologies to commercialize nanotechnology by creating intelligent batteries. Thought to be available within the next 12 to 18 months, these will last a very long time.

Normally, regular batteries contain metal electrodes in a solution of chemicals called *electrolytes.* When a battery is turned on, the electrolytes react with each other and electrons move between the electrodes. After much use, the electrolytes continue to react (even when turned off), and the battery's energy is drained. The new nanotechnology batteries separate electrolytes from the metal electrodes and don't lose energy when not in use.

Additionally, normal batteries don't work well with semiconductor processing, are slow to power up, and can't be integrated with computer chips. The new, improved nanotechnology battery has millions of silicon nanotube electrodes, standing vertical like a group of straws with a droplet of electrolyte on top of each nanotube (straw). The droplets can exist without interacting. However, when a voltage change forces the droplets into the spaces between the tubes, they react and create an electrical current.

The advantage of this nanotechnology battery is that it only creates energy when needed. This gives it a storage life of many years. The new silicon-based batteries are inexpensive, compatible with semiconductor processes, and reach full power quickly. They also are suitable for miniaturization, and mass production, and have high energy density.

An added benefit is that nanotube electrodes neutralize potentially toxic electrolytes during disposal. The green impact is that they won't pollute the environment or become a disposal problem.

Konarka Technologies of Lowell, Massachusetts makes plastic panels that absorb sunlight and indoor light and convert them into electricity. With the thickness and flexibility of wax paper, the panels can be introduced into fabrics and roofs. Created using nanoscale titanium dioxide particles coated in photovoltaic dyes, the panels produce electricity when light hits the dye. Lightweight and more flexible than earlier solar cells, Konarka's panels have wide applicability for power generation nationwide.

International Policy

An intergovernmental organization pledged to energy supply, economic growth, and environmental sustainability through energy policy collaboration is the International Energy Association (IEA). The 26 member countries are listed in Table 16-2.

International Energy Association countries acknowledge the world's growing energy interdependence. IEA members want to strengthen their energy segments to create the highest sustainable impact, as well as for public and environmental benefit. While making energy policies, free and open markets are key, along with energy security, maintenance, and environmental protection.

Table 16-2 The diversity of the IEA's membership points to global interest in green energy.

International Energy Association Members		
Australia	Greece	New Zealand
Austria	Hungary	Norway
Belgium	Ireland	Portugal
Canada	Italy	Spain
Czech Republic	Japan	Sweden
Denmark	Republic of Korea	Switzerland
Finland		Turkey
France	Luxembourg	United Kingdom
Germany	The Netherlands	United States

To further its goals, the IEA intends on adding diversity, efficiency, and flexibility to current energy generation programs. Green power energy sources will be encouraged and developed. They view clean and efficient use of fossil fuels as essential, but the development of nonfossil sources is also a priority.

Currently, nuclear and hydroelectric power are the main nonfossil fuel additions to the energy makeup of IEA countries. Since nuclear energy doesn't give off carbon dioxide, several IEA member countries plan to develop nuclear power at even higher safety standards than currently used. Renewable sources will also have an increasingly important contribution to make.

Improved global energy efficiency improves the environment and energy security in a big way. Ways to increase energy efficiency are possible throughout the energy cycle from production to consumption. A unified effort by individuals, cities, states, industry, and governments is critical.

Energy research, development, and marketing all become important as everyone turns and moves in the green power direction. Collaboration between energy consumers and providers worldwide will spread information and understanding, as well as encourage the development of environmentally adaptable power systems.

Individual countries' efforts are important and will help get the process moving faster. France has started including European companies in its innovative energy research funding in attempt to improve its own and the European Union's technological expertise.

France has also discussed joint funding with Germany for cutting-edge projects in high-technology industries, such as nanotechnology, biotechnology, solar cells, and environmentally friendly cars. The French government plans to increase high-tech energy investment in health, transportation, and energy. Their cooperative approach will involve industry, government, public laboratories, and potential buyers to focus critical research selection.

North America

North America is one of the world's most important energy regions, providing around 25% of global energy supply and using about 30% of the world's commercial energy. To address their joint energy needs and goals, U.S. President Bush, Canadian Prime Minister Chretien, and Mexican President Fox established the North America Energy Working Group (NAEWG) in 2001.

The NAEWG goals are to strengthen communication and cooperation among the three governments and energy sectors on common energy-related issues. While designing sustainable North American energy trade policy, the NAEWG also takes into account the domestic policies, jurisdictional authority, and existing trade obligations of each country.

To accomplish this, the NAEWG exchanges information on key factors affecting North American energy, like market trends, supply sources, technical specifications, and environmental science and technology development.

With all the progress being made in the United States and across the globe, the future looks bright. Consumer and industrial incentives at all levels are being implemented. The possibilities are growing as the reality of declining fossil fuel resources is slowly being understood.

The outlook for sustainable and well-stewarded Earth resources is coming into focus.

Quiz

1. Renewable energy certificates describe
   (a) coupons from fast food restaurants
   (b) electricity obtained from fossil fuels
   (c) only investments in solar power
   (d) environmental characteristics of power from renewable energy projects

2. The study of compounds at the single atom level or 10^{-9}-meter scale is called
   (a) chemical engineering
   (b) nanotechnology
   (c) microbiology
   (d) kinesiology

3. "Going green" means choosing alternative processes and products that
   (a) are used once and then thrown away
   (b) allow the environment to sustain itself
   (c) involve artificial turf/grass in sports stadiums
   (d) emit a small amount of radiation

4. The Geysers, a geothermal resource in California, generates what percentage of power demand to San Francisco and Oregon?
   (a) 35%
   (b) 50%
   (c) 70%
   (d) 90%

5. What country plans to fund innovative research projects in energy, a bid to boost their own and the European Union's competitiveness?
   (a) Holland
   (b) Italy
   (c) Belgium
   (d) France

6. One of the first states to enact environmental standards and a strict definition of green power was
   (a) California
   (b) North Dakota
   (c) Texas
   (d) Iowa

7. General Motors Corporation, Johnson & Johnson, Delphi Corporation, Dow Chemical USA, DuPont, and FedEx, among others
   (a) pay their employees in U.S. currency only
   (b) all have divisions in Madagascar
   (c) use only solar power for electricity generation
   (d) established a corporate partnership interested in using green power

8. For a cleaner electricity blend, consumers will have to pay as much as
   (a) 1–2% more per month
   (b) 5–8% more per month
   (c) 10–20% more per month
   (d) 25–40% more per month

9. The strength, electrical, and thermal properties of nanomaterials comes primarily from their
   (a) slippery skins
   (b) knotted ends
   (c) large surface area
   (d) color

10. Electricity production is considered green energy if it is created
    (a) without causing any harmful environmental impact
    (b) in the Rocky Mountains
    (c) only at night
    (d) by cutting down old growth forests

Part Four Test

1. What is produced by slicing thin wafers from a high-purity single crystal?
   (a) Monocrystalline silicon
   (b) Heterocrystalline sandstone
   (c) Diamonds
   (d) Graphite

2. Small brown-black blobs of oil that stick to your feet when you go to the beach are commonly known as
   (a) gumballs
   (b) snowballs
   (c) racquetballs
   (d) tar balls

3. Photovoltaic arrays follow the sun's path through the sky during the day using
   (a) single-axis tracking
   (b) two-axis tracking
   (c) three-axis tracking
   (d) disposable generators

4. Internal geothermal heat, fueled by core processes, supplies energy for all of the following except
   (a) plate tectonics
   (b) earthquakes
   (c) volcanic eruptions
   (d) beach erosion

5. Where plates collide and one plate is forced under another, it is known as
   (a) abduction
   (b) subduction
   (c) construction
   (d) liposuction

6. Buildings designed for passive solar and day lighting use features like
   (a) granite countertops
   (b) few windows or doors
   (c) large, south-facing windows and construction materials that absorb and slowly release the sun's heat
   (d) hardwood floors

7. The way the Earth's natural atmospheric gases decrease the amount of heat released from the atmosphere is called the
   (a) ventilation cycle
   (b) greenhouse effect
   (c) calcium cycle
   (d) lighthouse effect

8. Droughts affect hydroelectric plant operations because, when water is low or absent,
   (a) fish can't swim upstream
   (b) irrigation needs less water
   (c) plants can't make electricity
   (d) turbines turn faster

9. The uranium oxide mineral uraninite is commonly called
   (a) houndstooth
   (b) molasses
   (c) pitchblende
   (d) vermillion

10. The ERT's Clean Power Program includes all of the following except
    (a) verifying specific energy blocks as green
    (b) creating a Power Scorecard Rating System
    (c) developing an EcoPower[SM] ticket program that guarantees purchased power
    (d) investing in large quantities of foreign coal, oil, and gas

11. Aerodynamic lift is used in what type of power generation?
    (a) Solar
    (b) Hydroelectric
    (c) Wind
    (d) Biomass

12. Who is often called the Father of the Petroleum Industry?
    (a) Jed Clampett
    (b) Abraham Gesner
    (c) Howard Hughes
    (d) Douglas Williams

13. Solar cells
    (a) function more efficiently under focused light
    (b) function less efficiently under focused light
    (c) cannot function under focused light
    (d) only use focused light to generate heat

14. The most common transportation fuels in the United States today are
    (a) whale oil and animal fat
    (b) coal and solar power
    (c) gasoline and diesel fuel
    (d) ethanol and propane

15. In what year is the Yucca Mountain, Nevada, nuclear waste depository supposed to open?
    (a) 2005
    (b) 2008
    (c) 2010
    (d) 2015

16. What was the calculation called that determined the oil well extraction rate in the United States (lower 48 states) would peak around 1970, and then drop?
    (a) Pike's Peak
    (b) Hummer's Peak
    (c) Downward Spiral
    (d) Hubbert's Peak

17. The temperature gradient of the earth's crust is
    (a) 1–2°C per km of depth
    (b) 3–8°C per km of depth
    (c) 10–15°C per km of depth
    (d) 17–30°C per km of depth

18. By the year 2005, how many megawatts (MW) of wind energy will be generated worldwide?
    (a) nearly 20,000 MW
    (b) nearly 35,000 MW
    (c) nearly 50,000 MW
    (d) nearly 75,000 MW

19. The Three Mile Island nuclear power plant in Pennsylvania was made up of how many units?
    (a) 1
    (b) 2
    (c) 3
    (d) Three Mile Island was a solar power plant

20. Roughly what percentage of the world's nuclear power plants is found in the United States?
    (a) 15%
    (b) 25%
    (c) 40%
    (d) 55%

21. Which nonrenewable energy source takes millions of years to form?
    (a) Wind
    (b) Petroleum
    (c) Sewage
    (d) Solar

22. What is the main vehicular fuel in Brazil?
    (a) Ethanol
    (b) Propanol
    (c) Vinegar
    (d) Gas hydrates

23. Hydroelectric power plants affect water quality and flow and can cause
    (a) overgrowth of lily pads
    (b) fish to spawn prematurely
    (c) brownouts in metropolitan areas
    (d) low dissolved oxygen levels in the water

24. Coal provides nearly what fraction of the world's energy?
    (a) $^1/_8$
    (b) $^1/_4$
    (c) $^1/_3$
    (d) $^1/_2$

25. All of the following are used to clean up oil spills except
    (a) brooms
    (b) skimmers
    (c) sorbents
    (d) vacuum trucks

26. In January 2002, Yucca Mountain, Nevada was named as the
    (a) newest vacation retreat west of the Rockies
    (b) gambling capital of the world
    (c) nation's permanent nuclear waste depository
    (d) best glow-in-the-dark spa in the Western Hemisphere

27. In the United States, there are around 80,000 dams. However, only
    (a) 1500 produce power
    (b) 2400 produce power
    (c) 2800 produce power
    (d) 3200 produce power

28. Choosing alternative processes and products that allow the environment to sustain itself is called
    (a) being thrifty
    (b) natural selection

(c) going green
(d) fuel economy

29. What cutting-edge technology shows great potential in the areas of energy storage and transmission?
(a) Nanotechnology
(b) Biosensors
(c) Holographic imaging
(d) Advanced robotics

30. The uranium oxide mineral uraninite is found in
(a) sandstone
(b) granite and other volcanic rocks
(c) clays
(d) sodium chloride

31. In the western United States, how many communities are close enough to geothermal reservoirs to use geothermal district heating?
(a) over 50
(b) over 135
(c) over 275
(d) over 350

32. Green energy products must get a large percentage of their power from a mixture of
(a) blue-green algae
(b) nuclear isotopes
(c) biomass and battery recycling
(d) wind, solar, geothermal, biomass, and hydroelectric power

33. NOAA recommends that people sensitive to chemicals, including crude oil and petroleum products, avoid beaches with
(a) sand crabs
(b) tar balls
(c) small children
(d) jelly fish

34. In 1857, Michael Dietz invented a
(a) water-flushing toilet
(b) carbon-filament light bulb
(c) clean-burning kerosene lamp
(d) robotic vacuum cleaner

35. This most common type of hydroelectric power plant that uses a dam to store river water in a reservoir is called a(n)
    (a) diversion plant
    (b) impoundment plant
    (c) pumped storage plant
    (d) compoundment plant

36. NRC is the acronym for what energy agency?
    (a) National Rubber Committee
    (b) Nuclear Registration Committee
    (c) National Registered Cranes
    (d) Nuclear Regulatory Commission

37. If the individual percentages of electricity provided by geothermal energy in Utah, California and northern Nevada were added together, they would total
    (a) 12%
    (b) 15%
    (c) 18%
    (d) 25%

38. A working fluid, like isobutene, boils and flashes to a gas at
    (a) a lower temperature than water
    (b) the speed of light
    (c) a higher temperature than water
    (d) –273°C

39. When core temperatures escalate out of control in a nuclear power plant, causing a core meltdown, it is known as
    (a) The French Syndrome
    (b) Murphy's Law
    (c) The Law of Inverse Proportions
    (d) The China Syndrome

40. The Pacific Ocean's Ring of Fire is
    (a) low in geothermal potential
    (b) a film about hobbits
    (c) high in geothermal potential
    (d) a tale from Greek mythology

Final Exam

1. At sea level, pure water boils at 100°C and freezes at
   (a) 0°C
   (b) 4°C
   (c) 10°C
   (d) 32°C

2. Carbon dioxide, methane, and nitrous oxide are all
   (a) used in carbonated drinks
   (b) produced by gaseous cattle
   (c) greenhouse gases
   (d) used by dentists

3. Water has the highest surface tension of any liquid except
   (a) mercury
   (b) split pea soup
   (c) iodine
   (d) hot chocolate

4. The crust is
   (a) something on top of pies
   (b) the thinnest of the Earth's layers

(c) found around the edge of volcanoes
(d) the thickest of the Earth's layers

5. Greenhouse gases are
   (a) mostly released from gardening sheds
   (b) primarily argon and xenon
   (c) a natural part of the atmosphere
   (d) always found on the moon

6. A stream that flows into another stream is called a
   (a) brook
   (b) river
   (c) tributary
   (d) creek

7. Pitchblende is another name for
   (a) calcium chloride
   (b) uranium oxide
   (c) bismuth
   (d) magnesium sulfate

8. The biggest human-supplied gas to the greenhouse effect is
   (a) methane
   (b) carbon dioxide
   (c) sulfur dioxide
   (d) ammonia

9. What percentage of the world's population lives within 100 km of a coastline?
   (a) 25%
   (b) 40%
   (c) 60%
   (d) 72%

10. Since heat moves from hotter areas to colder areas, the Earth's heat moves from its fiery center towards its
    (a) North Pole
    (b) moon
    (c) core
    (d) surface

11. The average amount of time an element, like carbon or calcium, spends in a geological reservoir is known as
    (a) pool time
    (b) cosmic time
    (c) residence time
    (d) vacation time

12. Water held within plants returns to the atmosphere as a vapor through a process called
    (a) resuscitation
    (b) evacuation
    (c) transpiration
    (d) evolution

13. The upper mantle is also called the
    (a) rind
    (b) asthenosphere
    (c) ionosphere
    (d) coresphere

14. Where fresh river water joins salty ocean water, it is known as
    (a) polluted
    (b) pure
    (c) silt
    (d) brackish

15. Reykjavik, Iceland's capital, is
    (a) sooty and polluted
    (b) heated completely by geothermal energy from volcanic sources
    (c) an ice-covered wasteland
    (d) balmy and temperate

16. Marine biologist Rachel Carson, wrote what eye-opening book in 1962?
    (a) *The Silent Stream*
    (b) *The Silent Spring*
    (c) *The Noisy Stream*
    (d) *The Polluted Spring*

17. The length of time that water spends in the groundwater portion of the hydrologic cycle is known as the
    (a) carbon cycle
    (b) Krebs cycle
    (c) Cenozoic era
    (d) residence time

18. Africa, Antarctica, Australia, Eurasia, North America, and South America make up the
    (a) hot continents
    (b) prime vacation spots for families
    (c) six major landmasses
    (d) six wonders of the world

19. Nitrogen's residence time in the environment is approximately
    (a) 40 years
    (b) 4000 years
    (c) 4 million years
    (d) 400 million years

20. Hazardous waste is all of the following except
    (a) highly flammable
    (b) corrosive
    (c) nontoxic
    (d) reactive

21. The 1979 Hollywood movie called *The China Syndrome* was about
    (a) nuclear power
    (b) geothermal power
    (c) solar power
    (d) wind power

22. The middle layer of the three ocean density layers is known as the
    (a) pycnocline
    (b) thermocline
    (c) deep layer
    (d) halocline

23. When water soaks into the ground, it's called
    (a) evaporation
    (b) transpiration

(c) infiltration
(d) precipitation

24. What simple mode of power generation was once used to grind grain and pump water?
 (a) Gnat power
 (b) Nuclear power
 (c) Windmills
 (d) Hamster power

25. Fair weather cumulus clouds look like
 (a) Elvis
 (b) horses' tails
 (c) floating cotton balls
 (d) hanging fruit

26. Water is
 (a) lighter than air
 (b) heavier than air
 (c) drier than air
 (d) about the same as air

27. The layer of rock that drifts slowly over the supporting, malleable upper mantle layer is called
 (a) magmic rock
 (b) a guyout
 (c) a geological plate
 (d) bedrock

28. What is thought to create the most greenhouse gases?
 (a) Cattle burping
 (b) Jet contrails
 (c) Fertilizers
 (d) Burning of fossil fuels

29. When tributary streams link up with a larger stream at 90° angles, it is known as
 (a) dendritic drainage
 (b) radial drainage
 (c) rectangular drainage
 (d) trellis drainage

30. What is it called when rainfall or snow melt has no time to evaporate, transpire, or move into groundwater reserves?
    (a) No problem
    (b) Runoff
    (c) Infiltration
    (d) Permeation

31. When plants absorb carbon dioxide and sunlight to make glucose and other sugars for building cellular structures, it is part of the
    (a) biological carbon cycle
    (b) astrological cycle
    (c) unicycle
    (d) remediation cycle

32. On a per capita basis, which country is the world's largest nuclear power producer?
    (a) Argentina
    (b) Spain
    (c) India
    (d) France

33. Pollutants that come from agriculture, storm runoff, lawn fertilizers, and sewer overflows are
    (a) point source
    (b) open source
    (c) non–point source
    (d) easy to trace

34. The middle layer separating the lower stratosphere from the thermosphere is called the
    (a) troposphere
    (b) ionosphere
    (c) mesosphere
    (d) outer space

35. When marine animals create light through a chemical reaction, it is called
    (a) ozone
    (b) bioluminescence
    (c) radioactivity
    (d) photosynthesis

36. All of the Earth's frozen water, found in colder latitudes and at higher elevations in the form of snow and ice, is known as the
    (a) thermosphere
    (b) atmosphere
    (c) cryosphere
    (d) ionosphere

37. Which gas is the second-biggest additive to the greenhouse effect at around 20%?
    (a) Methane
    (b) Propane
    (c) Carbon dioxide
    (d) Neon

38. When flowing water creates energy that can be turned into electricity, it is called
    (a) rafting power
    (b) hydroelectric power
    (c) solar power
    (d) dispersion power

39. Infectious, pathological, and chemotherapy waste are types of
    (a) fossil fuel waste
    (b) agricultural waste
    (c) biological waste
    (d) radioactive waste

40. Who invented SCUBA (self-contained underwater breathing apparatus) equipment?
    (a) Ralph Nader
    (b) Jack Showers
    (c) Jacques Cousteau
    (d) Richard Smalley

41. Which of the following is a layer of the atmosphere?
    (a) Blue sphere
    (b) Cryosphere
    (c) Lithosphere
    (d) Stratosphere

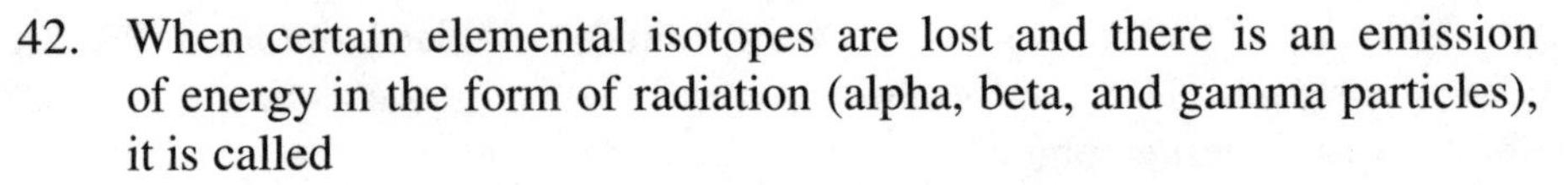

42. When certain elemental isotopes are lost and there is an emission of energy in the form of radiation (alpha, beta, and gamma particles), it is called
    (a) a solar flare
    (b) fireworks
    (c) chemical erosion
    (d) radioactive decay

43. How many oxygen atoms does ozone have?
    (a) 1
    (b) 2
    (c) 3
    (d) 4

44. The chemical 2,3,7,8-tetrachlorodibenzo-paradioxin (TCDD) comes from a family of toxic chemicals called
    (a) hydrogenated fats
    (b) dioxins
    (c) surfactants
    (d) dimethyl butanes

45. Roughly what percentage of chlorine in the stratosphere comes from manmade sources?
    (a) 22%
    (b) 45%
    (c) 75%
    (d) 84%

46. All of the following are water pollution contaminants except
    (a) heat
    (b) radioactive waste
    (c) rainbow trout
    (d) oil spills

47. Made of semiconducting materials, these convert sunlight directly into electricity:
    (a) magnifying lenses
    (b) photovoltaic cells
    (c) nuclear fuel rods
    (d) mood rings

48. Which of the following absorbs the greenhouse gas carbon dioxide in a big way?
    (a) Cattle
    (b) Forests
    (c) Champagne
    (d) Streams

49. Most streams follow a branching drainage pattern, which is known as
    (a) dendritic drainage
    (b) radial drainage
    (c) rectangular drainage
    (d) composite drainage

50. What are hurricanes called in the northwestern Pacific Ocean and Philippines?
    (a) Tsunamis
    (b) Blue northerns
    (c) Really big blows
    (d) Typhoons

51. When harnessing wind power, propeller-like turbine blades are used to generate
    (a) tourism
    (b) electricity
    (c) oxygen
    (d) heat

52. The Aurora Borealis and Aurora Australis are found in which atmospheric layer?
    (a) Troposphere
    (b) Stratosphere
    (c) Mesosphere
    (d) Thermosphere

53. A seamount is
    (a) an extinct undersea volcano
    (b) a tall coral reef
    (c) another name for a seahorse
    (d) a cliff overlooking the ocean

54. The amount of open space in the soil is called soil
    (a) aridity
    (b) porosity
    (c) crystallization
    (d) lithification

55. Radioactive waste materials, created by mining, includes
    (a) biomedical waste
    (b) boat sewage
    (c) oil spills
    (d) uranium dust

56. In 1755, Immanuel Kant offered the idea that the solar system was formed from a rotating cloud of gas and thin dust called
    (a) nebulae
    (b) Gaia
    (c) Pangaea
    (d) the South Atlantic Anomaly

57. Rainfall and streams on the east side of the Rocky Mountains drain to the Atlantic Ocean, while flowing water from the Rocky Mountain's western slopes runs to the
    (a) Gulf of Mexico
    (b) North Sea
    (c) Pacific Ocean
    (d) Indian Ocean

58. Global warming works a lot like
    (a) a guesthouse
    (b) a greenhouse
    (c) an electric blanket
    (d) a waffle maker

59. A place where water is stored for some period of time is called a
    (a) swimming pool
    (b) reservoir
    (c) pitcher
    (d) hose

60. When rainfall flows down from the top of a mountain peak, it is called
    (a) trellis drainage
    (b) rectangular drainage

(c) radial drainage
(d) dendritic drainage

61. Denudation takes place when
    (a) all the seeds of a dandelion are blown away
    (b) mercury is extracted from cinnabar
    (c) permafrost melts
    (d) rock disintegrates and is removed from the surface of continents

62. What is found at the upper edge of the zone of saturation and the bottom edge of the zone of aeration?
    (a) Bedrock
    (b) Water table
    (c) Mantle
    (d) Epicenter

63. A first-order stream has how many tributaries flowing into it?
    (a) 0
    (b) 2
    (c) 5
    (d) 10

64. The Tonga Trench drops the ocean depth nearly
    (a) 1000 meters in depth
    (b) 5000 meters in depth
    (c) 11,000 meters in depth
    (d) 14,000 meters in depth

65. Which atmospheric layer can reach temperatures of nearly 2000°C?
    (a) Lithosphere
    (b) Mesosphere
    (c) Thermosphere
    (d) Troposphere

66. In 1998, the United Nations declared
    (a) the war on poverty
    (b) International Year of the Fruit Fly
    (c) that all meetings must end after one hour
    (d) International Year of the Ocean

67. The measurement of the number of hydrogen ions in water is called
    (a) pH
    (b) hydrogenation

(c) total carbon content
(d) dissolved oxygen

68. The ban on what group of compounds in 1996 turned the tide on falling ozone levels?
(a) Hydrocarbons
(b) Chlorofluorocarbons
(c) Narcotics
(d) Analgesics

69. A major air pollutant formed in the atmosphere from nitrogen oxides and volatile organic compounds is
(a) argon
(b) water vapor
(c) benzene
(d) ozone

70. The specific focus of radiology that uses minute amounts of radioactive materials to study organ function and structure is called
(a) sedimentary medicine
(b) solar medicine
(c) metamorphic medicine
(d) nuclear medicine

71. When high levels of nitrates and phosphates cause the overgrowth of aquatic plants and algae, high dissolved oxygen consumption, and fish death, it is called
(a) aeration
(b) calcification
(c) eutrophication
(d) fossilization

72. What disinfectant is used in many water treatment plants?
(a) Toluene
(b) Chlorine
(c) Vinegar
(d) Arsenic

73. When chemical reactions dissolve limestone over a long period of time, creating a cave or cavern, the area is called a
(a) hot spot
(b) karst

(c) point source
(d) mud slide

74. Which of the following is the energy and heat source for the ocean's food chain?
(a) Chlorine
(b) Cold fusion
(c) Sunlight
(d) Wind

75. The categories *extremely arid, arid,* and *semiarid* describe
(a) deserts
(b) oceans
(c) tornadoes
(d) glaciers

76. What measures the rate of heat loss from exposed skin to that of surrounding air temperatures?
(a) A thermometer
(b) Wind chill factor
(c) Wind shear
(d) Your mother

77. Residence time is the
(a) time you spend at home after school
(b) amount of time water spends in the groundwater part of the hydrologic cycle
(c) growth time of kelp beds
(d) time magma spends in the mantle

78. Circulating air and water patterns affected by the Earth's rotation are known as the
(a) geomorphic cycle
(b) carbonization effect
(c) geosynclinal cycle
(d) Coriolis effect

79. What weather process has become the "bad boy" of the world's weather shifts?
(a) Nimbus clouds
(b) Glaciation

(c) El Niño
(d) High humidity

80. Aquifers that lie beneath layers of impermeable clay are known as
(a) unconfined aquifers
(b) confined aquifers
(c) dry aquifers
(d) vase aquifers

81. Ozone has a distinctive odor and is what color?
(a) Yellow
(b) Tan
(c) Blue
(d) Purple

82. Groundwater is stored in large, underground water reservoirs called
(a) swimming pools
(b) aquariums
(c) karma
(d) aquifers

83. Pathogenic microorganisms cause
(a) disease
(b) bad traffic
(c) diabetes
(d) hiking trails

84. Species that live exclusively in the total darkness of caves are called
(a) calcites
(b) stalagtites
(c) troglobites
(d) trilobites

85. Which of the following is not known to cause water turbidity?
(a) Clay
(b) Solar flares
(c) Silt
(d) Plankton

86. Before the 1970s, photovoltaic cells were mostly used in
(a) flashlight batteries
(b) rock concert lighting

(c) providing electricity to orbiting satellites
(d) cell phones

87. Water covers roughly what percentage of the Earth's surface?
(a) 25%
(b) 45%
(c) 60%
(d) 70%

88. Evaporation is when
(a) water freezes
(b) water changes from a liquid to a gas or vapor
(c) plants turn toward the sun
(d) rainfall leads to runoff

89. Seventy percent of greenhouse gases are made up of
(a) methane
(b) carbon dioxide
(c) nitrous oxide
(d) carbon monoxide

90. The displacement of sand and particles by the wind is called
(a) a real problem for contact lens wearers
(b) the Coriolis effect
(c) metamorphism
(d) eolian movement

91. The boundary between two watersheds is called a
(a) divide
(b) continental margin
(c) water table
(d) soggy place

92. Groundwater is found in the
(a) saturation zone
(b) aeration zone
(c) ablation zone
(d) beach zone

93. Total organic carbon is used by hydrologists to
(a) standardize No. 2 pencils
(b) size metamorphic rock particles

(c) check the health of fresh water
(d) see if their fire-roasted marshmallows are ready

94. Solar cycles and changes in the sun's radiation levels naturally increase the Earth's
(a) acidity
(b) water levels
(c) radius
(d) temperature

95. Carbon in the lithosphere is stored in
(a) colas
(b) both inorganic and organic forms
(c) pencils
(d) the Aurora Borealis

96. Rain shadow deserts are formed near
(a) cities
(b) puddles
(c) mountains
(d) golf courses

97. Fecal coliform bacteria are naturally found in
(a) iron mines
(b) glaciers
(c) petrified wood
(d) feces and intestinal tracts of humans and warm-blooded animals

98. Which of the following factors is not used to classify deserts?
(a) Volcanism
(b) Humidity
(c) Total rainfall
(d) Wind

99. Which greenhouse gas is about 145% higher now than it was in the 1800s?
(a) Butane
(b) Octane
(c) Methane
(d) Nitrogen

100. Which of the following is a significant marker of a lake's ability to support aquatic life?
   (a) Ozone
   (b) Gravel bottom
   (c) Local fish hatchery
   (d) Dissolved oxygen

Answers to Quiz, Test, and Exam Questions

CHAPTER 1

1. C	2. D	3. C	4. B	5. A
6. B	7. B	8. C	9. A	10. D

CHAPTER 2

1. D	2. B	3. C	4. B	5. D
6. D	7. B	8. A	9. A	10. C

CHAPTER 3

1. C	2. B	3. D	4. C	5. C
6. B	7. D	8. B	9. A	10. B

CHAPTER 4

1. B	2. A	3. B	4. D	5. D
6. C	7. C	8. B	9. A	10. C

PART ONE TEST

1. D	2. B	3. C	4. B	5. C
6. B	7. C	8. B	9. A	10. C
11. A	12. A	13. C	14. D	15. B
16. C	17. D	18. A	19. B	20. A
21. D	22. C	23. B	24. D	25. A
26. C	27. B	28. D	29. C	30. B
31. C	32. D	33. A	34. D	35. A
36. B	37. B	38. C	39. B	40. D

CHAPTER 5

1. B	2. D	3. A	4. B	5. C
6. D	7. C	8. C	9. D	10. A

CHAPTER 6

1. D	2. C	3. A	4. B	5. D
6. A	7. B	8. D	9. D	10. C

CHAPTER 7

1. D	2. B	3. C	4. D	5. C
6. D	7. C	8. D	9. B	10. A

CHAPTER 8

1. C	2. C	3. D	4. A	5. D
6. D	7. B	8. A	9. B	10. C

PART TWO TEST

1. C	2. B	3. D	4. A	5. B
6. D	7. B	8. D	9. C	10. B
11. B	12. C	13. B	14. A	15. A
16. B	17. A	18. D	19. C	20. C
21. C	22. B	23. B	24. D	25. C
26. B	27. C	28. A	29. D	30. B
31. C	32. A	33. C	34. B	35. D
36. B	37. D	38. B	39. A	40. C

CHAPTER 9

1. C	2. B	3. A	4. A	5. D
6. B	7. D	8. C	9. B	10. C

CHAPTER 10

1. C	2. A	3. C	4. D	5. B
6. B	7. B	8. C	9. A	10. D

CHAPTER 11

1. D	2. C	3. D	4. D	5. B
6. B	7. A	8. B	9. A	10. C

CHAPTER 12

1. C	2. D	3. A	4. B	5. C
6. B	7. A	8. B	9. C	10. D

PART THREE TEST

1. A	2. D	3. B	4. A	5. C
6. D	7. C	8. C	9. B	10. B
11. B	12. D	13. B	14. C	15. C
16. D	17. C	18. C	19. A	20. D
21. B	22. D	23. A	24. D	25. A
26. D	27. B	28. C	29. B	30. B
31. A	32. D	33. B	34. C	35. A
36. B	37. D	38. C	39. A	40. B

CHAPTER 13

1. D	2. A	3. B	4. D	5. C
6. A	7. D	8. A	9. B	10. C

CHAPTER 14

1. C	2. D	3. B	4. A	5. C
6. A	7. C	8. B	9. D	10. D

CHAPTER 15

1. C	2. A	3. D	4. B	5. B
6. A	7. C	8. D	9. C	10. D

CHAPTER 16

1. D	2. B	3. B	4. C	5. D
6. A	7. D	8. C	9. C	10. A

PART FOUR

1. A	2. D	3. A	4. D	5. B
6. C	7. B	8. C	9. C	10. D
11. C	12. B	13. A	14. C	15. C
16. D	17. D	18. C	19. B	20. B
21. B	22. A	23. D	24. B	25. A
26. C	27. B	28. C	29. A	30. B
31. C	32. D	33. B	34. C	35. B
36. D	37. C	38. A	39. D	40. C

FINAL EXAM

1. A	2. C	3. A	4. B	5. C
6. C	7. B	8. B	9. C	10. D
11. C	12. C	13. B	14. D	15. B
16. B	17. D	18. C	19. D	20. C
21. A	22. A	23. C	24. C	25. C
26. B	27. C	28. D	29. C	30. B
31. A	32. D	33. C	34. C	35. B
36. C	37. A	38. B	39. C	40. C
41. D	42. D	43. C	44. B	45. D
46. C	47. B	48. B	49. A	50. D
51. B	52. D	53. A	54. B	55. D
56. A	57. C	58. B	59. B	60. C
61. D	62. B	63. A	64. C	65. C
66. D	67. A	68. B	69. D	70. D
71. C	72. B	73. B	74. C	75. A
76. B	77. B	78. D	79. C	80. B
81. C	82. D	83. A	84. C	85. B
86. C	87. D	88. B	89. B	90. D
91. A	92. A	93. C	94. D	95. B
96. C	97. D	98. A	99. C	100. D

APPENDIX I

Terms and Organizations

Acronym	Description
AA	Attainment area
ACL	Alternative concentration limit
AO	Agent orange—an herbicide/defoliant containing dioxin used in Vietnam by the United States
AMS	American Meteorological Society
API	American Petroleum Institute
AQCR	Air quality control region
ARAR	Applicable relevant and appropriate requirements (cleanup standards)
ASTM	American Society for Testing and Materials
ATSDR	Agency for Toxic Substances and Disease Registry

Acronym	Description
AWEA	American Wind Energy Association
BACM	Best available control measure
BACT	Best available control technology
BAT	Best available technology
BATEA	Best available technology economically achievable
BCT	Best conventional technology
BDAT	Best demonstrational achievable technology (also best demonstrational available technology)
BDT	Best demonstrational technology
BEJ	Best engineering judgment
BIF	Boiler and industrial furnace
BMP	Best management practice
BOD	Biological oxygen demand—index of amount of oxygen used by bacteria to decompose organic waste
BPJ	Best professional judgment
BPCT	Best practical control technology
BRS	Biennial reporting system
BTU	British thermal unit—energy required to raise 1 pound of water 1°F.
C&D	Construction and demolition
CAA	Clean Air Act
CBI	Confidential business information
CAIR	Comprehensive Assessment Information Rule
CAMU	Corrective action management unit
CAS	Chemical Abstract Service
CCP	Commercial chemical product
CDD	Chlorodibenzodioxin

Acronym	Description
CDF	Chlorodibenzofuran
CEM	Continuous emission monitoring
CEQ	Council on Environmental Quality
CERCLA	Comprehensive Environmental Response, Compensation, and Liability Act of 1980 (amended 1984)
CESQG	Conditionally exempt small-quantity generator (of hazardous wastes)
CFC	Chlorofluorocarbon—an ozone-depleting refrigerant
CFR	Code of Federal Regulations
CHEMTREC	Chemical Transportation Emergency Center
CHIPS	Chemical Hazards Information Profiles
CIESIN	Center for International Earth Science Information Network
CNG	Compressed natural gas
COD	Chemical oxygen demand
CPSC	Consumer Product Safety Commission (16 CFR)
CTG	Control Techniques Guidelines
CWA	Clean Water Act
DCO	Delayed compliance order
DDT	Dichlorodiphenyltrichloroethane—a toxic pesticide
DEP	Department of Environmental Protection
DMR	Discharge monitoring reports
DNA	Deoxyribonucleic acid—made of phosphates, sugars, purines, and pyrimidines; helix shape carries genetic information in cell nuclei
DO	Dissolved oxygen
DOD	Department of Defense
DOE	Department of Energy
DOJ	Department of Justice

Acronym	Description
DOT	Department of Transportation
DRE	Destruction and removal efficiency
EERE	Energy efficiency and renewable energy
EIS	Environmental Impact Statement
ELF	Extremely low frequency electromagnetic wave (<300 Hz)—emitted by electrical power lines
EMS	Environmental management system (also see ISO14000)
EP	Extraction procedure
EPA	Environmental Protection Agency
EPR	Extended product responsibility
EREF	Environmental Research and Education Foundation
ERT	Environmental Resources Trust, Inc.
ESA	Environmental site assessment
ESI	Environmental sustainability index
ESP	Electrostatic precipitor
EWEA	European Wind Energy Association
FDA	Food and Drug Administration (21 CFR)
FFCA	Federal Facility Compliance Act
FIFRA	Federal Insecticide, Fungicide, and Rodenticide Act
First Third	August 17, 1988, Federal Register (53 FR 31138)—the first of the hazardous waste land disposal restrictions
FR	Federal Register
GATT	General Agreement on Tariffs and Trade—since 1947; over 100 member countries
GCM	Global climate model
GLP	Good laboratory practices
GMER	Green Mountain Energy Resources

Acronym	Description
GMO	Genetically modified organism
GPO	Government printing office
GMP	Good manufacturing procedures
GRAS	Generally recognized as safe
HazWoper	29 CFR 1910.120—the OSHA/EPA requirement to have all employees trained if they will be handling, managing, or shipping hazardous wastes.
HRS	Hazard ranking system
HHW	Household hazardous waste
HazMat	Hazardous material
HW	Hazardous waste
HSWA	Hazardous and Solid Waste Amendments—1984
HWM	Hazardous waste management
Hz	Hertz—frequency with which alternating current changes direction
ID	Hazardous waste identification number assigned to RCRA generators, transporters, and TSDFs
IEA	International Energy Association
INPO	Institute of Nuclear Power Operations
INUR	Inventory Update Rule
IPM	Integrated pest management
ISB	In-situ burning
ITC	Interagency Testing Committee
kWh	Kilowatt-hour
LAER	Lowest achievable emission rate
LCA	Life cycle analysis/assessment
LDR	Land Disposal Restrictions (40 CFR Part 268)
LED	Light-emitting diode
LEPC	Local Emergency Planning Committee

Acronym	Description
LNG	Liquid natural gas
LOEL	Lowest observed effect level
LPG	Liquid petroleum gas (or propane)
LQG	Large-quantity generator (of hazardous wastes)—this term has a specific definition under RCRA
LUST	Leaking underground storage tanks
MACT	Maximum achievable control technology
Maglev movement	|Magnetic levitation train—using magnetic forces for high speed
MARPOL	International Convention on the Prevention of Pollution from Ships
MCL	Maximum concentration limit (or level)
MRT	Mean residence time—the amount of time a water molecule spends in a reservoir before moving on
MSDS	Material safety data sheet (under OSHA)
MSW	Municipal solid waste (trash and nonhazardous waste)
MW	Mega-watt—1,000 kilowatts (1 million watts)
NAA	Non-attainment area
NAAQS	National Ambient Air Quality Standards
NAEWG	North America Energy Working Group
NAFTA	North American Free Trade Agreement—Canada, Mexico and the United States
NCAR	National Center for Atmospheric Research
NCP	National Contingency Plan
NEPA	National Environmental Policy Act
NESHAP	National Emissions Standard for Hazardous Air Pollutants
NIOSH	National Institute of Occupational Safety and Health
NIPDWR	National Interim Primary Drinking Water Regulation

Acronym	Description
NGO	Nongovernmental organizations—over 10,000 organizations worldwide linked by ECONET
NPDES	National Pollutants Discharge Elimination System
NPL	National Priorities List—list of Superfund sites
NRC	Nuclear Regulatory Commission
NSDWR	National Secondary Drinking Water Regulation
NSPS	New Source Performance Standards
NSR	New Source Review
ODC	Ozone-depleting chemical
OGF	Old-growth forest—high-biodiversity ecosystem with trees 300–1,000 years old
OH	Hydroxyl radicals—atmospheric molecule that reduces methane (CH_4), carbon monoxide (CO), and ozone
OPM	Operation and maintenance
ORD	Office of Research and Development
OSHA	Occupational Safety and Health Administration
OSW	Office of Solid Waste
OSWER	Office of Solid Waste and Emergency Response
OTG	Off-the-grid power generation independent of a major power plant
PAIR	Preliminary Assessment Information Rule
PCB	Polychlorinated biphenyl—used in dyes, paints, light bulbs, transformers and capacitors
PSP	Point source pollution
PEL	Permissible exposure limit
pH	Logarithmic scale that measures acidity (pH 0) and alkalinity (pH 14); pH 7 is neutral
PM-10	Particulate matter < 10 micrometers
PNA	Polynuclear aromatic compounds

Acronym	Description
PNIN	Premanufacture notification
POP	Publicly owned treatment works
PPB	Parts per billion
PPM	Parts per million
PRP	potentially responsible parties
PSD	Prevention of significant deterioration
PV	Photovoltaic device—generates electricity through semiconducting material
PVC	Petrochemical formed from toxic gas vinyl chloride and used as a base in plastics.
QA/QC	Quality assurance/quality control
R	Richter scale—logarithmic scale (0–9) used to measure the strength of an earthquake
rad	Radiation absorbed dose—amount of radiation energy absorbed in 1 gram of human tissue
R&D	Research and development
RACM	Reasonably available control measure
RACT	Reasonably available control technique
rDNA	Recombinant DNA—new mix of genes spliced together on a DNA strand; (a.k.a. biotechnology)
RCRA	Resource Conservation and Recovery Act of 1976—resulted in hazardous waste regulations
rem	R-roentgen equivalent man—biological effect of a given radiation at sea level is 1 rem.
RI/FS	Remedial investigation/feasibility study
RNA	Ribonucleic acid—formed on DNA and involved in protein synthesis
RPCC	Release prevention, control and countermeasure
RQ	Reportable quantity
RUST	RCRA underground storage tanks

Acronym	Description
SARA	Superfund Amendments and Reauthorization Act
SBS	Sick-Building Syndrome
S&BA	Slash and burn agriculture
SDWA	Safe Drinking Water Act of 1974
Second Third	June 23, 1989, Federal Register—hazardous waste land disposal restrictions
SERC	State Emergency Response Commission
SIC	Standard industrial classification
SIP	State Implementation Plan
SMCL	Secondary maximum contamination level
SOC	Schedule of compliance
SPCC	Spill prevention control and countermeasures
SPDES	State pollutant discharge elimination
SQG	Small-quantity generator (of hazardous wastes)
SW-846	Test methods for evaluating solid waste, physical/chemical methods
SWMU	Solid waste management unit
TCDD	Tetrachlorodibenzodioxin
TCE	Tetrachloroethylene, perchloroethylene
TCLP	Toxic Characteristic Leaching Procedure
Third Third	June 1, 1990, Federal Register—refers to hazardous wastes "land ban" land disposal restrictions
TOC	Total organic carbon
TRU	Transuranic wastes
TSCA	Toxic Sustances Control Act of 1976—regulates asbestos, PCBs, new chemicals being deveoped for sale and other chemicals
TSDF	Treatment, storage, or disposal facility (permitted hazardous waste facility)
TSS	Total suspended solids

Acronym	Description
TWA	Time weighted average
TWC	Third-world countries
UIC	Underground injection control
USCG	United States Coast Guard
USDA	United States Department of Agriculture
USDW	Underground source of drinking water
USGS	United States Geological Survey—manages LandSat, which images the environment via satellite
USPS	United States Postal Service
UV	Ultraviolet radiation from the sun (UVA, UVB types)
VOC	Volatile Organic Compound—carbon-containing compounds that evaporate easily at low temperatures
VOME	Vegetable oil methyl ester—biodiesel derived from the reaction of methanol with vegetable oil
W	Watt—unit of electrical power
WAP	Waste analysis plan
WB	World Bank—owned by governments of 160 countries and funds hydroelectric plants and encourages ecotourism; funds no nuclear energy
WCU	World Conservation Union
WGI	World Glacier Inventory
WHS	World Heritage Site—natural or cultural site recognized as globally important and deserving international protection
WRI	World Resources Institute

APPENDIX

Conversion Factors

LENGTH

1 centimeter	0.3937 inch
1 inch	2.5400 centimeters
1 meter	3.2808 feet; 1.0936 yards
1 foot	0.3048 meter
1 yard	0.9144 meter
1 kilometer	0.6214 mile (statute); 3281 feet
1 mile (statute)	1.6093 kilometers
1 mile (nautical)	1.8531 kilometers
1 fathom	6 feet; 1.8288 meters
1 micrometer (micron)	0.0001 centimeter
1 nanometer	1×10^{-7} centimeter
1 angstrom	1×10^{-8} centimeter

VELOCITY

1 kilometer/hour	27.78 centimeters/second
1 mile/hour	17.60 inches/second

AREA

1 square centimeter	0.1550 square inch
1 square inch	6.452 square centimeters
1 square meter	10.764 square feet; 1.1960 square yards
1 square foot	0.0929 square meter
1 square kilometer	0.3861 square mile
1 square mile	2.590 square kilometers
1 acre (U.S.A.)	4840 square yards

VOLUME

1 cubic centimeter	0.0610 cubic inch
1 cubic inch	16.3872 cubic centimeters
1 cubic meter	35.314 cubic feet
1 cubic foot	0.02832 cubic meter
1 cubic meter	1.3079 cubic yards
1 cubic yard	0.7646 cubic meter
1 liter	1000 cubic centimeters
1 gallon (U.S.A.)	3.7853 liters

MASS

1 gram	0.03527 ounce
1 ounce	28.3496 grams
1 kilograms	2.20462 pounds
1 pound	0.45359 kilogram

PRESSURE

1 kilogram/ square centimeter	0.96784 atmosphere; 14.2233 pounds per square inch; 0.98067 bar
1 bar	0.98692 atmosphere; 10^5 pascals

References

Ball, P., *Life's Matrix: A Biography of Water,* New York, NY: Farrar, Straus and Giroux, 2000.

Campbell, Colin J., and Jean Laherrere, "The End of Cheap Oil," *Scientific American,* 278 (3), 78–83 (1998).

Deffeyes, Kenneth S., *Hubbert's Peak: The Impending World Oil Shortage,* Princeton, NJ: Princeton University Press, 2001.

Energy Information Administration, Office of Integrated Analysis and Forecasting, *International Energy Outlook 2004,* rep. no. DOE/EIA-0484(2004), Washington, D.C.: U.S. Department of Energy, 2004.

Environmental Protection Agency, *In Brief: The U.S. Greenhouse Gas Inventory*, Washington, D.C.: Office of Air and Radiation (EPA 430-F-02-008), 2002.

Gallant, R. A., *Structure: Exploring Earth's Interior*, New York, NY: Benchmark Books, 2003.

Gohau, G., *A History of Geology*, New Brunswick, NJ: Rutgers University Press, 1990.

Goodstein, David, *Out of Gas: The End of the Age of Oil*, New York, NY: W.W. Norton & Company, 2004.

Hambrey, M., and J. Alean, *Glaciers*, New York, NY: Press Syndicate of the University of Cambridge, 1992.

Holmes, G., et al., *Handbook of Environmental Management and Technology*, New York, NY: John Wiley & Sons, Inc., 1993.

Jackson, J. L., ed., *Dictionary of Geological Terms,* 3rd ed., National Academy of Sciences for the American Geological Institute, New York, NY: Knopf Group, 1984.

Jensen, J., *Remote Sensing of the Environment: An Earth Resource Perspective*, Upper Saddle River, NJ: Prentice Hall, 2000.

Jones, R. L., *Soft Machines: Nanotechnology and Life*, Oxford, United Kingdom: Oxford University Press, 2004.

Kump, L., et al., *The Earth System*, Upper Saddle River, NJ: Prentice Hall, 1999.

Lambert, D., *The Field Guide to Geology*, Facts on File, Inc., New York, NY: Diagram Visual Information, Inc., 1997.

Lovelock, J., *GAIA: A New Look at Life on Earth*, Oxford, United Kingdom: Oxford University Press, 2000.

Lüsted, M., and G. Lüsted, *A Nuclear Power Plant*, New York, NY: Lucent Books, 2005.

Montgomery, C., *Fundamentals of Geology,* 3rd ed. Chicago, IL: William C. Brown Publishers, 1997.

Murphy, B., and D. Nance, *Earth Science Today*, Pacific Grove, CA: Brooks/Cole Publishing Co., 1999.

Phillips, D., et al., *Blame It on the Weather*, San Diego, CA: Advantage Publishers Group, 2002.

Press, F., and R. Siever, *Understanding Earth,* 2nd ed., New York, NY: W. H. Freeman and Company, 1998.

Reynolds, R., *Guide to the Weather*, New York, NY: Cambridge University Press, 2000.

Saign, G., *Green Essentials: What You Need to Know about the Environment*, San Francisco, CA: Mercury House, 1994.

Shellenberger, M., and T. Nordhaus, *The Death of Environmentalism: Global Warming Politics in a Post-Environmental World*, 2004, essay presented at the October 2004 meeting of the Environmental Grantmakers Association.

Williams, J., *The Weather Book*, New York, NY: Vintage Books, 1992.

Internet References

ATMOSPHERE

- http://osrl.uoregon.edu/projects/globalwarm/report/

BIODIVERSITY

- http://www.biodiv.org/default.aspx
- http://www.biodiversityhotspots.org/xp/Hotspots/hotspotsScience/

EARTH'S FORMATION

- http://hubble.nasa.gov
- http://science.msfc.nasa.gov
- www.nasa.gov/home/index.html
- http://hubblesite.org/newscenter

ECOLOGY

- www.itopf.com/stats.html
- http://earthtrends.wri.org/

FOSSIL FUELS

- http://www.exxonmobil.com/corporate/files/corporate/CCR2002_energy.pdf
- http://greenwood.cr.usgs.gov/energy/WorldEnergy/DDS-60

GEOTHERMAL

- http://www.sandia.gov/geothermal
- http://geology.er.usgs.gov/eastern/tectonic.html

GLACIERS

- www.glacier.rice.edu/

- http://nsidc.org/cryosphere/index.html
- www.coolantarctica.com/toc.htm
- http://nsidc.org/data/g01130.html

GLOBAL WARMING

- www.epa.gov/globalwarming
- http://www.ipcc.ch
- www.unfccc.de

GREEN ENERGY

- www.ert.net
- www.eere.energy.gov

NANOTECHNOLOGY

- http://www.cnst.rice.edu

OCEANS

- www.usgs.gov
- www.nws.noaa.gov
- http://www.mcbi.org
- www.epa.gov/owow/nps/prevent.html
- http://clean-water.uwex.edu/
- http://sciencenow.sciencemag.org/cgi/content/full/2004/715/1

SOLAR

- http://www.eere.energy.gov/

SPACE

- www.nasa.gov/home/index.html
- http://nssdc.gsfc.nasa.gov/photo_gallery/

- http://earth.jsc.nasa.gov/sseop/efs/
- http://www.noaa.gov/satellites.html

WASTES

- http://www.epa.gov/osw/
- http://www.epa.gov/superfund/
- http://www.epa.gov/superfund/sites/npl/

WATER POLLUTION

- http://clean-water.uwex.edu/wav/

WEATHER

- www.weather.com
- www.theweathernetwork.com
- www.wunderground.com/

WIND

- www.awea.org
- www.britishwindenergy.co.uk/ref/noise.html
- http://www.eere.energy.gov/windandhydro/windpoweringamerica/
- http://www.iii.org/media/facts/statsbyissue/hurricanes/?table_sort_738608=4

INDEX

A
Ablation zone, 155–156
Abyssal plains, 132, 134
Accumulation zone, 155–156
Acid/base water contaminants, 179
Acid deposition, 184–186
Acid precipitation, 182
Acid rain, 182–186
  treatment, 185–186
Acid shock, 184
Acquifer recharge zone, 117
Acronyms, 383–392
Adélie penguin, 154
Aerosols, 75
Agassiz, Louis, 158
Agrology, 5
Air pressure, 50
Albertite, 287
Alpine glaciers, 149–150
Alternative energy sources. *See* Future policy and alternatives
Altocumulus clouds, 57–58
Alum, 177–178
Alzheimer's disease, 185
Antarctica, 153–154
Antarctic Circumpolar current, 133
Aphotic zone, 130
Aquifers, 115–118
Arctic Ocean, 155
Aristotle, 4
Arroyos, 234
ASTER (Advanced Spaceborne Thermal Emission and Reflection Radiometer), 36–37
Asthenosphere, 16
Atlantic Ocean, 134–135
Atmosphere
  clouds. *See* Clouds
  composition. *See* Atmosphere, composition of
  convection, 53–54
  ecosystems and biodiversity. *See* Ecosystems and biodiversity
  fronts, 54–55
  greenhouse effect and global warming. *See* Greenhouse effect and global warming
  Internet reference, 397
  introduction, 7–8, 41–42
  jet stream, 49–50
  of other planets, 8
  pollution, 67–69
  pressure of, 50–51
  relative humidity, 52–53
  wind, 51–52

Atmosphere, composition of
introduction, 42
mesosphere, 48
ozone. *See* Ozone
stratosphere, 44–45
thermosphere, 48–49
troposphere, 43–44
Aurora Australis (Southern Lights), 49
Aurora Borealis (Northern Lights), 49

B
Barchan dunes, 231–232
Barometer, 50
Beaufort, Admiral Sir Francis, 51
Beaufort wind scale, 51–52
Benthic life, 131
Bergschrund, 161
Billows clouds, 57, 62–63
Binary geothermal power plants, 330–331
Biochemical oxygen demand (BOD), 181
Biodegradable, definition of, 267
Biodiversity, 23–25, 236
endemic species, 24–25
Internet references, 397
Biodiversity Early Warning System, 35
Bioengineering, 5
Biological and biohazardous wastes, 265–266
Biological carbon cycle, 252–253
Biological weathering, 208–209
Bioluminescence, 131
Biomass, 294
Biomineralization, 245
Biosphere
crust, 13–15
definition of, 11, 22
hydrosphere, 12
introduction, 11, 22–23
lithosphere, 12–13
Black smokers, 130
Blowout dunes, 231–232
BOD (biochemical oxygen demand), 181
Botany, 5
Bottom trawling, 137, 139
Brackish water, 127
Breeder reactors, 305
Bycatch, 139

C
Calcium, 244–248
Calcrete, 233
Caliche, 233
Calving (glaciers, icebergs), 156
Carbohydrates, 267
Carbon
balance, 253–254
biological carbon cycle, 252–253
geological carbon cycle, 250–252
greenhouse effect and global warming, 82–83
importance of, 249, 254–255
introduction, 249–250
Carbon cycle, 250, 254
Carbon dioxide, 76–77
cascade, 255
Carbonic acid, 209
Carson, Rachel, 262
Center for Applied Biodiversity Science at Conservation International, 35
Chemical wastes, 180, 266–268
Chemical weathering, 209–210
Chemosynthesis, 130
Chernobyl, 307–308
China Syndrome, 308
Chlorination, 178
Chlorofluorocarbons (CFCs), 46–47
CI (Conservation International), 32, 35
Cirrostratus clouds, 56–57
Cirrus clouds, 56–58
Clean Air Act, 179, 299
Clean fuels, 290
Clean Water Act, 179
Climate change, 83–85. *See also* Global warming
Clouds
high-level clouds, 56–58
introduction, 55–56, 106

mid-level clouds, 58–59
speciality clouds, 60–62
tornadoes, 62–64
types of, 57
vertical clouds, 59–60
Coagulation, 177
Coal, 293
Coastal deserts, 226–227
Co-firing method, 294
Cold front, 54
Cold glacier, 153
Columbia Ice Field, 151
Composite dunes, 231–233
Composting, 274
Condensation, 105–106, 108
Confined aquifers, 116–117
Conifers, 30
Conservation biology, 5
Conservation International (CI), 32, 35
Contaminants, water, 179–182
Continental crust, 14
Continental glaciers, 149–150
Continental margin, 131
Continental plates, 17
Continental rise, 132
Continental shelf, 14, 131–132
Continental slope, 14, 131–132
Contrail clouds, 57, 61
Convection, 53–54
Conversion factors, 393–394
Core, 16–17
Coriolis effect, 225
Costeau, Jacques, 125–126
Creep, 213, 216
Crevasses (glacial), 160–161
Crust, 13–15
continental crust, 14
elements in, 13
introduction, 13
oceanic crust, 14–15
Cryosphere, 12, 149
Crystalline silicon solar cells, 310–311
Cumulonimbus clouds, 57, 60
Cumulus clouds, 57, 59
Currents, ocean, 15, 133–134
Cyclones, 66

D

DDT (dichlorodiphenyltrichloroethane), 262
Decay rates, radioactive, 268–269
de Coriolis, Gaspard Gustave, 225
Deforestation, 29–32
Dendritic drainage, 113–114
Density, ocean, 128
Denudation, 204–205
Desertification
degradation of dry lands, 236
economic factors, 237–238
introduction, 235
land overuse, 236
population density, 237
rainfall, 236
soil and vegetation, 237
trends, 238–239
Deserts
coastal deserts, 226–227
desertification. *See* Desertification
dunes, 230–233
hot and dry deserts, 224
introduction, 221–223
midlatitude deserts, 225
monsoon deserts, 228–229
oases, 230
paleodeserts, 239
plants, 233–234
polar deserts, 229
rain shadow deserts, 227–228
semiarid deserts, 225–226
soils, 233
trade wind deserts, 224–225
types of, 223–229
water, 234–235
Differential weathering, 205

Dinoflagellate blooms, 174
Dioxins, 267
Discharge, definition of, 116
Dissolution, 183, 210
Dissolved oxygen, 175–176
Distributed power, 315
Diversion hydroelectric plant, 324–325
Dolinen, 256
Draas, 233
Drainage basin, 112
Drift, 160
Dry steam geothermal power plants, 330
Dunes, 230–233
  sand dunes, 230–233
    barchan, 231–232
    blowouts, 231–232
    composite, 231–233
    linear, 231–232
    transverse, 231–232
  erosion of, 216–217
    blowouts, 216
    transgressive (creeping), 216
Dursban insecticide, 189
Dust bowl, 233
Dynamite fishing, 138
Dysphotic zone, 130

E

Earth
  atmosphere. *See* Atmosphere
  biosphere. *See* Biosphere
  core, 16–17
  formation, 4, 6, 397
  Gaia hypothesis, 8–11
  introduction, 3–4
  layers of, 6
  magnetism, 17–18
  mantle, 16
  place in the galaxy, 6–7
  size and shape, 4
Earth Remote Sensing Data Analysis Center, 36
East Wind Drift, 133
Ecological niche, 23
Ecology, 5, 397
Ecosystems and biodiversity
  biodiversity, 23–25
  biosphere, 22
  deforestation, 29–32
  ecosystems, 21–22
  endangered species, 26–27
  endemic species, 24–25
  habitat, 25–29
  hotspots. *See* Hotspots
  introduction, 21–22
  wetlands, 27–29
Ecotourism, 38
Electricity, 292
Electricity grid, 314
Electrochemical solar cells, 311–312
El Niño, 136–137
Emperor penguin, 154
Endangered species, 26–27, 141
Endangered Species List, 25–26
Endemic species, 24–25
Energy, 289–290
Energy resources, 10
Enhanced greenhouse effect, 81–82
Environmental biology, 3
Environmental geology, 5
Environmental Protection Agency (EPA), 80, 85, 179, 262–263
Environmental Resources Trust (ERT), 341
Environmental science
  atmosphere. *See* Atmosphere
  definition of, 3
  fields of study, 5
  land. *See* land
  water. *See* Water
  what can be done. *See* Fossil fuels; Future policy and alternatives; Geothermal energy; Hydroelectric power; Nuclear energy; Solar energy; Wind, energy
Eolian movement, 229

EPA (Environmental Protection Agency), 80, 85, 179, 262–263
Epikarst, 256
Equilibrium line, 156
Ergs, 233
Erosion
  glaciers, 163–164
  land. *See* Weathering and erosion
Erratics, 161
ERT (Environmental Resources Trust), 341
Ethanol, 292–293
Euphotic zone, 129–130
European Space Agency's Envisat Earth observation satellite, 47, 143
Evaporation, 105–108
Evapotranspiration, 106
Exploration geophysics, 5
Extinction, 26, 32
Exxon Valdez oil spill, 173–174, 186

F

Fecal coliform bacteria, 181
Feely, Richard, 83
Field sampling, 36
Filtration, 178
Final exam, 359–375
Firn, 148
Fisheries, 138–141
  non-native (alien) species, 139–140
Fission, 270
Flashed steam geothermal power plants, 330
Flocculation, 178
Flooding, 118–120
Flowing artesian well, 116
Food web of the sea, 129
Forestry, 5
Fossil fuels
  biomass, 294
  coal, 293
  definition of, 288
  electricity, 292
  energy, 289–290
  ethanol, 292–293
  first oil use, 287–288
  Hubbert's peak, 290–292
  Internet references, 397
  introduction, 287
  methanol, 293
  natural gas, 294
  new gasoline types, 295
  oil demand, 290
  oil spills. *See* Oil spills
  oil wells, 288–289
  propane, 294
Fractured aquifers, 115
Francis turbine, 326
Fronts, atmospheric, 54–55
Frost wedging, 208
Fujita, T. Theodore, 63
Fujita wind damage scale, 63–64
Future policy and alternatives
  green power. *See* Green power
  green practices, 343–344
  international policy, 346–348
  introduction, 335
  nanotechnology, 344–346
  North America, 348
  sustainability, 335–337

G

Gaia hypothesis, 8–11
Gases, greenhouse. *See* Greenhouse gases
Gasohol, 292
Gasoline, new types of, 295
Geochemical cycling
  calcium, 244–248
  carbon. *See* Carbon
  introduction, 243–244
  karst formation, 255–258
  nitrogen, 248–249
Geochemistry, 5
Geological carbon cycle, 250–252
Geomorphology, 5
Geophysics, 5

Geothermal energy
core heat, 327–328
disadvantages, 332
geothermal power generation, 329–330
Internet references, 397
introduction, 327
miscellaneous uses, 331–332
types of geothermal plants, 330–331
water's part, 328–329
Gesner, Abraham, 287–288
Gill netting, 139
Glaciers
Antarctica, 153–154
Arctic Ocean, 155
crevasses, 160–161
erosion, 163–164
future of, 168
global warming and, 166–167
icebergs, 156–158
ice caps and sheets, 150–152
Internet references, 398
introduction, 147–148
meltwater streams, 153
permafrost, 164–165
Pleistocene Era, 148–150
speed and movement, 158–160
surge, 158–160
temperatures, 152–153
till and moraines, 161–163
Vostok ice-core, 167
zones, 155–156
Glaciology, 5
Global warming, 75, 83–85
glaciers and, 166–167
Internet references, 398
Going green, 336–337, 343
Green certificates, 338–340
Green energy, 321, 337
Internet references, 398
Green-E Renewable Branding Program, 342–343
Greenhouse effect and global warming
carbon, 82–83
climate change, 83–85
enhanced greenhouse effect, 81–82
greenhouse gases. *See* Greenhouse gases
introduction, 73–75
ozone and greenhouse effect, 81
Greenhouse gases
carbon dioxide, 76–77
formation of, 79–81
halocarbons, 79
introduction, 75–76
inventories, 85–86
methane, 78–79
nitrogen oxides, 77–78
reducing, 86–87
Green power
Environmental Resources Trust (ERT), 341
Green-E Renewable Branding Program, 342–343
introduction, 337–338
renewable energy certificates (RECs), 338–340
Green practices, 343–344
Grid-connected systems, 314
Ground moraines, 162–163
Groundwater, 110–111, 116
Gulf Stream, 133–134
Gullies, 217
Gyre, 15

H

Habitat, 25–29
loss of, 25
Halocarbons, 79
Halocline layer, 127
Halons, 48
Halophytes, 234
Halos, 57
Hardpan, 235
Hard water, 245
Headward erosion gullies, 218
High-level clouds, 56–58
High-pressure system, 50

Hot and dry deserts, 224
Hotspots
  conservation, 35
  fragility of, 34
  introduction, 32–34
  list of, 33
  remote sensing, 36–38
  solutions, 38–39
Hubbert, M. King, 290–291
Hubbert's peak, 290–292
Hubble space telescope, 6
Humidity, 108, 221–222
Hurricanes, 65–67
  categories, 67
  naming of, 66
  tropical depression, 65
  tropical storm, 65
  wind shear, 65
Hybrid geothermal power plants, 330–331
Hydroelectric power
  disadvantages, 327
  introduction, 323–324
  turbines, 326
  types of hydroelectric plants, 324–325
Hydrologic cycle
  aquifers, 115–118
  condensation, 108
  description of, 105–107
  evaporation, 107–108
  flooding, 118–120
  future, 121
  groundwater, 110–111
  introduction, 101–103
  precipitation, 109
  properties of water, 103–104
  runoff, 118
  transpiration, 110
  transport, 108–109
  watershed, 112–114
  water use and quality, 120–121
Hydrology, 5, 101
Hydrolysis, 210
Hydrosphere, 12
Hydrothermal reservoirs, 328

I

Icebergs, 156–158
Ice caps and sheets, 150–152
Icefalls, 160
Ice fields, 150–151
IEA (International Energy Association), 346–347
Impoundment hydroelectric plant, 324–325
Infiltration, 117
Inorganic water contaminants, 179
In-situ burning, 296
In-situ data collection, 36
Institute of Nuclear Power Operations (INPO), 308
Intergovernmental Panel on Climate Change (IPCC), 86
International Energy Association (IEA), 346–347
International Oil Tanker Owner's Pollution Federation (IOTPF), 173
International policy, 346–348
Internet references, 397–399
Ionosphere, 42, 49
Isobars, 51

J

Jet stream, 49–50
Joints (in rocks), 207–208

K

Kant, Immanuel, 4
Kaplan propeller turbine, 326
Karst, 115, 255–258
Kerosene, 288
Kuroshio current, 133
Kyoto Protocol, 84

L

Land
  deserts. *See* Deserts
  geochemical cycling. *See* Geochemical cycling

Land *(Continued)*
  solid and hazardous wastes. *See* Wastes, solid and hazardous
  weathering and erosion. *See* Weathering and erosion
Landslides, 212
Lateral moraines, 162–163
Laterite soil, 210–211
Laughing gas, 77
Lead (crack in ice), 152
Lenticular clouds, 62–63
Light, oceans and, 129–131
Linear dunes, 231–232
Lithosphere, 12–13
Littoral zone, 133
Loess, 233
Longlining, 139
Love Canal, 263
Lovelock, James, 8–9
Low-pressure system, 50
Lysocline, 245

M
Magnetism, 17–18
Mammatus clouds, 57, 61–62
Mantle, 16
Marine Conservation Biology Institute, 142
Marsh gas, 78
Mass movement, 212–214
Mass wasting, 204, 209
Meltwater, 153, 159
Mesosphere, 16, 42–43, 48
Meteorologist, definition of, 41
Methane, 78–79
Methanol, 293
Metric conversions, 393–394
Midlatitude deserts, 225
Mid-level clouds, 58–59
Milky Way, 6–7
Millibars, 50
Mineralogy, 5
Monsoon deserts, 228–229
Moraines, 161–163
MTBE (methyl tertiary-butyl ether), 179
Muro ami, 138
Myers, Norman, 32

N
NAEWG (North America Energy Working Group), 348
Nanotechnology, 344–346
  Internet reference, 398
National Aeronautics and Space Administration (NASA), 6, 36, 78, 143, 238, 313
National Center for Atmospheric Research (NCAR), 239
National Oceanic and Atmospheric Administration (NOAA), 136, 143–144
National Research Council (NRC), 297
Natural gas, 294
Nebulae, 6
Nebular hypothesis, 4
Nick points, 218
Nilas, 152
Nimbostratus clouds, 57, 59
Nitrogen, 180, 248–249
Nitrogen cycle, 249
Nitrogen oxides, 77–78, 184
NOAA (National Oceanic and Atmospheric Administration), 136, 143–144
Nonhazardous waste, 272–273
Non-native fish species, 139–140
Non–point source pollution, 1212
North America Energy Working Group (NAEWG), 348
North Equatorial current, 133
North Pole, 155
Nuclear energy
  accidents, 307–308
  disadvantages, 306–308
  introduction, 303–306
  regulations, 308–309
Nuclear medicine, 186

Nuclear waste, 268–271
Nunataks, 152

O
Oases, 230
Occluded front, 55
Oceanic crust, 14–15
Oceanography, 5
Oceans
  Atlantic Ocean, 134–135
  currents, 133–134
  density, 128
  El Niño, 136–137
  future of, 141–143
  Internet references, 398
  introduction, 125–126
  light, 129–131
  Pacific Ocean, 135–136
  physical changes, 137–138
  pressure, 128–129
  salinity, 127
  shorelines, 132–133
  temperature, 127–128
  tsunamis, 143–144
  zones, 131–132
Oil
  demand, 290
  first oil, 287–288
  slicks, 186, 296
  spills. *See* Oil spills
  weathering, 299
  wells, 288–289
Oil spills, 173, 295–299
  list of, 174
  run-off and leakage, 297
  tar balls, 289–299
  wildlife impact, 297–298
Old-growth forests, 30–31
Organic matter
  definition of, 249
  in water, 179, 181
Organic waste, 265
Organizations and terms, 383–392
Orographic clouds, 57, 61–62
Outlet glacier, 151
Outwash, 120
Oxidation, 209
Oxygen solubility, 176
Ozone, 80
  depletion of, 46–48
  greenhouse effect and, 81
  introduction, 45–46
  layer, 43
  oxidation (surface water treatment), 178

P
Pacific Ocean, 135–136
Pack ice, 152
Paleodeserts, 239
Passive solar technology, 312
Pathogens in water, 181–182
Pedalfer soil, 210
Pedocal soil, 210
Pedology, 5
Pelton turbine, 326
Penguins of Antarctica, 154
Perched water table, 116
Periglacial areas, 165
Permafrost, 164–165
Peru current, 133
Petroleum, 287. *See also* Oil
Petrology, 5
pH of water, 175
Phosphorus, 180
Photosynthesis, 129
Photovoltaic (PV) reaction, 310, 312–314
Physical weathering, 207–208
Phytoplanktons, 174
Pinatubo effect, 68
Pitchblende, 304
Plants, 11, 22–23
  desert, 233–234
Pleistocene Era, 148–150
Plutonium, 269, 305

Point source pollutant, 121
Polar deserts, 229
Polar stratospheric clouds, 46
Polar vortex, 46
Pollution, air, 67–69
  sources, 69
Pollution, water
  acid deposition, 184–186
  acid rain, 182–183
  chemicals, 180
  contaminants, 179–182
  definition of, 171
  dissolved oxygen, 175–176
  future, 188–189
  Internet reference, 399
  introduction, 171–174
  natural water, 172
  oil slicks, 186
  organic matter, 181
  pathogens, 181–182
  pH, 175
  radioactivity, 186–187
  surface water treatment, 177–178
  thermal pollution, 187–188
  turbidity, 176–177
  waste water treatment, 178–179
Population density, desertification and, 237
Porous media aquifers, 115
Precipitation, 105–107, 109
Pressure
  air, 50–51
  ocean, 128–129
Pressure ridge, 152
Propane, 294
Propeller turbine, 326
Pumped storage hydroelectric plant, 324–325
PV (photovoltaic) reaction, 310, 312–314
Pycnocline layer, 128

R

Radial drainage, 113–114
Radioactive tracers, 187
Radioactive waste, 179, 186–187, 268–271
Radiopharmaceuticals, 186
Rain shadow deserts, 227–228
Range, 24
Recharge, definition of, 117
Rectangular drainage, 113–114
Recycling, 273
Redwoods, 31
References, 395–399
  Internet references, 397–399
Relative humidity, 52–53
Remote sensing, 36–38
Renewable energy certificates (RECs), 338–340
Reservoir, water, 103, 106
Residence time, 110, 244
Reuse, 273
Rills, 217
Rock, 120
  weathering speed, 206
Rockfalls and slides, 215
Rock flour, 162
Rock shift, 204
Runoff
  oil, 297
  water, 118

S

Saguaro cacti, 234
Salinity, 127
Saltation, 216
Scour gullies, 217
Seamounts (extinct volcanos), 132, 134
Sedimentation, 178
Semiarid deserts, 225–226
Séracs, 160
Shearing stress, 213–214
Shear strength, 214
Sheen, oil, 296
Shorelines, ocean, 132–133
Silent Spring, 262a
Silicon solar cells, 310–311
Sinkholes, 255–256

Slip face movement, 230–231
Sluicing, 217
Smoking paddocks, 215
Snout, 153
Soil
  calcium in, 247
  chemical waste in, 268
  in desert, 233
  erosion, 211–217
  horizons, 211–212
  permeability, 111
  porosity, 111
  types of, 210–211
  water and, 120
Solar energy
  common uses, 314–316
  crystalline silicon solar cells, 310–311
  electrochemical solar cells, 311–312
  Internet reference, 398
  introduction, 304, 309–310, 313–314
  solar technology types, 312
  thin solar cells, 311
Solubility, 209
Somali current, 134
South Equatorial current, 133
Space, 399
Speciality clouds, 60–62
Species, 22
Spent radioactive fuel, 270
Squall line, 60
Staebler-Wronski Effect, 311
Stationary fronts, 55
Stenciling, storm drain, 188–189
Steppes, 223
Stratocumulus clouds, 57, 59
Stratopause, 45
Stratosphere, 42–45
Stream gage, 112
Structural geology, 5
Styrofoam, 267
Subduction, 17
Sulphur dioxide, 184
Super-catalysts, 345
Supercells, 60
Superfund sites, 263–264
Super Outbreak, 62
Surface tension, 104
Surface water treatment, 177–178
Sustainability, 22, 335–337
Swamp gas, 78

T

Tar balls, 289–299
Temperate glacier, 152
Temperate rain forest, 30
Temperature, ocean, 127–128
Temperature inversion, 44
Terminal moraines, 162–163
Terms and organizations, 383–392
Texture (weathering factor), 205
Thermals, 54, 60
Thermal water pollution, 187–188
Thermocline layer, 128
Thermoregulation, 298
Thermosphere, 42–43, 48–49
Thin solar cells, 311
THM (trihalomethane), 178
Three Mile Island, 307–308
Till, glacial, 120, 161–162
Tillite, 162
Tornadoes, 62–64
  Fujita wind damage scale, 63–64
Total organic carbon (TOC), 253
Trade wind deserts, 224–225
Transgressive (creeping) dunes, 216
Transpiration, 105–106, 110
Transport (water), 108–109
Transuranic waste, 270–271
Transverse dunes, 231–232
Trellis drainage, 113–114
Trenches, deep-sea, 132, 136
Tributary, 112
Trihalomethane (THM), 178
Troglobites, 257

Troglophiles, 257
Tropical depression, 65
Tropical dry forest, 30
Tropical Prediction Center, 66
Tropical rain forest, 30
Tropical storm, 65
Tropic of Cancer, 226
Tropic of Capricorn, 226
Tropopause, 44
Troposphere, 42–44
Tsunamis, 143–144
Turbidity, 176
Turbidity currents, 15, 131
Turbines
  hydroelectric, 326
  wind, 320
Typhoons, 66

U
Ultraviolet (UV) radiation, 45
Unconfined aquifers, 116–117
United States Geological Survey (USGS), 112
Uranium, 269, 304

V
Vertical clouds, 59–60
Vienna Convention, 47
Volcanology, 5
Vostok ice-core, 167

W
Wadis, 234
Warm front, 55
Wastes, solid and hazardous
  biological and biohazardous wastes, 265–266
  chemical wastes, 266–268
  Internet references, 399
  introduction, 261–264
  making a difference, 273–274
  nonhazardous waste, 272–273
  radioactive wastes, 268–271
Waste water treatment, 178–179
Water
  chemical waste in, 268
  on deserts, 234–235
  erosion, 217–218
  fisheries. *See* Fisheries
  geothermal energy and, 328–329
  glaciers. *See* Glaciers
  hydrologic cycle. *See* Hydrologic cycle
  natural water, 172
  oceans. *See* Oceans
  pollution. *See* Pollution, water
  properties of, 103–104
  sources of, 102–103
  table, 111
  treatment, 177–179. *See also* Pollution, water
  universal solvent, 104
  use and quality, 120–121
Watershed, 112–114
Wave energy, 216
Weather, 399
Weathering and erosion
  biological weathering, 208–209
  chemical weathering, 209–210
  denudation, 204–205
  dune erosion, 216–217
  introduction, 203–207
  physical weathering, 207–208
  rockfalls and slides, 215
  soil erosion, 211–217
  soil types, 210–211
  water erosion, 217–218
  wind erosion, 214–215
West Wind Drift, 133
Wetlands, 27–29
  impact of human activities on, 28
WGI (World Glacier Inventory), 168
White smokers, 130
Wildlife biology, 5
Wildlife impact of oil spills, 297–298

Wind, 51–52
  energy, 319–322
  erosion, 214–216
  Internet references, 399
  shear, 65
Wind chill factor, 51, 53
Working fluid, 330–331
World Conservation Union, 26, 238
World Glacier Inventory (WGI), 168
World Wildlife Fund, 25

Y

Yucca Mountain radioactive waste site, 309

Z

Zones
  ablation, 155–156
  accumulation, 155–156
  acquifer recharge, 117
  aeration, 111
  aphotic, 130
  dysphotic, 130
  euphotic, 129–130
  glaciers, 155–156
  littoral, 133
  oceans, 131–132
  saturation, 111
Zoology, 5

ABOUT THE AUTHOR

Linda D. Williams is a nonfiction writer with specialties in science, medicine, and space. A resident of Houston, Texas, Ms. Williams' work has ranged from biochemistry and microbiology to genetics and human enzyme research. She has worked as a lead scientist and technical writer for NASA and McDonnell Douglas Space Systems, and served as a science speaker for the Medical Sciences Division at NASA-Johnson Space Center. Currently, Ms. Williams works in the Weiss School of Natural Sciences at Rice University, Houston, Texas.